T0199807

Spintronic Materials and Technology

Series in Materials Science and Engineering

Series Editors: **B Cantor,** University of York, UK
A Eades, Lehigh University, Bethlehem, Pa., USA
M J Goringe, University of Surrey, UK
E Ma, Johns Hopkins University, Baltimore, Md, USA

Series in Materials Science and Engineering

Spintronic Materials and Technology

Y B Xu
The University of York, UK

S M Thompson
The University of York, UK

CRC Press
Taylor & Francis Group
Boca Raton London New York

CRC Press is an imprint of the
Taylor & Francis Group, an **informa** business
A TAYLOR & FRANCIS BOOK

CRC Press
Taylor & Francis Group
6000 Broken Sound Parkway NW, Suite 300
Boca Raton, FL 33487-2742

First issued in paperback 2019

ISBN-13: 978-0-8493-9299-3 (hbk)
ISBN-13: 978-0-367-39007-5 (pbk)

Library of Congress Cataloging-in-Publication Data

Spintronic materials and technology / edited by Yongbing Xu, Sarah Thompson.
    p. cm.
  Includes bibliographical references and index.
  ISBN 0-8493-9299-3 (alk. paper)
    1. Microelectromechanical systems. 2. Nanotechnology. 3. Nuclear spin. I. Xu, Yongbing. II. Thompson, Sarah.

TK7875.S65 2006
621.381--dc22

2006044660

Visit the Taylor & Francis Web site at
http://www.taylorandfrancis.com

and the CRC Press Web site at
http://www.crcpress.com

Preface

Spintronics is an exciting and rapidly expanding new field of microelectronics and nanoelectronics, which is based on exploiting the fact that electrons have spin as well as charge. For decades, the information technology (IT) industry has followed Moore's law, that the number of transistors on a chip doubles about every 2 years, but conventional solid state electronics may soon reach a limit due to the increasing heat dissipation challenges of charge current and quantum size effects in small devices. Spintronics, molecular electronics, and carbon-based electronics along with advances in nanotechnology are expected to ensure continued adherence to Moore's law in the future. Within the context of spintronics, the electrons' spins, not just their electrical charge, are manipulated within electronic circuits. These spintronic devices, combining the extra degree of freedom offered by magnetic materials and the versatility of active semiconductor devices, are anticipated to be nonvolatile, versatile, fast, and capable of simultaneous data storage and processing, while consuming less energy. They already play an increasingly significant role in high-density data storage, microelectronics, magnetic sensors, quantum computing, bio-medical applications, and so on.

It was the discovery of giant magnetoresistance (GMR) in the early 1980s that initiated spintronic research and resulted in the first generation of a spintronic device in the form of the spin valve. Today, spintronic devices are ubiquitous on the desktop as spin valves play their role as the active element in the read head of most hard disk storage devices. GMR was followed rapidly by the discovery of tunneling magnetoresistance leading to the magnetic tunnel junction that has been utilized in developing the next generation of memory known as magnetic random access memory — an example of another spintronic device. Second-generation spintronic devices will integrate magnetic materials and semiconductor devices to create new flexible devices such as spin transistors and spin logic. These second-generation spintronic devices will not just improve the existing capabilities of electronic transistors, but will have new functionalities enabling future computers to run faster, but consume less power, and have the potential to revolutionize the IT industry as did the development of the transistor 50 years ago.

This book is divided into three parts, covering the main research challenges in spintronics. Part I, "Spintronic Material and Characterization," deals with the properties, desirable features, and the search for new spintronic materials ranging from magnetic oxides and metallic GMR materials to dilute magnetic semiconductors. Their magnetic, structural, and spin-dependent transport properties are characterized using many different conventional laboratory-based techniques and more novel synchrotron radiation-based measurements. In Part II, "Spin Torque and Domain Wall Magnetoresistance," issues concerning the operation of spintronic devices are addressed including the new principle for the operation of future spintronic devices using spin-polarized current rather than conventional applied magnetic fields. This promises to enable the switching of the individual spin components of the device while avoiding cross talk at the nanoscale. In Part III, "Spin Injection and Spin Devices," complete device ideas are explored describing both Si and III-V semiconductor-based spin transistors and the integration of spin technology with photonics.

The book provides background, introduction, the latest research results, and an extensive list of references in each chapter. The textbook style is intended to satisfy the needs of graduate students and young researchers, who have little knowledge of this new area. The collection of this material in one book enables the challenges and progress toward the ultimate goal of spin-controlled devices to be seen in context, and the individual chapters are designed to be self-contained to aid the researcher concerned with one particular area. Due to the multidisciplinary nature of spintronics research, the book covers a wide range of topics in materials science, physics, device fabrication, characterization, and operation. All the authors are active researchers and leading experts in these areas, ensuring that the book provides an excellent insight into the current development and future of spintronics.

The editors wish to thank the authors in writing their chapters so that this very first book in spintronics could be accessible to a general audience. The editors thank Professor Brian Cantor, FREng, for advice in preparing this book. The editors would also like to thank John Navas, Amber Donley, and Gail Renard from the Taylor & Francis Group, LLC.

Yongbing Xu
Sarah Thompson

The Editors

Yongbing Xu is an anniversary reader, which is a senior professorship in nanotechnology at the University of York, UK, and heads the Spintronics Laboratory in Electronics. Before moving to York in 2000, he was a senior research fellow at the Cavendish Laboratory, University of Cambridge. He was awarded Ph.D.s by Nanjing University, China, in 1993, and the University of Leeds, UK, in 1996. His research interests are in the areas of spintronics, magnetic nanomaterial, nanodevices, and nanofabrication. He has published more than 120 refereed papers in leading academic journals and given many invited talks at international conferences. He is the coordinator and chair of the steering committee of the WUN (Worldwide University Network) Grand Challenge Project "Spintronics," which includes 20 partners from 12 leading universities in the UK, USA, and China. He is the section editor in spintronics of the journal *Current Opinion in Solid State and Materials Science* published by Elsevier. In 2000, he was awarded the prestigious EPSRC advanced fellowship. He is the stream leader of the York MEng/BEng program in nanotechnology and teaches introductory nanotechnology, solid state devices, nanoelectronics, and advanced information storage.

Sarah Thompson is a senior lecturer at the University of York, York, UK, and heads the Magnetic Thin Films Research Group in the physics department. She has 15 years' research experience in the fabrication, structural, magnetic, and transport properties of magnetic thin films and multilayers concentrating on spin dependent transport, recently in spin electronics. Since 1999, Dr. Thompson has pioneered the development of infrared spectroscopy in a magnetic field to probe the spin dependent transport in magnetic materials, and combining these two areas of expertise, she is the champion for the spintronics Flagship Project for the Development of the 4GLS at Daresbury Laboratory, UK.

Contributors

S. Abe
Department of Materials Science
Tohoku University
Sendai, Japan

William Allen
Clarendon Laboratory
University of Oxford
Oxford, United Kingdom

A.J. Behan
Department of Physics and
  Astronomy
University of Sheffield
Sheffield, United Kingdom

J.A.C. Bland
Cavendish Laboratory
University of Cambridge
Cambridge, United Kingdom

H.J. Blythe
Department of Physics and Astronomy
University of Sheffield
Sheffield, United Kingdom

Gustaaf Borghs
Imec
Leuven, Belgium

C.D. Damsgaard
Department of Physics
Denmark Technical University
Lyngby, Denmark

Jo De Boeck
Imec
Leuven, Belgium

Cindi L. Dennis
National Institute of Standards
  and Technology
Gaithersburg, Maryland

D.M. Edwards
Department of Mathematics
Imperial College
London, United Kingdom

P.E. Falloon
School of Physics
University of Western Australia
Crawley, Australia
Institut de Physique et Chimie des
  Matériaux de Strasbourg
Centre National de la Recherche
  Scientifique, Université Louis Pasteur
  (CNRS-ULP)
Strasbourg, France

I. Farrer
Cavendish Laboratory
University of Cambridge
Cambridge, United Kingdom

F. Federici
NEST-CNR-INFM and Classe
  di Scienze
Scuola Normale Superiore
Pisa, Italy

A.M. Fox
Department of Physics and Astronomy
University of Sheffield
Sheffield, United Kingdom

G.A. Gehring
Department of Physics and Astronomy
University of Sheffield
Sheffield, United Kingdom

V. Gopar
Institut de Physique et Chimie des
  Matériaux de Strasbourg
Centre National de la Recherche
  Scientifique, Université Louis Pasteur
  (CNRS-ULP)
Strasbourg, France
Max-Planck-Institut für Physik
  komplexer Systeme
Dresden, Germany

John F. Gregg
Clarendon Laboratory
University of Oxford
Oxford, United Kingdom

J.B. Hansen
Department of Physics
Denmark Technical University
Lyngby, Denmark

M. Hickey
Cavendish Laboratory
University of Cambridge
Cambridge, United Kingdom

A. Hirohata
Quantum Nano-Scale Magnetics
  Laboratory
Frontier Research System, RIKEN
Wako, Japan

S.N. Holmes
Toshiba Research Europe Limited
Cambridge Research Laboratory
Cambridge, United Kingdom

A. Husmann
Toshiba Research Europe Limited
Cambridge Research Laboratory
Cambridge, United Kingdom

R.M. Ibrahim
Department of Physics and Astronomy
University of Sheffield
Sheffield, United Kingdom

K. Inomata
Department of Materials Science
Tohoku University and CREST-JST
Sendai, Japan

C.S. Jacobsen
Department of Physics
Denmark Technical University
Lyngby, Denmark

R.A. Jalabert
Institut de Physique et Chimie des
  Matériaux de Strasbourg
Centre National de la Recherche
  Scientifique, Université Louis Pasteur
  (CNRS-ULP)
Strasbourg, France

R. Jansen
MESA+ Institute for Nanotechnology
University of Twente
Enschede, The Netherlands

Y. Jiang
CREST-JST
Sendai, Japan

G.A.C. Jones
Cavendish Laboratory
University of Cambridge
Cambridge, United Kingdom

H. Kohno
Graduate School of Engineering
  Science
Osaka University
Toyonaka, Japan

R.F. Lee
Cavendish Laboratory
University of Cambridge
Cambridge, United Kingdom

Serban Lepadatu
Department of Electronics
The University of York
York, United Kingdom

Contributors

S. Abe
Department of Materials Science
Tohoku University
Sendai, Japan

William Allen
Clarendon Laboratory
University of Oxford
Oxford, United Kingdom

A.J. Behan
Department of Physics and
  Astronomy
University of Sheffield
Sheffield, United Kingdom

J.A.C. Bland
Cavendish Laboratory
University of Cambridge
Cambridge, United Kingdom

H.J. Blythe
Department of Physics and Astronomy
University of Sheffield
Sheffield, United Kingdom

Gustaaf Borghs
Imec
Leuven, Belgium

C.D. Damsgaard
Department of Physics
Denmark Technical University
Lyngby, Denmark

Jo De Boeck
Imec
Leuven, Belgium

Cindi L. Dennis
National Institute of Standards
  and Technology
Gaithersburg, Maryland

D.M. Edwards
Department of Mathematics
Imperial College
London, United Kingdom

P.E. Falloon
School of Physics
University of Western Australia
Crawley, Australia
Institut de Physique et Chimie des
  Matériaux de Strasbourg
Centre National de la Recherche
  Scientifique, Université Louis Pasteur
  (CNRS-ULP)
Strasbourg, France

I. Farrer
Cavendish Laboratory
University of Cambridge
Cambridge, United Kingdom

F. Federici
NEST-CNR-INFM and Classe
  di Scienze
Scuola Normale Superiore
Pisa, Italy

A.M. Fox
Department of Physics and Astronomy
University of Sheffield
Sheffield, United Kingdom

G.A. Gehring
Department of Physics and Astronomy
University of Sheffield
Sheffield, United Kingdom

V. Gopar
Institut de Physique et Chimie des
  Matériaux de Strasbourg
Centre National de la Recherche
  Scientifique, Université Louis Pasteur
  (CNRS-ULP)
Strasbourg, France
Max-Planck-Institut für Physik
  komplexer Systeme
Dresden, Germany

John F. Gregg
Clarendon Laboratory
University of Oxford
Oxford, United Kingdom

J.B. Hansen
Department of Physics
Denmark Technical University
Lyngby, Denmark

M. Hickey
Cavendish Laboratory
University of Cambridge
Cambridge, United Kingdom

A. Hirohata
Quantum Nano-Scale Magnetics
  Laboratory
Frontier Research System, RIKEN
Wako, Japan

S.N. Holmes
Toshiba Research Europe Limited
Cambridge Research Laboratory
Cambridge, United Kingdom

A. Husmann
Toshiba Research Europe Limited
Cambridge Research Laboratory
Cambridge, United Kingdom

R.M. Ibrahim
Department of Physics and Astronomy
University of Sheffield
Sheffield, United Kingdom

K. Inomata
Department of Materials Science
Tohoku University and CREST-JST
Sendai, Japan

C.S. Jacobsen
Department of Physics
Denmark Technical University
Lyngby, Denmark

R.A. Jalabert
Institut de Physique et Chimie des
  Matériaux de Strasbourg
Centre National de la Recherche
  Scientifique, Université Louis Pasteur
  (CNRS-ULP)
Strasbourg, France

R. Jansen
MESA+ Institute for Nanotechnology
University of Twente
Enschede, The Netherlands

Y. Jiang
CREST-JST
Sendai, Japan

G.A.C. Jones
Cavendish Laboratory
University of Cambridge
Cambridge, United Kingdom

H. Kohno
Graduate School of Engineering
  Science
Osaka University
Toyonaka, Japan

R.F. Lee
Cavendish Laboratory
University of Cambridge
Cambridge, United Kingdom

Serban Lepadatu
Department of Electronics
The University of York
York, United Kingdom

Duncan Loraine
Department of Physics
University of York
York, United Kingdom

A. Mokhtari
Department of Physics
and Astronomy
University of Sheffield
Sheffield, United Kingdom

Vasyl F. Motsnyi
Imec
Leuven, Belgium

J.R. Neal
Department of Physics
and Astronomy
University of Sheffield
Sheffield, United Kingdom

M. Pepper
Cavendish Laboratory
University of Cambridge
Cambridge, United Kingdom
Toshiba Research Europe Limited
Cambridge Research Laboratory
Cambridge, United Kingdom

David Pugh
Department of Physics
University of York
York, United Kingdom

D.A. Ritchie
Cavendish Laboratory
University of Cambridge
Cambridge, United Kingdom

Maciej Sawicki
Institute of Physics
Polish Academy of Sciences
Warsaw, Poland

Chitnarong Sirisathitkul
Magnet Laboratory
Walailak University
Thammarat, Thailand

R.L. Stamps
School of Physics
University of Western Australia
Crawley, Australia

S.J. Steinmuller
Cavendish Laboratory
University of Cambridge
Cambridge, United Kingdom

G. Tatara
Department of Physics
Tokyo Metropolitan University
Tokyo, Japan

N. Tezuka
Department of Materials Science
Tohoku University and CREST-JST
Sendai, Japan

Sarah M. Thompson
Department of Physics
University of York
York, United Kingdom

Gerrit van der Laan
Magnetic Spectroscopy Group
Daresbury Laboratory
Warrington, United Kingdom

Pol Van Dorpe
Imec
Leuven, Belgium

Willem Van Roy
Imec
Leuven, Belgium

D. Weinmann
Institut de Physique et Chimie des
Matériaux de Strasbourg
Centre National de la Recherche
Scientifique, Université Louis Pasteur
(CNRS-ULP)
Strasbourg, France

Ping Kwan J. Wong
Department of Electronics
The University of York
York, United Kingdom

Ke Xia
Institute of Physics
Chinese Academy of Sciences
Beijing, China

Xiangqian Xiu
Key Laboratory of Advanced Photonic
  and Electronic Materials
Department of Physics
Nanjing University
Nanjing, China

Yongbing Xu
Department of Electronics
The University of York
York, United Kingdom

Rong Zhang
Key Laboratory of Advanced Photonic
  and Electronic Materials
Department of Physics
Nanjing University
Nanjing, China

Table of Contents

Part I

Spintronic Materials and Characterizations

1 Magneto-Optical Studies of Magnetic Oxide Semiconductors

G.A. Gehring, A.J. Behan, H.J. Blythe, A.M. Fox, R.M. Ibrahim, A. Mokhtari, and J.R. Neal

CONTENTS

1.1 INTRODUCTION

Until a few years ago, the spin of electrons was ignored in current charge-based electronics. A new technology recently developed has been called *spintronics*, meaning *spin transport electronics* or *spin-based electronics*. In spintronics, it is not the electron's charge but the electron's spin that carries information.[1]

Spintronics is today one of the most rapidly growing fields in electronics[1–6] and combines both the spin and charge of the electrons to obtain devices with new functionality and increased performance. The advantages of these new devices are nonvolatility, increased data processing speed, decreased electric power consumption, and increased transistor density compared with conventional semiconductor devices.[7–9] Within this field of research, dilute magnetic semiconductors (DMSs), also known as semimagnetic semiconductors, are especially important materials. In these materials, rare earth or transition metal ions replace a considerable number of the native ions in the semiconductor's lattice.[10,11] This offers opportunities for a new generation of devices combining standard microelectronics with spin-dependent effects that arise from the interaction between the spin of the carrier and the magnetic properties of the material.[1]

Most of the devices in spintronics made so far are based on metallic thin films. Ferromagnetic metallic films with metallic spacer layers show giant magnetoresistance and have been used in read-heads in computer disk drives for several years. A new version with an insulating spacer layer shows tunneling magnetoresistance which is very much larger and is now being developed as a material for nonvolatile random access memory. Ferromagnetic metals have several advantages for use in devices:

- the transition temperatures are very well above room temperature
- the magnetic anisotropy can be tuned by alloying
- the fabrication routes are well established

However, it is less easy to couple them to conventional semiconductors because of the very large resistance mismatch. Thus, the hunt is now on for a ferromagnetic semiconductor that could be integrated with standard semiconductor technology. It is anticipated that if ferromagnetism and electronics transport could be coupled in semiconductors, the effect of magnetism would be significantly stronger than the phenomena observed in metals.[12]

The emerging field of semiconductor spintronics seeks to exploit the spin of charge carriers in semiconductors. It is widely expected that new functionalities for electronics and photonics can be derived if the injection, transfer, and detection of carrier spin can be controlled above room temperature. Among this new class of devices are spin transistors operating at very low power for mobile applications that rely on batteries, optical emitters with encoded information through their polarized light output, fast nonvolatile semiconductor memory, and integrated magnetic/electronic/photonic devices (electromagnetism-on-a-chip).

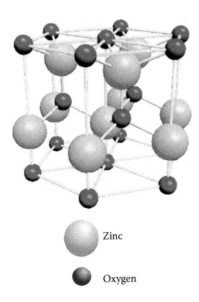

Zinc

Oxygen

FIGURE 1.1 The wurtzite structure.

1.2 DOPED ZnO

The interest for spintronics in doped ZnO applications stems from the seminal paper by Dietl et al.[17] in 2000, which predicted room temperature ferromagnetism in a range of p-doped semiconductors including ZnO containing 5% of manganese, as shown in Figure 1.2.

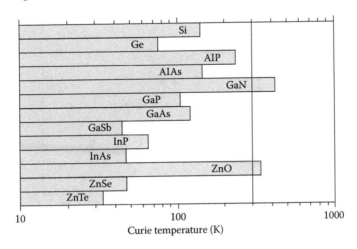

FIGURE 1.2 Theoretical predictions of the Curie temperature T_C for various p-type semiconductors containing 5% of Mn and 3.4×10^{20} holes per cm^3. (From FT. Dietl, H. Ohno, F. Matsukura, J. Cibert, and D. Ferrand, *Science*, 287, 1019, 2000. With permission.)

The reason why the prediction was made for p doping was that the interaction between magnetic d electrons on the manganese and the oxygen p electrons in the valence band was known to be much stronger than their interaction with the s electrons in the conduction band. It is necessary to have unequal filling of the magnetically split band to stabilize ferromagnetism, also following extensive work on GaMnAs, which is of the p type.

The experimental situation, however, is controversial, because there are numerous experimental reports both on the presence[18–23] and the absence[22,24] of ferromagnetism in ZnO (the case of Mn-doped samples is particularly controversial) flanked by theoretical controversies on the nature of the ferromagnetism of ZnO DMS.[17,25] The reasons why the data are greeted with amazement, and sometimes disbelief, is that the reported transition temperatures are above room temperature in samples with very low doping (2 to 5%), the magnetic moments per transition metal ion are low, the carrier concentration can also be low, and the doping is n-type rather than p-type.[17] However, there are now systematic studies[20] that show that increased magnetic moments only occur when there is a large electron doping. The nature of the doped states (i.e., whether these are localized or conducting) is still a central issue.

Bulk material has been produced that shows ferromagnetism, but the moments are usually very small.[21–23] For Mn doping, it is important to keep to low temperature processing.[23] In Figure 1.3, we show the influence of annealing temperature on a 2% Mn-doped ZnO that had been ground at room temperature. It is apparent that the magnetism is destroyed at high annealing temperatures.

Larger moments occur in thin films that can be made by sputtering,[26] chemical methods,[27] or (most commonly) pulsed laser deposition (PLD). In Table 1.1, we list the experiments that have used PLD. In this technique, it is important to specify the

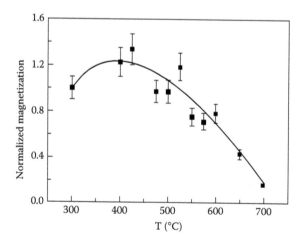

FIGURE 1.3 Influence of anneal temperature (T) on the magnetization measured at 1 T and 290 K for a $Zn_{0.98}Mn_{0.02}O$ powder mix after grinding for 4 h in a mortar and pestle; the data has been corrected for the paramagnetic contribution and normalized to the value before any anneal. Errors of 10% are shown. (From H.J. Blythe, R.M. Ibrahim, G.A. Gehring, J.R. Neal, and A.M. Fox, *J. Magn. Magn. Mater.*, 283, 117, 2004. With permission.)

TABLE 1.1
Summary of Transition-Metal-Doped ZnO Films Fabricated by the PLD Technique

Dopant	Transition Metal Content	Substrate	Growth Temperature (°C)	Oxygen Pressure (Torr)	T_C (K)	Behavior Observed	Reference
Co	0.02–0.5	c-Sapphire	300–700	10^{-6}–10^{-1}	~ 300	Spin-glass	28
Co	0.05	Sapphire	500–750	0.1	300	Ferromagnetic	29
Co	0.025	Sapphire	300–700	10^{-6}–10^{-1}	300	Ferromagnetic	30
Co	0–0.05	Sapphire	340	10^{-5}–10^{-4}	> 300	0.1–0.3 μ_B/Co	31
Co	0.25	Sapphire	400–700	10^{-5}	300	Ferromagnetic	32
Co	0.05	Sapphire R-cut	600	0.76–7.6×10^{-5}	> 300	2.6 μ_B/Co	33
Co, Mn, Cr, or Ni	0.05–0.25	Sapphire	350–600	2–4×10^{-5}	300	1.8 μ_B/Co	34
V	0.05	Sapphire	280–400	10^{-9}	> 350	0.5 μ_B/V	35
V	0.05	Sapphire	600–650	0.1	> 350	0.6 μ_B/V	36
V	0.05	Sapphire R-cut	600	0.76–7.6×10^{-5}	> 300	0.5 μ_B/V	33
Ni	0.01–0.25	Sapphire R-cut	300–700	1×10^{-5}		Superparamagnetic or ferromagnetic	33
Ni	0.05	Sapphire	600	0.76–7.6×10^{-5}	> 300	0.1–0.2 μ_B/Ni	33
Ti	0.05	Sapphire R-cut	600	0.76–7.6×10^{-5}	> 300	0.1–0.2 μ_B/Ti	33
Sc	0.05	Sapphire R-cut	600	0.76–7.6×10^{-5}	> 300	0.05–0.1 μ_B/Sc	33
Mn	0.01–0.36	Sapphire	610	5×10^{-5}		Paramagnetic	37
Mn	01–0.3	Sapphire		10^{-9}	> 30–45	0.15–0.17 μ_B/Mn	38
Mn	0.36	Sapphire	600	5×10^{-5}	N/A	Spin-glass	39
Mn	< 0.04	Fused quartz	400	228×10^{-3}	> 425	0.05 EMU/gm	40
Mn	≤ 0.35	Sapphire			N/A	Paramagnetic	41
Mn	0.065	Sapphire	500		> 375	0.013 EMU/gm	42
Mn	0.05	ZnO	200		250	0.5 μ_B/Mn	43

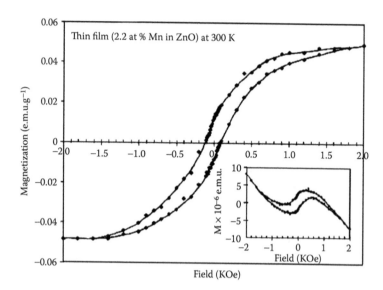

FIGURE 1.4 Magnetization at 300 K for a $Zn_{0.978}Mn_{0.022}O$ PLD thin film on fused quartz. The curve was obtained after subtracting the diamagnetic contribution from the substrate. Inset: The as-obtained data from the SQUID measurements including the diamagnetic contribution arising from the fused quartz substrate. (From P. Sharma, A. Gupta, K.V. Rao, F.J. Owens, R. Sharma, R. Ahuja, J.M.O. Guillen, B. Johansson, and G.A. Gehring, *Nat. Mater.*, 2, 673, 2003. With permission.)

amount of O_2 pressure in the deposition chamber. A stoichiometric film of ZnO is grown in 10 mtorr of O_2, and so films grown with lower pressure will have either oxygen vacancies or Zn interstitials, both of which make the film *n*-type. Table 1.1 gives an indication of the activity in this field and the diversity of the results obtained.

Even where films have been found to be ferromagnetic, there is still an active controversy as to whether there is a minority phase present that is ferromagnetic or if the ferromagnetism is really incorporated in the ZnO structure and thus mediated by the electron states that are characteristic of the ZnO.

The magnetic moments per transition metal ion are often rather less than the theoretical maximum, indicating that not all the transition metal ions are participating in the ferromagnetic structure. In some cases,[33,44] the magnetism is actually higher than the predicted maximum, indicating that another magnetic species is involved, possibly the oxygen defect states.

Typical hysteresis loops are shown in Figure 1.4 and Figure 1.5 for Mn and V-doped films. The coercive field is typically around 100 Oe.

Most superconducting quantum interference device (SQUID) measurements are made with the applied magnetic field in the plane of the film. However, we found very similar behavior when the field was perpendicular, unlike Venkatesan et al.[33] who, as shown in Figure 1.6, found different values for the saturation magnetization depending on the field direction. This is relevant here because all the Faraday rotation measurements were made with the field perpendicular to the film plane. Figure 1.6[33]

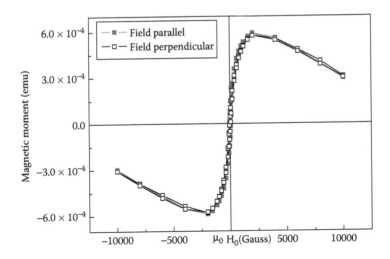

FIGURE 1.5 Magnetic moment data for a $Zn_{0.95}V_{0.05}O$ films measured at room temperature. The data have not been corrected for the diamagnetic contribution from the sapphire substrate. The closed squares are for a film mounted with its plane vertical (parallel to the field) and the open squares are for a film mounted with its plane horizontal (perpendicular to the field). (From R.M. Ibrahim et al., unpublished.)

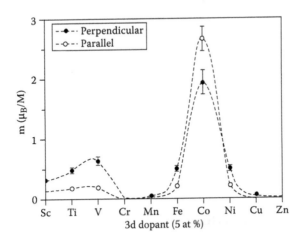

FIGURE 1.6 Magnetic moment of $(Zn_{1-x}M_x)O$ films for M = Sc, Ti, … , Cu, Zn measured at room temperature. Solid circles are for the field applied perpendicularly to the film plane and open circles are for the field applied in the plane of the film. The moment is expressed as μ_B/M. The trend measured at 5 K is similar. (From M. Venkatesan, C.B. Fitzgerald, J.G. Lunney, and J.M.D. Coey, *Phys. Rev. Lett.*, 93, 177206, 2004. With permission.)

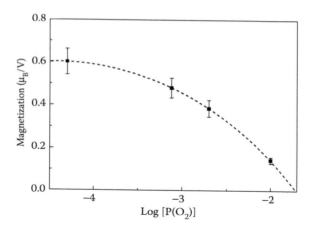

FIGURE 1.7 Magnetic moment per V atom for the $Zn_{0.94}V_{0.06}O$ films as a function of oxygen pressure measured at room temperature. The data is well-fitted by a second-order polynomial equation, with the magnet moment ($= -1.01 - 0.750x - 0.08x^2$) μ_B/V (see dotted line). (From R.M. Ibrahim, unpublished.)

was very important because it indicated strong similarities and trends between the different transition metal dopants.

The ZnO samples show the strongest ferromagnetism when they are electron-doped. It is interesting that what is measured is the magnetic moment per transition metal ion rather than the transition temperature. The reason for this is that the films are sensitive to heat treatment and heating them above room temperature to measure T_c may cause structural changes. It appears that at low doping, many of the transition ions are trapped in antiferromagnetic clusters. As the doping increases, more of them become aligned to the ferromagnetic moment. An example of the behavior of the moment with doping is given in Figure 1.7 for vanadium, which shows a decrease in the magnetization with increasing oxygen pressure, which itself causes a reduction in the doping density.

An important way to check if the ferromagnetism is associated with the ZnO band electrons is to look for magneto-optic effects at the frequency near the ZnO band edge. This is the main subject of this chapter.

1.3 OPTICAL PROPERTIES

In this section, we give a very brief introduction to the field of magneto-optics (MO). The refractive index and the absorption of light are caused by electric dipole transitions in the solid. We consider light traveling parallel to the direction of magnetization. Circularly polarized transitions occur between the magnetically quantized electronic states in which $\Delta m_j = \pm1$. If the spin-orbit coupling is weak, then the selection rules become $\Delta m_s = 0$ and $\Delta m_l = \pm1$, and some spin-orbit coupling is necessary in order that the orbital moment reflects information about the magnetism, which is associated with the electron spin. There are two effects observed in transmission. The refractive index can be different for left- and right-handed circularly polarized light — this is the Faraday effect, which causes a rotation of the plane of polarization and depends on a weighted average over all the transitions in the crystal. The other effect is the magnetic circular dichroism (MCD), which is the difference in absorption for left- and right-circularly polarized light at a particular frequency. This causes plane polarized light to become elliptically polarized.

The Faraday effect is very useful in applications because it is finite in spectral regions where the crystal is transparent. However, the MCD is of most use for determining the nature of the magnetic states. It is clear that the MCD is only nonzero where the crystal is absorbing. We shall see that the MCD spectrum is often much sharper than the absorption spectrum and so it gives better energy resolution as well as information on the wave functions.

Opaque samples are usually studied in reflection. The largest signals are obtained using the polar Kerr effect in which light is incident normally on the sample that is magnetized parallel to the propagation direction. Both Faraday and polar Kerr measurements require the magnetization to be parallel to the direction of the incident light.

The reflected light (the polar Kerr effect) or transmitted light (the Faraday effect) becomes elliptically polarized to a slight degree with the major axis of the ellipse rotated with respect to the incident polarization. Subsequently, the rotation and ellipticity of the reflected light or transmitted light is measured (Figure 1.8). The analysis of the magneto-optical activity of a substance is generally carried out in two stages. First, the measured quantities, that is, the Faraday or the Kerr rotation θ_F or θ_k and ellipticity η_F or η_k, are related to the dielectric tensor $\tilde{\varepsilon}$. In a second step, these functions are then interpreted in terms of crystal field states of magnetic ions or spin-polarized band-structure calculations. We shall see in Section 1.4 that the relationship between the Faraday and Kerr effects is a little surprising.[45]

It is useful to distinguish here between the uses of MO as a tool for magnetometry and in spectroscopy. In Kerr magnetometry, one wishes to obtain hysteresis loops in real time, often when the magnetism is in the plane of a thin film. This means that it is the longitudinal Kerr effect that is measured. In these measurements, the aim is to check the relative magnitude of the magnetization as a function of field.

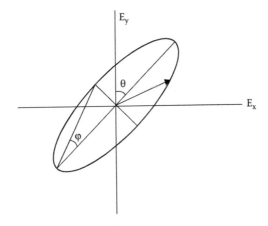

FIGURE 1.8 The Faraday or Kerr rotation (θ) and ellipticity ($\eta = \tan \varphi$).

This is done at a fixed frequency — usually from a laser that has high intensity and good spatial resolution and that enables the most accurate measurement techniques using compensators that work at laser frequency to be used. This technique can be used to measure either the real or the imaginary part of the Kerr angle.

MO spectroscopy is much less common and is the subject of this chapter. In this case, one wants absolute measurements of the MO coefficients over as wide a frequency range as possible to understand the nature of the occupied and excited electronic states. This means that a continuous light source is used with a spectrometer — this has a much lower intensity than a laser and, furthermore, a different measurement technique is used, which is described in detail in Section 1.5. This technique has the advantage that both the real and imaginary parts of the MO response can be measured together over a wide frequency range, but it is not as accurate as the single-frequency method using compensators. Thus it is more difficult to obtain hysteresis loops, although it can be done — usually by averaging the data many times.

1.4 MACROSCOPIC THEORY

1.4.1 MAGNETO-OPTIC THEORY IN TERMS OF THE DIELECTRIC TENSOR

At optical frequencies, the propagation of electromagnetic waves may be fully described by a complex dielectric tensor $\tilde{\varepsilon}$. In a cubic crystal (or a uniaxial crystal magnetized along its optic axis) with $M \| \hat{z}$, this tensor is of the form shown below,[46] where the off-diagonal components $\tilde{\varepsilon}_{xy}$ are nonzero only in magnetic samples, as will be shown later. We shall use the ~ notation to imply that the quantity is complex.

$$\tilde{\varepsilon} = \begin{bmatrix} \tilde{\varepsilon}_{xx} & -i\tilde{\varepsilon}_{xy} & 0 \\ i\tilde{\varepsilon}_{xy} & \tilde{\varepsilon}_{xx} & 0 \\ 0 & 0 & \tilde{\varepsilon}_{zz} \end{bmatrix} \tag{1.1}$$

For an absorbing medium, the components of the dielectric tensor are complex and are given by $\tilde{\varepsilon}_{ij} = \varepsilon'_{ij} + i\,\varepsilon''_{ij}$.

We consider the propagation of light in a medium with dielectric tensor given by Equation 1.1. The relevant Maxwell equations are given below, with any current incorporated in the complex part of $\tilde{\varepsilon}$:

$$\vec{\nabla} \times \vec{E} = -\frac{\partial \vec{B}}{\partial t} \tag{1.2a}$$

and

$$\vec{\nabla} \times \vec{H} = \frac{\partial \vec{D}}{\partial t} \tag{1.2b}$$

The constituent equations are $\vec{D} = \breve{\varepsilon}\varepsilon_o\vec{E}$ and $\vec{B} = \mu_o\vec{H}$. It has sometimes been argued that in ferromagnets the relation $\vec{M} = \chi\vec{H}$ does not hold and $\mu \neq 1$. We assume that at optical frequencies one should take $\mu = 1$ even for ferromagnets.[45]

We look for solutions at frequency ω, where the propagation is along z. These correspond to right- and left-circular polarized (RCP, LCP) light with propagation wave vectors $K_\pm$ along z.

$$\vec{E}_\pm = E_o(\hat{e}_x \pm i\hat{e}_y)\exp[i(\omega t - K_\pm z)] \tag{1.3}$$

where $\hat{e}_x$ and $\hat{e}_y$ are unit vectors. We define $K^2_\pm = \omega^2\tilde{n}^2_\pm/c^2$ and $c = 1/\sqrt{\varepsilon_o\mu_o}$ and solve for the complex refractive indices, $\tilde{n}_\pm$:

$$\frac{\omega^2\tilde{n}^2_\pm}{c^2}E_o(\hat{e}_x \pm i\hat{e}_y)\exp[i(\omega t - K_\pm z)]$$

$$= \varepsilon_o\mu_o\omega^2 E_o[(\tilde{\varepsilon}_{xx} \pm \tilde{\varepsilon}_{xy})\hat{e}_x + (i\tilde{\varepsilon}_{xy} \pm i\tilde{\varepsilon}_{xx})\hat{e}_y]\exp[i(\omega t - K_\pm z)]. \tag{1.4}$$

This may be solved for $\tilde{n}_\pm$. We find

$$\tilde{n}^2_\pm = \tilde{\varepsilon}_{xx} \pm \tilde{\varepsilon}_{xy}, \tag{1.5a}$$

that may be simplified to give

$$\tilde{n}_\pm = \sqrt{\tilde{\varepsilon}_{xx}} \pm \frac{\tilde{\varepsilon}_{xy}}{2\sqrt{\tilde{\varepsilon}_{xx}}}, \tag{1.5b}$$

provided that

$$\tilde{\varepsilon}_{xx} \gg \tilde{\varepsilon}_{xy}. \tag{1.5c}$$

Magneto-optic effects are given in terms of the difference between left- and right-circular polarizations and hence depend on $\tilde{n}_+ - \tilde{n}_- = \tilde{\varepsilon}_{xy}/\sqrt{\tilde{\varepsilon}_{xx}}$.

The Faraday rotation, θ_F, and MCD, η_F, in a sample of thickness l_o are given in terms of the real and imaginary parts of X_+, namely $n_\pm$ and $k_\pm$, by:

$$\theta_F = \text{Re}\left(\frac{\omega l_o}{2c}(\tilde{n}_+ - \tilde{n}_-)\right) = \frac{\omega l_o}{2c}(n_+ - n_-) = \frac{\omega l_o}{2c}\Delta n, \tag{1.6}$$

$$\eta_F = \tan\left[\text{Im}\left[\frac{\omega l_o}{2c}(\tilde{n}_+ - \tilde{n}_-)\right]\right] = \tan\left[\frac{\omega l_o}{2c}(k_+ - k_-)\right] \approx \frac{\omega l_o}{2c}\Delta k. \tag{1.7}$$

The small angle approximation used for η_F is almost always valid. The expression $\tilde{n}_+ - \tilde{n}_- = \tilde{\varepsilon}_{xy}/\sqrt{\tilde{\varepsilon}_{xx}}$ shows that, if we want to obtain the spectrum of $\tilde{\varepsilon}_{xy}$ from a measurement of $\tilde{n}_+ - \tilde{n}_-$, we also need to measure the spectrum of $\tilde{\epsilon}_{xx}$. Again assuming that $\tilde{\epsilon}_{xx} \gg \tilde{\epsilon}_{xy}$, we expect that $\tilde{\epsilon}_{xx}$ will be independent of the magnetization and write $\tilde{\epsilon}_{xx} = (n + ik)^2$.

Accordingly, the expression for the off-diagonal dielectric constant is given by

$$(\tilde{n}_+ - \tilde{n}_-)(n + ik) = \tilde{\epsilon}_{xy}. \tag{1.8}$$

This may be reexpressed in terms of the measured Faraday rotation and MCD effects

$$\epsilon'_{xy} = \frac{2c}{\omega l_o}(n\theta_F - k\eta_F), \tag{1.9}$$

$$\epsilon''_{xy} = \frac{2c}{\omega l_o}(k\theta_F + n\eta_F). \tag{1.10}$$

Only if the overall absorption is very small, $k \ll n$, are the real and imaginary part of $\tilde{\epsilon}_{xy}$ related simply to the Faraday rotation and MCD, respectively:

$$\epsilon'_{xy} \approx \frac{2c}{\omega l_o}n\theta_F, \tag{1.11}$$

$$\epsilon''_{xy} \approx \frac{2c}{\omega l_o}n\eta_F. \tag{1.12}$$

We now derive the corresponding expressions to relate the Kerr angles to the dielectric constant. In this case, we evaluate the change in amplitude and phase for the left- and right-circularly reflected components.[45] These can be written in terms of the complex refractive indices:

$$\tilde{r}_\pm = \frac{\tilde{n}_\pm - 1}{\tilde{n}_\pm + 1} = r_\pm e^{i\Delta_\pm}. \tag{1.13}$$

We may assume that the rotations, $\Delta_\pm$, are very small and also that $|r_+ - r_-/r_+ + r_-| \ll 1$. In this case, the Kerr rotation, θ_K, equals half the difference between the phase shifts of RCP and LCP light after reflection and can be shown to be given by the following expression:[45]

$$\theta_K = \frac{(\Delta_+ - \Delta_-)}{2} = \mathrm{Im}\frac{\tilde{r}_+ - \tilde{r}_-}{\tilde{r}_+ + \tilde{r}_-} = \mathrm{Im}\frac{\tilde{n}_+ - \tilde{n}_-}{\tilde{n}_+ \tilde{n}_- - 1} = \mathrm{Im}\frac{\tilde{\epsilon}_{xy}}{(\tilde{\epsilon}_{xx} - 1)\sqrt{\tilde{\epsilon}_{xx}}}. \tag{1.14}$$

The Kerr ellipticity is similarly given by:[45]

$$\eta_K = -\frac{r_+ - r_-}{r_+ + r_-} \approx -\mathrm{Re}\,\frac{\tilde{r}_+ - \tilde{r}_-}{\tilde{r}_+ + \tilde{r}_-} = -\mathrm{Re}\,\frac{\tilde{n}_+ - \tilde{n}_-}{\tilde{n}_+ \tilde{n}_- - 1} = -\mathrm{Re}\,\frac{\tilde{\varepsilon}_{xy}}{(\tilde{\varepsilon}_{xx} - 1)\sqrt{\tilde{\varepsilon}_{xx}}}. \quad (1.15)$$

These can be written in terms of n and k, where $\tilde{\varepsilon}_{xx} = (n + ik)^2 = n^2 - k^2 + 2ink$:

$$\varepsilon'_{xy} = -B\theta_K - A\eta_K, \quad (1.16)$$

$$\varepsilon''_{xy} = A\theta_K - B\eta_K, \quad (1.17)$$

with

$$A = n^3 - 3nk^2 - n, \quad (1.18a)$$

$$B = -k^3 + 3n^2 k - k. \quad (1.18b)$$

Reim and Schoenes[45] explained clearly the difference between the Kerr and Faraday effects. We summarize the equations below.

The Faraday rotation and MCD are given by

$$\theta_F = \mathrm{Re}\left(\frac{\omega l_o}{2c}\frac{\tilde{\varepsilon}_{xy}}{\sqrt{\tilde{\varepsilon}_{xx}}}\right), \quad (1.19a)$$

$$\eta_F = \tan\left(\mathrm{Im}\left[\frac{\omega l_o}{2c}\frac{\tilde{\varepsilon}_{xy}}{\sqrt{\tilde{\varepsilon}_{xx}}}\right]\right). \quad (1.19b)$$

The Kerr rotation and Kerr ellipticity are given by

$$\theta_K = \mathrm{Im}\,\frac{\tilde{\varepsilon}_{xy}}{(\tilde{\varepsilon}_{xx} - 1)\sqrt{\tilde{\varepsilon}_{xx}}}, \quad (1.19c)$$

$$\eta_K = -\mathrm{Re}\,\frac{\tilde{\varepsilon}_{xy}}{(\tilde{\varepsilon}_{xx} - 1)\sqrt{\tilde{\varepsilon}_{xx}}}. \quad (1.19d)$$

We note that the real and imaginary parts have been switched between the two effects, and also the diagonal component of the dielectric tensor $\tilde{\varepsilon}_{xx}$ enters the two expressions differently.

The Kerr effect depends on a different absorption between LCP and RCP light. Thus, the Kerr rotation is identically zero if there is no absorption. Whereas in Faraday geometry, the Faraday rotation is finite when there is no absorption, but the MCD vanishes.

We consider briefly the reasons for using either Faraday or Kerr geometry. If the sample is a film that is thinner or comparable with the skin depth, then the Faraday effect should be used. In this case, the light passes through the substrate and care has to be taken over accounting for multiple reflections. As Reim and Schoenes[45] show, the Faraday effect is larger than the Kerr effect and should be used, provided some light will pass through the sample. The Kerr effect is used for highly reflective thin films and for bulk samples.

The appearance of the term $(\tilde{\varepsilon}_{xx} - 1)$ in the denominator of the expression for the Kerr angles in Equation 1.19 is interesting. The real part of this quantity vanishes at the plasma edge and this gives enhanced Kerr angles at that frequency that are unrelated to any peak in $\tilde{\varepsilon}_{xy}$. This effect was first seen by Feil and Haas.[47] Note that if we wish to use the measurement of the MO angles to obtain the real and imaginary parts of $\tilde{\varepsilon}_{xy}$, then it is also necessary to measure the diagonal component.

1.4.2 MAGNETO-OPTIC THEORY IN TERMS OF THE OPTICAL CONDUCTIVITY

In this short section, we comment on the confusion that sometimes arises in the literature when the complex optical conductivity is used instead of the complex dielectric constant. The reason for this is that there is a different definition between the two tensors. The dielectric constant is defined in the way it is because this arises naturally when we consider transitions involving circularly polarized light. However, the conductivity tensor is used to link up with a low frequency when the off-diagonal terms are just the Hall conductivity, and in this case, it is more natural to keep everything real:

$$\tilde{\varepsilon} = \begin{bmatrix} \tilde{\varepsilon}_{xx} & -i\tilde{\varepsilon}_{xy} & 0 \\ i\tilde{\varepsilon}_{xy} & \tilde{\varepsilon}_{xx} & 0 \\ 0 & 0 & \tilde{\varepsilon}_{zz} \end{bmatrix}, \tag{1.20a}$$

and

$$\tilde{\sigma} = \begin{bmatrix} \tilde{\sigma}_{xx} & \tilde{\sigma}_{xy} & 0 \\ -\tilde{\sigma}_{xy} & \tilde{\sigma}_{xx} & 0 \\ 0 & 0 & \tilde{\sigma}_{zz} \end{bmatrix}, \tag{1.20b}$$

The relationships between the diagonal components $\tilde{\varepsilon}_{xx}$ and $\tilde{\sigma}_{xx}$ are obtained in a straightforward manner using Maxwell's equations.[16] These two tensors are not independent but describe the same phenomena. The two equivalent forms of the fourth Maxwell equation are

$$\vec{\nabla} \times \vec{H} = \frac{\partial \vec{D}}{\partial t} = i\omega\varepsilon_0 \tilde{\varepsilon} \, \vec{E},$$
(1.21)

and

$$\vec{\nabla} \times \vec{H} = \tilde{\sigma}\vec{E},$$
(1.22)

implying

$$\tilde{\sigma} = i\varepsilon_0 \omega \tilde{\varepsilon}.$$
(1.23)

This gives a simple relationship between the diagonal elements of the dielectric tensor and optical conductivity tensors. However, care needs to be exercised in the evaluation of the off-diagonal terms because of the extra i in the definition of the dielectric tensor. This means that the conductivity tensor can be written in terms of the dielectric tensor as follows

$$\tilde{\sigma} = \varepsilon_0 \omega \begin{bmatrix} i\tilde{\varepsilon}_{xx} & \tilde{\varepsilon}_{xy} & 0 \\ -\tilde{\varepsilon}_{xy} & i\tilde{\varepsilon}_{xx} & 0 \\ 0 & 0 & i\tilde{\varepsilon}_{zz} \end{bmatrix},$$
(1.24a)

with

$$\tilde{\sigma}_{xx} = i\varepsilon_0 \omega \tilde{\varepsilon}_{xx},$$
(1.24b)

and

$$\tilde{\sigma}_{xy} = \varepsilon_0 \omega \tilde{\varepsilon}_{xy}.$$
(1.24c)

The theory in Section 1.3 can now be worked through using the conductivity tensor instead of the dielectric tensor.

1.5 MEASUREMENT OF THE OFF-DIAGONAL COMPONENTS OF THE DIELECTRIC TENSOR

In this section, we discuss the methods that are used to measure the dielectric tensor. This involves measuring both a MO response and the diagonal component of the dielectric tensor.

1.5.1 MEASUREMENT OF MAGNETO-OPTIC SPECTRA

The methods used to measure Faraday and Kerr angles are similar — there is just a different light path. In both cases, the magnetic field is perpendicular to the plane of the surface. In Faraday rotation and MCD geometry, the light travels through the sample (and the substrate). Sometimes the substrate is made slightly wedge-shaped in order to avoid multiple reflections. In Kerr geometry, the polar Kerr configuration is used, so light is incident on the sample at a small angle, typically about 5°, and the reflected light is analyzed. (One cannot use a half-reflecting plate to separate the incident and reflected beams as this would induce some phase changes.) The Kerr effect can also be measured with the field in the plane of the sample, but the effect is smaller and so this is not done in spectroscopy unless the magnetic field needed to pull the magnetization out of plane is too high.

The techniques most often used for measuring MO Kerr and Faraday effect are (i) the Faraday cell, which has very high sensitivity (0.0001° at 500 nm) but is limited to a small range of wavelengths and is also rather sensitive to stray magnetic fields, and (ii) the method that was pioneered by Sato,[48] and further improved by van Drent,[49] using a photoelastic modulator (PEM). This latter method has the advantage that it works over a wide wavelength range and that the orientation and the ellipticity of the light are measured simultaneously. Its accuracy is not as high for a single pass as for the Faraday cell, so the published spectra are the averages of many scans. In this mode, an accuracy of 0.0001° can also be achieved.

As this is the method that is used almost universally, in Figure 1.9 we give a diagram of the apparatus employed for Kerr geometry; it is the same as in Faraday geometry, except that the analyzer and the detector are behind the sample. Experimentally, one can change from one geometry to the other by using flipper mirrors.

Assuming that light propagating along the $\hat{z}$ direction is polarized along $\hat{x}$ by the initial polarizer and then modulated by a retardation, $\delta = \delta_0 \sin 2\pi pt$ by the PEM. This is expressed in terms of the two circular components (RCP and LCP) $\hat{r} = (\hat{x} + i\hat{y})/\sqrt{2}$ and $\hat{l} = (\hat{x} - i\hat{y})/\sqrt{2}$. After passing through the sample, the amplitude and phase of the two components are changed by factors $\hat{r}_+ = r_+ e^{i\theta+}$ and $\hat{r}_- = r_- e^{i\theta-}$,

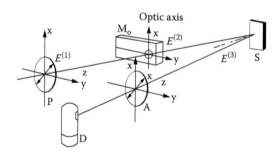

FIGURE 1.9 A schematic diagram of the PEM method for measuring the MO spectrum. P — polarizer, M_o — modulator, S — sample, A — analyzer, D — detector. (From K. Sato, *Jpn. J. Appl. Phys.*, 20, 2403, 1981. With permission.)

respectively. The Kerr rotation and ellipticity are given by $\varphi_K = -(1/2)\Delta\theta$ and $\eta_K = (1/4)(\Delta R / R)$, where $\Delta R = (r_+^2 - r_-^2)$ and $R = (1/2)(r_+^2 + r_-^2)$.

The polarization is measured after the light passes through the analyzer, which is set at an angle ϕ to the $\hat{x}$ axis; this angle is chosen so as to optimize the accuracy.[49] The intensity at the detector is given by

$$I = \frac{\varepsilon E^2}{16\pi}\left\{R + \frac{\Delta R}{2}\sin\delta + R\sin(\Delta\theta + 2\phi)\cos\delta\right\}. \qquad (1.25)$$

This is simplified by using the standard expansions in terms of Bessel functions

$$\sin(\delta_0 \sin 2\pi pt) = 2J_1(\delta_0)\sin 2\pi pt + ... \qquad (1.26a)$$

and

$$\cos(\delta_0 \sin 2\pi pt) = J_0(\delta_0) + 2J_2(\delta_0)\sin 4\pi pt + ... \qquad (1.26b)$$

A lock-in detector allows us to measure the ratio of the intensities at frequencies, 0, p, and $2p$ and hence measure the values of $\Delta\theta$ and ΔR directly:

$$\frac{I_1}{I_0} = A\frac{J_1(\delta_0)\Delta R/R}{1 + J_0(\delta_0)\sin(\Delta\theta + 2\phi)} = AJ_1(\delta_0)\frac{\Delta R}{R}, \quad for\ \phi = 0, \qquad (1.27)$$

$$\frac{I_2}{I_0} = B\frac{2J_2(\delta_0)\sin(\Delta\theta + 2\phi)}{1 + J_0(\delta_0)\sin(\Delta\theta + 2\phi)} = BJ_2(\delta_0)2\Delta\theta, \quad for\ \phi = 0. \qquad (1.28)$$

The choice of $\phi = 0$ is often made, but other choices are possible.[45,46]

1.5.2 MEASUREMENT OF $\tilde{\varepsilon}_{xx}$

The electronic properties of a magnetic material are reflected in the complex functions $\tilde{\varepsilon}_{xx}(\omega)$ and $\tilde{\varepsilon}_{xy}(\omega)$. There are three different ways to measure $\tilde{\varepsilon}_{xx}(\omega)$. These are (i) measurement of $R(\omega)$ and $T(\omega)$ for thin films,[50,51] (ii) measurement of $R(\omega)$ alone over a wide energy range and using Kramers–Kronig relations,[52–54] and (iii) by ellipsometry.[55,56] We compare the three methods to obtain $\tilde{\varepsilon}_{xx}$, n, and k.

Obtaining results for films using $R(\omega)$ and $T(\omega)$ is the simplest experimental method, but calculations do not have high accuracy (method i) because the transmitted light has to cross three separate surfaces.

The use of reflectivity alone (method ii), together with the Kramers–Kronig relation, requires a wide range of frequencies, but when it can be used, it is experimentally simpler than ellipsometry. Careful calculation is needed.

The ellipsometry method (method iii) is more difficult than the other two methods. However, in ellipsometry, because we measure a ratio of two values of the reflection, we have highly accurate and reproducible results (even in low light levels). Furthermore, no reference sample is necessary and also it is not as susceptible to scatter or lamp fluctuations.[55] In the experiments described in Section 1.7, literature values based on method ii are employed to analyze data.

1.6 CLASSICAL AND QUANTUM THEORY OF ABSORPTION

In this section, we review the theory of absorption and show how this is affected by magnetism.

A material is envisaged as an assembly of a large number of polarized atoms, each of which is small (in comparison to the wavelength of light) and close to its neighbors. When a light wave is incident on such a medium, each atom can be thought of as a classical forced oscillator, being driven by a time-varying electric field, $\vec{E}(t)$, which here is assumed to be applied in the $\hat{z}$ direction. The force exerted on an electron of charge q_e by the $\vec{E}(t)$ field of the harmonic wave at frequency, ω, is of the form

$$\vec{F}_E = q_e \vec{E}(t) = \hat{z} q_e E_0 e^{i\omega t}. \tag{1.29}$$

Newton's second law provides the equation of motion:

$$q_e E_0\, e^{i\omega t} = m_e \frac{d^2 z}{dt^2} - m_e \gamma \frac{dz}{dt} + m_e \omega_0^2 z. \tag{1.30}$$

where γ is the damping constant and ω_0 is the natural frequency. We have the solution:

$$z(t) = z_0 e^{i\omega t}, \tag{1.31a}$$

or

$$z(t) = \frac{q_e/m_e}{\omega_0^2 - \omega^2 + i\gamma\omega} E_0 e^{i\omega t}. \tag{1.31b}$$

This is the relative displacement between the negative cloud and the positive nucleus. Without a driving force, the oscillator will vibrate at its resonant frequency, ω_0.

The dipole moment is equal to the product of the charge q_e and its displacement, and if there are N contributing electrons per unit volume, the electric polarization, P (i.e., the density of dipole moments) is

$$P = q_e z N. \tag{1.32}$$

Hence

$$P = \frac{q_e^2 NE / m_e}{\omega_0^2 - \omega^2 + i\gamma\omega}. \tag{1.33}$$

Using the fact that $\tilde{n}^2 = \tilde{\varepsilon}_{xx}$, we can arrive at an expression for $\tilde{n}$ as a function of ω, which is known as a dispersion equation:

$$\tilde{\varepsilon}_{xx}(\omega) = 1 + \frac{P(t)}{\varepsilon_0 E(t)} = 1 + \frac{q_e^2 N / m_e}{\varepsilon_0 (\omega_0^2 - \omega^2 + i\gamma\omega)} = \tilde{n}^2(\omega). \tag{1.34}$$

The quantum mechanical expression is obtained from this classical result by assuming that there are N molecules per unit volume, and that each oscillator has a range of natural frequencies ω_{oj}, each with oscillator strength f_j. In this case:

$$\tilde{\varepsilon}_{xx}(\omega) = 1 + \frac{Nq_e^2}{\varepsilon_0 m_e} \sum_j \left(\frac{f_j}{\omega_{0j}^2 - \omega^2 + i\gamma_j\omega} \right). \tag{1.35}$$

Assuming that the average field experienced by an electron moving about within a conductor is just the applied field, $\vec{E}\ (t)$, we can extend the dispersion equation of a rare medium to read

$$\tilde{\varepsilon}_{xx}(\omega) = 1 + \frac{Nq_e^2}{\varepsilon_0 m_e} \left[\frac{f_e}{-\omega^2 + i\gamma_e\omega} + \sum_j \frac{f_j}{\omega_{0j}^2 - \omega^2 + i\gamma_j\omega} \right]. \tag{1.36}$$

The first bracketed term is the contribution from the free electrons, wherein N is the number of atoms per unit volume. Each of these has f_e conduction electrons, which have no natural frequencies. The second term arises from the bound electrons and is identical to the diagonal element of the dielectric tensor:

$$\tilde{\varepsilon}_{xx}(\omega) = 1 + \frac{Nq_e^2}{\varepsilon_0 m_e} \sum_j \frac{f_j}{\omega_0^2 - \omega^2 + i\gamma_j\omega}. \tag{1.37}$$

We can get a rough idea of the response of metals to light by making a few simplifying assumptions. Accordingly, neglecting the bound electron contribution and assuming that γ_e is also negligible for very large ω, we obtain:

$$n^2(\omega) = 1 - \frac{Nq_e^2}{\varepsilon_0 m_e \omega^2} = 1 - \left(\frac{\omega_p}{\omega} \right)^2, \tag{1.38}$$

where ω_p is the plasma frequency. The plasma frequency serves as a critical value below which the index is complex and the penetrating wave decays exponentially from the boundary. At frequencies above ω_p, n is real, absorption is small, and the conductor is transparent. In the latter circumstance, n is less than 1, as it was for dielectrics at very high frequencies.[16]

For electric dipole transitions, the off-diagonal element is given by a sum over all initial states, i, and final states, j

$$\tilde{\varepsilon}_{xy} = 1 + \frac{Nq_e^2}{2\varepsilon_o m_e}\left[\sum_{ij}\frac{f_{ij+}}{\omega_{ij+}^2 - \omega^2 + i\gamma_{ij}\omega} - \sum_{ij}\frac{f_{ij-}}{\omega_{ij-}^2 - \omega^2 + i\gamma_{ij}\omega}\right] \tag{1.39}$$

where $f_{ij\pm}$ are the oscillator strengths for transitions caused by RCP and LCP light. The theory of the off-diagonal tensor is given clearly by Kahn and Pershan.[46]

There are two reasons why there might be a MO response. First, the two oscillator strengths may be different, $f_{ij+} \neq f_{ij-}$ because of different populations, but $\omega_{ij+} \approx \omega_{ij-}$. (This means that the splitting in the ground state is large compared with the thermal energy so that a population difference occurs, but small compared with the optical linewidths or experimental resolution.) In this case, both the real and imaginary components of $\tilde{\varepsilon}_{xy}$ have the same frequency dependence as a normal oscillator, but the magnitude is reduced because of the difference in populations. This gives a so-called paramagnetic line shape, in which the imaginary part has a peak and the real part shows a dispersive shape as shown in Figure 1.10B. Since a paramagnetic signal depends on population differences, it is temperature dependent.

The other reason why one can get a signal is because the two absorption frequencies are different due to splitting in either the ground state or excited state, or both, so $\omega_{ij+} \neq \omega_{ij-}$, but the oscillator strengths are the same, $f_{ij+} = f_{ij-}$. In this case, the subtraction of the two absorptions gives rise to a dispersive feature in the absorption as shown in Figure 1.10A. Using the Kramers–Kronig analysis, one expects to observe a peak in the real part, and this is confirmed by calculation.[46] We note that in this case, because of the sign in the inverse Kramers–Kronig transform,

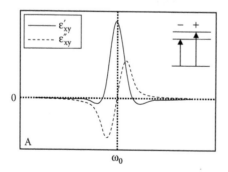
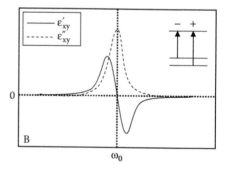

FIGURE 1.10 ε'_{xy} and ε''_{xy} and energy level scheme, (A) diamagnetic and (B) paramagnetic line shapes.

the absorption and dispersion have opposite polarities. The line shape in this case is called diamagnetic.

We note that one cannot have a population difference without having some energy splitting between the states of the ground state manifold, and so this separation is in part idealized. However, the paramagnetic line shape will be observed if the spectral resolution is too broad to pick up the very small splitting that causes the population differences.

There is an important difference between the real and imaginary parts of the dielectric constant. The imaginary part at frequency ω is sensitive to the differential absorption at that frequency; whereas, there may be contributions to the real part from other regions of the spectrum. Hence, MO spectroscopy concentrates on the imaginary part.

If there is a purely diamagnetic effect arising from a splitting of the levels of either the ground or the excited state, then ε''_{xy} is a purely dispersive feature that has equal weight in its positive and negative parts. If the transition is stronger for one polarization than the other, then this usually indicates unequal populations of two states with different orbital quantum numbers. This can be in the ground state in which the transition is stronger than the state with the higher population, or in the excited state, in which case the transition to the state with a higher population has a reduced strength. The temperature dependence of the paramagnetic component gives an indication of the degree of ordering of the magnetic states. It is also important to stress that the MO response occurs from electric dipole transitions that are sensitive to the spin ordering only because of spin-orbit coupling. We now discuss the theory of the MO behavior of the II-VI materials.

1.7 MAGNETO-OPTICAL RESPONSE OF II-VI DILUTE MAGNETIC SEMICONDUCTORS

1.7.1 Cubic Semiconductors

The understanding of the semiconductors with the zinc-blende structure, for example, CdTe and GaAs, is very advanced. The theory of the MO effects are given clearly in Ando's reviews;[57,58] here we summarize the important points. There is a direct gap at the Γ point between the conduction band Γ_6 formed from s electrons and the valence band that arises from the p electrons. The valence band is split at the Γ point by the spin-orbit coupling into the $p_{3/2}$ states, which are in a Γ_8 state and the $p_{1/2}$ states, which are in a Γ_7 state. Away from the Γ point, the Γ_8 states split into the heavy and light hole states. When a magnetic transition metal is doped at concentration x, there is a splitting of the bands as shown in Figure 1.11. The value of $<S_z>$ is induced by the external field. The values of the splitting parameters, $N_0\alpha$ and $N_0\beta$, depend on both the dopant and the semiconductor. The notation used here is used universally.

The α and β values have been measured and calculated. Typically the value of $N_0\alpha$ is of order 0.2 eV in CdTe; it is positive and rather insensitive to the host and dopant. The value of $N_0\beta$ is larger and negative for Mn and Co; $N_0\beta$ is of order -0.88 eV for Mn in CdTe. The large value of $N_0\beta$ indicates that the II-VI semiconductors would

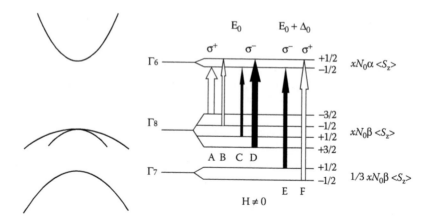

FIGURE 1.11 The Zeeman splittings and the σ^+ and σ^- transitions at the Γ point in $Cd_{1-x}Mn_xTe$. The width of the arrow indicates the strength of the transition. (From K. Ando, *Magneto-optics,* Springer Series in Solid State Science, 128, 211, 2000. With permission.)

be ferromagnetic if they were hole-doped.[17] The theory described here can be used to explain the MO spectrum of $Cd_{1-x}Mn_xTe$; an important feature that is observed is the opposite sign between the MO signal originating from the Γ_8 and the Γ_7 levels.

1.7.2 WURTZITE SEMICONDUCTORS

ZnO grows in the wurtzite structure, which is similar to zinc blende, but with a hexagonal distortion. This produces a crystal field energy that splits and mixes the Γ_8 and the Γ_7 states. The theory was worked out in detail[59] in terms of two parameters: the crystal field splitting Δ_1 and the spin-orbit coupling Δ_2. It is found that there are three doubly degenerate states, *A*, *B*, and *C* with energies and wave functions given below

$$E_A = \Delta_1 + \Delta_2; \quad \left|+\frac{3}{2}\right\rangle = |1,\uparrow\rangle, \quad \left|-\frac{3}{2}\right\rangle = |-1,\downarrow\rangle. \tag{1.40a}$$

$$E_B = \frac{\Delta_1 - \Delta_2}{2} + \sqrt{\left(\frac{\Delta_1 - \Delta_2}{2}\right)^2 + 2\Delta_2^2};$$

$$\left|+\frac{1}{2}\right\rangle = \sqrt{1-b^2}\,|1,\downarrow\rangle + b\,|0\uparrow\rangle, \quad \left|-\frac{1}{2}\right\rangle = -\sqrt{1-b^2}\,|-1,\uparrow\rangle + b\,|0\downarrow\rangle. \tag{1.40b}$$

$$E_C = \frac{\Delta_1 - \Delta_2}{2} - \sqrt{\left(\frac{\Delta_1 - \Delta_2}{2}\right)^2 + 2\Delta_2^2};$$

$$\left|+\frac{1}{2}\right\rangle = b\,|1,\downarrow\rangle - \sqrt{1-b^2}\,|0\uparrow\rangle, \quad \left|-\frac{1}{2}\right\rangle = b\,|-1,\uparrow\rangle + \sqrt{1-b^2}\,|0\downarrow\rangle. \tag{1.40c}$$

The mixing parameter b is given by:

$$b^2 = \frac{1}{2}\left[1 - \frac{\Delta_1 - \Delta_2}{2\sqrt{\left(\frac{\Delta_1 - \Delta_2}{2}\right)^2 + 2\Delta_2^2}}\right]. \tag{1.41}$$

If we set the crystal field term Δ_1 equal to zero, then we recover the results for the zinc-blende structure:

$$E_A = E_B = \Delta_2, \tag{1.42a}$$

$$E_c = -2\Delta_2, \tag{1.42b}$$

and

$$b^2 = 2/3. \tag{1.42c}$$

The parameters have been measured in pure ZnO to be $\Delta_1 = 0.043$ eV and $\Delta_2 = 0.005$ eV.[60,61] These values lead to a rather small value of b: $b^2 = 0.031$.

The theory is extended to include a coupling to a magnetic ion by including a term $B = 1/2 N_0 \beta x <S_z>$ as was done for the zinc-blende materials:

$$E_A^{\pm} = \Delta_1 + \Delta_2 \pm B, \tag{1.43a}$$

$$E_B^{\pm} = \frac{\Delta_1 - \Delta_2}{2} + E_{\mp}, \tag{1.43b}$$

$$E_C^{\pm} = \frac{\Delta_1 - \Delta_2}{2} - E_{\mp}, \tag{1.43c}$$

and

$$E_{\mp} = \sqrt{\left(\frac{\Delta_1 - \Delta_2 \mp 2B}{2}\right)^2 + 2\Delta_2^2}. \tag{1.43d}$$

It is instructive to examine the case where the value of Δ_2 is so small that it can be neglected. The energies and the b values are given below for the case where $\Delta_2 = 0$:

$$E_A^{\pm} = \Delta_1 \pm B, \tag{1.44a}$$

$$E_B^{\pm} = \Delta_1 \mp B, \tag{1.44b}$$

$$E_C^{\pm} = \pm B, \tag{1.44c}$$

$$b^2 = 0. \tag{1.44d}$$

For experimental values, namely $\Delta_1 = 43$ meV and $\Delta_2 = 5$ meV, we find (linear terms in B only):

$$E_A^{\pm} = 48 \pm B \quad \text{meV}, \tag{1.45a}$$

$$E_B^{\pm} = 39 \mp 0.94B \quad \text{meV}, \tag{1.45b}$$

$$E_C^{\pm} = -1.3 \pm 0.94B \quad \text{meV}, \tag{1.45c}$$

$$b^2 = 0.03. \tag{1.45d}$$

The experimentally observed splitting between the A and B bands is 9.5 meV, which is rather small. Thus the crystal-field splitting plays an important role in quenching the orbital angular momentum. We see that if the spin-orbit coupling can be neglected, then the energies of the A and B states are coincident, when the direction of the magnetic field is reversed. This has the effect that the MCD of the A and B states cancel exactly.[60] The mixing parameter b is also important. If $\Delta_1 = 0$, then $b^2 = 2/3$, and this causes the intensity ratio of the σ transitions from the heavy and light holes to be 3:1 and the intensity from Γ_7 state to be 2/3 that of the heavy holes. If $b = 0$, the intensities of the A and B states are equal, and the state C is now pure $m_l = 0$. Therefore, there are no σ polarized transitions between the state C and the conduction band. In the absence of spin-orbit coupling, the spin index is conserved in an electric dipole transition. This is shown in Figure 1.12.

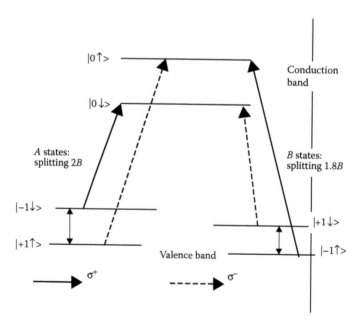

FIGURE 1.12 Schematic demonstration of the relevant transitions for MCD in ZnO using parameters taken from Reynolds et al. (see Reference 61). The center of the A and B levels differ by only 9.5 meV and the Zeeman splittings of the A and B levels are very nearly equal and opposite.

In general, there are two types of transition that may be observed. There are band transitions and transitions to exciton states. In CdMnTe, band transitions are observed from the heavy and light hole band and also from the Γ_7 level as described by Ando.[57,58] In ZnO, we do not expect to see such transitions because of the cancellation described above. ZnO has particularly strongly bound excitons, the binding energy is ~60 meV.[61] This binding energy is sufficiently large that the excitons from the A, B, and C states all lie below the smallest direct band gap. However, it is the lowest exciton from the A state that is observed most strongly.

It is interesting to note that the magnetic response of the ZnO valence band should be very anisotropic because the energies, as given in Equation 1.43, were for a magnetic moment of the localized spin along the hexagonal axis. In an orthogonal direction, the A and B states would mix and the magnetism would be reduced. This effect should be important if the valence band is partially occupied. It is intriguing to note that films have been reported to show a strong anisotropy[33] and that, in some cases, it is known that the easy direction is along the hexagonal axis,[62] and yet the films are n-type, so that the valence bands should be fully occupied.

1.8 MAGNETO-OPTIC DATA ON DOPED ZnO

In this section, we describe the measurements of the coupling parameters α and β and also present MO studies on a range of doped ZnO samples at different temperatures.

1.8.1 MEASUREMENTS OF α AND β

We remind ourselves that the spin splitting of the valence band is given by $N_0\beta x <S_z>$ and that of the conduction band by $N_0\alpha x <S_z>$. The MO spectrum depends on combinations of α and β and so at least two excitons must be measured to obtain both coefficients. Typically $\beta > \alpha$ and it is the β term that can drive a magnetic transition in a hole-doped material. ZnO is magnetic when it is electron-doped, and so the term that is responsible for the magnetism must come either from the donor levels or the conduction band.

So far there have been two estimates of the quantity $N_0(\beta - \alpha)$ for doped ZnO. Ferrand et al.[63] measured the reflectance of a very low doped sample (0.4%) of Co using circularly polarized light in a magnetic field at low temperatures. At this doping they could assume that the Co ions were isolated from each other. They were able to see excitons on both the A and B levels. In circularly polarized light, they measured only one spin component for each exciton and saw that the two g values were equal and opposite as expected from theory. They found the zero-field splitting of the excitons to be 7 meV, in good agreement with the value of 9 meV found earlier.[61] They reported a value of $N_0|\beta - \alpha| = 0.4$ eV. This is actually smaller than for other semiconductors.

The other measurement was a little less direct. Schwartz et al.[64] measured the change in the band gap that occurs as a function of a magnetic field at low temperatures. They assumed a typical value for $N_0\alpha = 0.2$ eV and used the observed band gap shift to estimate that $N_0\beta$ is −2.3 eV for 1.7% Co doping and −4.5 eV for 1.5% Ni doping. These are certainly large values and comparable with what has been observed in other semiconductors.

1.8.2 Magnetic Circular Dichroism in Ferromagnetic Samples

Following on from the seminal work of Ando et al.,[57] there has been much effort to see what effects the dopant transition metals have on the MCD at the band edge. This is important because there is plenty of evidence of SQUID loops showing room temperature ferromagnetism. The question is whether this is due to a minority impurity phase[63] or due to the transition metal ions that are incorporated into the ZnO lattice and that the magnetism is indeed due to the exchange interaction mediated by ZnO conduction electrons. An additional question that is pertinent to the use of these materials in devices is the extent to which the mobile electrons are polarized. As we shall see, MCD can give useful information on both of these questions.

An important energy is the ZnO band edge that occurs near 3.4 eV at 4 K but varies with both temperature and doping. Oxygen vacancies give rise to a deep level at 2.3 eV and a Zn interstitial acts as a shallow donor at ~10 meV.[65] The A-B splitting is only about 9 meV and the Γ_7 level is about 49 meV below the A level. The Co and Mn ions have characteristic absorption spectra that are observable below the band edge. Theory suggests that conduction band effects are too weak to cause the observed ferromagnetism[33] and the exchange should be caused by electrons in an impurity band. The oscillator strength of the impurity band will be low (although it should be strongly spin-polarized) but it has been observed weakly in MCD.[65]

Most of the films are grown on sapphire substrates. The films are usually about 200 nm thick, so that some light will pass through, even above the ZnO band gap. This means that the ratio of the thickness of the substrate to the film is enormous. Sapphire is chosen because it is transparent in the region of the band gap at ~3.4 eV, but some substrates do have a MCD that is considerably larger than the film — this can actually be modeled as the tail of the absorption that occurs near 7 eV. The MCD of a blank substrate is shown in Figure 1.13. It is very weakly temperature dependent and varies linearly with the external field. In the plots below, the substrate contribution has been subtracted off. If the MCD from the substrate is not subtracted correctly, then a spurious signal is added to the data that depends linearly on the applied field.

MCD measurements have been made on Co-, Mn-, V-, and Ti-doped samples. All the Sheffield data was measured in a field of 0.5 T. In a pure ZnO crystal, the A exciton can be seen up to room temperature; however, this is not the case in doped crystals. In particular, increasing the number of interstitial Zn or oxygen vacancies quenches the exciton, even at low temperatures. We observe that the MCD response near the ZnO band edge is rather similar in all of these materials, pointing to a single mechanism to explain the magnetism. The common features are that there is often a negative excitonic response at low temperatures and then, as this is quenched, a high temperature positive hump appears a little above the band edge. This is observed in all doped samples, but not in pure ZnO. Theory tells us that the α value is similar, and positive, for all transition metals. We suggest that the positive peak arises from the differential occupation of the spin components of the conduction band.

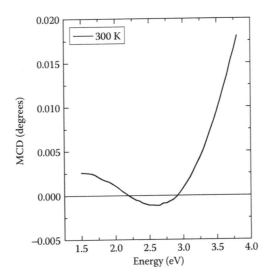

FIGURE 1.13 The MCD of a sapphire C-cut substrate in a field of 0.5 T. (Data from J.R. Neal et al., *Phys. Rev. Lett.*, 96, 197208, 2006.)

1.8.2.1 ZnO:Co

This is the material that has been studied by the most groups. The first experiments were by Ando et al.[58,60] At low temperatures there are two main features in this spectrum. There is a characteristic feature near 2 eV due to the transition $^4A_2 \rightarrow {}^4T_1$ from all magnetic Co^{2+} ions and also the exciton line near to the band edge that was first observed by Ando et al.[60] The negative sign of the MCD at the exciton line is expected from the large and negative value of $N_0\beta$ that is predicted by theory and observed in other semiconductors containing Co. The peak reduces as the temperature is raised. In Ando et al.'s data at 1.6% Co, the exciton peak persists up to 200 K; in our samples that are more heavily doped, it disappears after about 100 K. More recently, a hysteresis loop has been obtained at 10 K.[66] These authors also used the measured shift of the band gap to confirm ferromagnetism.[67] We show in Figure 1.14 some of our data taken at 10 K and 290 K. SQUID data showed that this film has a saturated moment of ~$0.5\mu_B$ per cobalt, which indicates that some 15% of the Co ions are participating in the ferromagnetic state. The peak due to the *d-d* transition at 2 eV peak is sensitive to *all* the Co ions. The variation of the intensity of this peak is consistent with a model in which essentially all the Co ions contribute at 10 K and the intensity of the paramagnetic fraction falls as $1/T$, while the ferromagnetic fraction remains at room temperature. The decrease in the exciton peak occurs rapidly as the temperature is raised as the exciton becomes quenched. The positive ferromagnetic peak discussed above becomes apparent at high temperature and is only weakly temperature dependent.

We see that, as the negative peak at the band edge loses intensity, we have a large positive peak. This is much more intense than the corresponding peak observed by Ando et al.[60] It is weakly dependent on temperature. We note that it has a paramagnetic

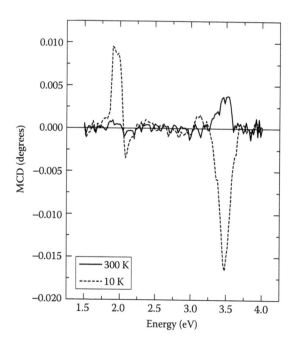

FIGURE 1.14 MCD spectra for 2% Co in ZnO grown by PLD in 10 mtorr of O_2 on C-cut sapphire. (Data from J.R. Neal et al., *Phys. Rev. Lett.*, 96, 197208, 2006.)

line shape that can only arise from an imbalance of population. As this material is *n*-type, we can assume that the only place where unequal populations can occur that would give a MCD at this energy is in the conduction band. We therefore make a tentative identification of this feature with the polarization of the carriers in the conduction band. It is weak because most of the states in the conduction band are empty, and hence, the effect of a spin polarization on the occupied states is a weak effect. In addition, the conduction band is mostly *s*-like, and hence, the spin-orbit coupling acts only through the admixtures of other states. The fact that its intensity increases with temperature between 100 K and 300 K is due to the increased number of carriers. (It also becomes more noticeable as the exciton peak, which is negative, weakens as the temperature is raised). Similar effects were observed by Kittilstved et al.[65] who also showed a dramatic change in both the optical and magnetic properties after exposure of the films to Zn vapor, indicating the importance of Zn interstitials. There is some indication of structure just below the band edge at ~3 eV, which may correspond to transitions to the impurity band.

1.8.2.2 ZnO:Mn

Figure 1.15 shows the MCD of $Zn_{0.98}Mn_{0.02}O$ at 10 K and 300 K. At 10 K, there are characteristic Mn features at ~3 eV from *d-d* transitions on the Mn.[68] It is also possible that there are minority phases of Mn_2O_3 or Mn_3O_4,[69] which have band gaps around where we observe absorption. It is interesting that these low energy peaks disappear rapidly, so they cannot be seen at 100 K, which is consistent with the known ordering temperatures of oxides of Mn.

Ando et al.[60] showed that there were MO features at the band edge. We have found that the spectra for Mn in this region is rather similar to that for Co. There is a sharp negative peak at low temperatures that gives way to a positive peak whose height is relatively independent of temperature between 100 K and 300 K. In this case, there is also an indication of absorption to an impurity band at ~3.1 eV at low temperatures. As shown in Figure 1.15, our low and room temperature data taken at the same value of the magnetic field demonstrate that the MCD changes shape as a function of temperature, and that the overall variation does not scale with $1/T$.

1.8.2.3 ZnO:Ni

This has been shown to have a small MO effect at the band edge and the shift in the band edge also indicates a strong *sp-d* exchange.[64]

1.8.2.4 ZnO:V and ZnO:Ti

There have been earlier reports that no MO effects exist in V- or Ti-doped samples,[60] and although a signal was obtained, it did not show a hysteresis loop at low temperature.[66]

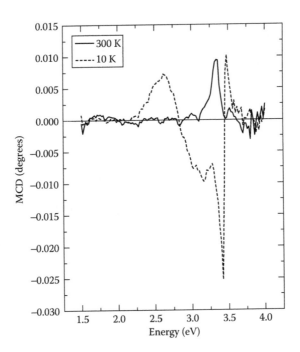

FIGURE 1.15 MCD spectra for 2% Mn in ZnO grown by PLD in 10 mtorr of O_2 on C-cut sapphire. (Data from J.R. Neal et al., *Phys. Rev. Lett.*, 96, 197208, 2006.)

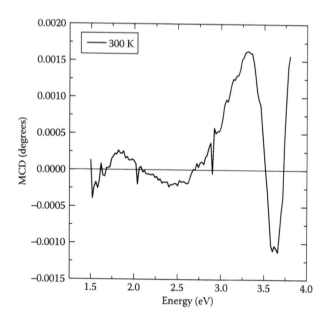

FIGURE 1.16 MCD for films of ZnO with 5% V_2O_5 grown by PLD in 10 mtorr of O_2 on C-cut sapphire. (From J.R. Neal et al., unpublished.)

However, in our sample, we do observe a strong MO signal at the band edge even at room temperature for both V and Ti as shown in Figure 1.16 and Figure 1.17. Both of these dopants are showing evidence of the ferromagnetism in the MCD at the band edge for ZnO.

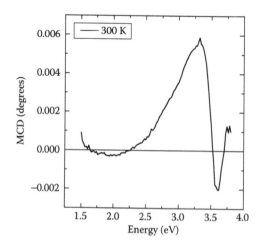

FIGURE 1.17 Room temperature MCD spectra for 2% Ti in ZnO grown in 10 mtorr of O_2 on C-cut sapphire. (From J.R. Neal et al., unpublished.)

The value of the *sp-d* mixing has been calculated for both V and Ti.[70] Ti is expected to have ferromagnetic coupling, but the level splitting is more complex for V. Again the temperature dependence of the MCD is incompatible with a paramagnetic line shape.[71]

1.9 CONCLUSIONS

In this chapter, we have reviewed the spectroscopic experiments made on doped ZnO materials. There are still papers that suggest that the magnetism is not intrinsic to ZnO. However, there is an increasing body of work showing that there is a convincing pattern of behavior for different dopants. We explained that magneto-optic measurements are a powerful technique to measure the intrinsic magnetic properties of doped ZnO, because we can choose to study the effects that occur at the ZnO band edge that are characteristic of bulk ZnO.

We have reviewed the theory of MO measurements and also given a discussion of the band energies of ZnO and explained why the expected MCD at the band edge should not be observed in non-magnetic ZnO.

The existence of a high temperature positive peak at the band edge in the MCD is good evidence for the underlying ferromagnetism. The magnitude of the hump is expected to depend on the underlying magnetism of the magnetic ions, the number of conduction electrons and the polarization of the conduction electrons. It is weakly temperature dependent, indicating that it depends on the ferromagnetic moments of the doped ions. Because the strength of this peak is predicted to depend on the doping level, it is expected to vary between samples. The interesting aspect of this peak is that it is clear evidence for the polarization of mobile carriers. This will be very important if these materials are to have applications in electronic devices.

ACKNOWLEDGMENTS

We would like thank Professor M. Kucera for much help in the initial stages of setting up our laboratory and the Engineering and Physical Sciences Research Council and the Royal Society for financial support.

REFERENCES

1. S.A. Chambers, *Mater. Today*, 2, 34, 2002.
2. S. Das Sarma, J. Fabian, X. Hu, and Z. Zutic, *IEEE Trans. Magn.*, 36, 2821, 2000.
3. H. Ohno, D. Chibu, F. Matsukura, T. Omiya, E. Abe, T. Dietl, Y. Ohno, and K. Ohtani, *Nature (London)*, 408, 944, 2000.
4. K. Sato and H. Katayama-Yoshida, *Jpn. J. Appl. Phys., Part 1*, 40, 485, 2001.
5. D.D. Awschalom and J.M. Kikkawa, *Phys. Today*, 52, 33, 1999.
6. S.A. Wolf, D.D. Awschalom, R.A. Buhrman, J.M. Daughton, S. von Molnar, M.L. Roukes, A.Y. Chtchelkanova, and D.M. Treger, *Science*, 294, 1488, 2001.
7. H. Munekata, H. Ohno, S. von Molnar, Armin Segmüller, L.L. Chang, and L. Esaki, *Phys. Rev. Lett.*, 63, 1849, 1989.

8. J.K. Furdyna, *J. Appl. Phys.*, 64, R29, 1998.

9. G. Prinz, *J. Magn. Magn. Mater.*, 200, 57, 1999.

10. S. Das Sarma, J. Fabian, X. Hu, and I. Zutic, *Solid State Commun.*, 119, 207, 2001.

11. J. De Boeck, W. Van Roy, J. Das, V. Motsnyi, Z. Liu, L. Lagae, H. Boeve, K. Dessein, and G. Borghs, *Semicond. Sci. Technol.*, 17, 342, 2002.

12. S.J. Pearton, W.H. Heo, M. Ivill, D.P. Norton, and T. Steiner, *Semicond. Sci. Technol.*, 19, R59, 2004.

13. P. Sharma, K. Sreenivas, and K.V. Rao, *J. Appl. Phys.*, 93, 3963, 2003.

14. D.C. Look, *Mater. Sci. Eng. B.*, 80,383, 2001.

15. E. Derenzoa, M.J. Webera, and M.K. Klintenberg, *Nucl. Instrum. Methods Phys. Res. A*, 486, 214, 2002.

16. N.W. Ashcroft and N.D. Mermin, *Solid State Physics*, Holt, Reinhart, and Winston, Orlando, FL, 1976.

17. T. Dietl, H. Ohno, F. Matsukura, J. Cibert, and D. Ferrand, *Science*, 287, 1019, 2000.

18. S.J. Pearton, W.H. Heo, M. Ivill, D.P. Norton, and T. Steiner, *Semicond. Sci. Technol.*, 19, R59, 2004.

19. W. Prellier, A. Fouchet, Ch. Simon, and B. Mercey, *Mat. Sci. and Eng. B.*, 109, 192, 2004.

20. T. Andrearczyk, J. Jaroszynski, G. Grabecki, T. Dietl, T. Fukumura, and M. Kawasaki, cond-mat/0502574, unpublished.

21. H.J. Blythe, R.M. Ibrahim, G.A. Gehring, J.R. Neal, and A.M. Fox, *J. Magn. Magn. Mater.*, 283, 117, 2004.

22. J.L. Costa-Krämer, F. Briones, J.F. Fernández, A.C. Caballero, M. Villegas, M. Diaz, M.A. García, and A. Hernando, *Nanotechnology*, 16, 214, 2005.

23. P. Sharma, A. Gupta, K.V. Rao, F.J. Owens, R. Sharma, R. Ahuja, J.M.O. Guillen, B. Johansson, and G.A. Gehring, *Nat. Mater.*, 2, 673, 2003.

24. G. Lawes, A.S. Risbud, A.P. Ramirez, and R. Seshadri, *Phys. Rev. B*, 71, 045201, 2005.

25. N.A. Spaldin, *Phys. Rev. B*, 69, 125201, 2004.

26. K.J. Kim and Y.R. Park, *Appl. Phys. Lett.*, 81, 1420, 2002.

27. K.R. Kittilstved and D.R. Gamelin, *J. Am. Chem. Soc.*, 127, 5292, 2005.

28. J.H. Kim, H. Kim, D. Kim, Y.E. Ihm, and W.K. Choo, *J. Appl. Phys.*, 92, 6066, 2002.

29. W. Prellier, A. Fouchet, B. Mercey, Ch. Simon, and B. Raveau, *Appl. Phys. Lett.*, 82, 3490, 2003.

30. J.H. Kim, J.B. Lee, H. Kim, D. Kim, Y. Ihm, and W.K. Choo, *IEEE Trans. Magn.*, 38(5), 2880, 2002.

31. H. Saeki, H. Matsui, T. Kawai, and H. Tabata, *J. Phys. Condens. Matter*, 16, S5533, 2004.

32. J.H. Kim, H. Kim, D. Kim, Y. Ihm, and W.K. Choo, *J. Eur. Ceram. Soc.*, 24, 1847, 2004.

33. M. Venkatesan, C.B. Fitzgerald, J.G. Lunney, and J.M.D. Coey, *Phys. Rev. Lett.*, 93, 177206, 2004.

34. K. Ueda, H. Tabata, and T. Kawai, *Appl. Phys. Lett.*, 79,988, 2001.

35. H. Saeki, H. Tabata, and T. Kawai, *Solid State Commun.*, 120, 439, 2001.

36. N.H. Hong, J. Sakai, and A. Hassini, *J. Phys. Condens. Matter*, 17, 199, 2005.

37. S.W. Jung, S.J. An, G.C. Yi, C.U. Jung, S.I. Lee, and S. Cho, *Appl. Phys. Lett.*, 80, 4561, 2002.

38. T. Fukumura, Z. Jin, M. Kawasaki, T. Shono, T. Hasegawa, S. Koshihara, and H. Koinuma, *Appl. Phys. Lett.*, 78, 958, 2001.

39. J.H. Kim, J.B. Lee, H. Kim, W.K. Choo, Y. Ihm, and D. Kim, *Ferroelectrics*, 273, 71, 2002.

40. Y.Q. Chang, D.B. Wang, X.H. Luo, X.Y. Xu, X.H. Chen, L. Li, C.P. Chen, R.M. Wang, J. Xu, and D.P. Yu, *Appl. Phys. Lett.*, 83, 4020, 2003.

41. Y.M. Kim, M. Yoon, I.W. Park, Y.J. Park, and J.H. Lyou, *Solid State Commun.*, 129,175, 2004.
42. K.W. Nielsen, J.B. Philipp, M. Opel, A. Erb, J. Simon, L. Alff, and R. Gross, *Superlattices and Microstructures*, 37, 327, 2005.
43. D.P. Norton, M.E. Overberg, S.J. Pearton, K. Pruessner, J.D. Budai, L.A. Boatner, M.F. Chisholm, J.S. Lee, Z.G. Khim, Y.D. Park, and R.G. Wilson, *Appl. Phys. Lett.*, 83, 5488, 2003.
44. T. Wakano, N. Fujimura, Y. Morinaga, N. Abe, A. Ashida, and T. Ito, *Physica E*, 10, 260, 2001.
45. W. Reims and J. Schoenes, in *Handbook of Ferromagnetic Materials*, E.P. Wohlfarth and K.H.J. Buschow, Eds., North Holland, Amsterdam, 1990.
46. F.J. Kahn and P.S. Pershan, *Phys. Rev.*, 186, 89, 1969.
47. H. Feil and C. Hass, *Phys. Rev. Lett.*, 58, 65, 1987.
48. K. Sato, *Jpn. J. Appl. Phys.*, 20, 2403, 1981.
49. W.P. Van Drent and Suzuki, *J. Mag. Magn. Mater.*, 175, 53, 1997.
50. G.Q. Di and S. Uchiyama, *Phys. Rev. B.*, 53, 3327, 1996.
51. Y. Okimoto, Y. Konishi, M. Izumi, T. Manako, M. Kawasaki, and Y. Tokura, *J. Phys. Soc. Jpn*, 71, 613, 2002.
52. L. Degiorgi, I. Blatter-Morke, and P. Wachter, *Phys. Rev. B.*, 35, 5421, 1987.
53. S.K. Park, T. Ishikawa, and Y. Tokura, *Phys. Rev. B.*, 58, 3717, 1998.
54. E. Saitoh, A. Asamitsu, Y. Okimoto, and Y. Tokura, *J. Phys. Soc. Jpn.*, 69, 3614, 2000.
55. D.E. Aspnes and A.A. Studna, *Phys. Rev. B.*, 27, 985, 1983.
56. W.F.J. Fontijn, P.J. van der Zaag, M.A.C. Devillers, and V.A.M. Brabers, *Phys. Rev. B.*, 56, 5432, 1997.
57. K. Ando, *Magneto-optics*, Springer Series in Solid State Science, 128, 211, 2000.
58. K. Ando, H. Saito, V. Zayets, and M.C. Debnath, *J. Phys. Condens. Matter*, 16, S5541, 2004.
59. M. Arciszewska and M. Nawrocki, *J. Phys. Chem. Sol.*, 47, 309, 1986.
60. K. Ando, H. Saito, Z. Jin, T. Fukumura, M. Kawasaki, Y. Matsumoto, and H. Koinuma, *J. Appl. Phys.*, 89, 7284, 2001.
61. D.C. Reynolds, D.C. Look, B. Jogai, C.W. Litton, G. Cantwell, and W.C. Harsch, *Phys. Rev. B*, 60, 2340, 1999.
62. A. Dinia, G. Schmerber, C. Meny, V. Pierron-Bohnes, and E. Beaurepaire, *J. Appl. Phys.*, 97, 123908, 2005.
63. D. Ferrand, S. Marcet, W. Pacuski, E. Gheereart, P. Kossacki, J.A. Gaj, J. Cibert, C. Deparis, H. Mariette, and C. Morhain, *J. Super.*, 18, 15, 2005.
64. D.A. Schwartz, N.S. Norberg, Q.P. Nguyen, J.M. Parker, and D.R. Gamelin, *J. Am. Chem. Soc.*, 125, 13205, 2003.
65. K.V. Kittilstved, J.Z. Zhao, W.K. Lui, J.D. Bryan, and D.R. Gamelin, cond-mat/050712, unpublished.
66. H. Saeki, H. Matsui, T. Kawai, and H. Tabata, *J. Phys. Condens. Matter*, 16, S5533, 2004.
67. S. Alexander, *Phys. Rev. B.*, 13, 304, 1975.
68. Z.W. Jin et al., *Appl. Phys. Lett.*, 83, 39, 2003.
69. C.D. Kundalia, S.B. Ogale, S.E. Lofland, S. Dhar, C.J. Metting, S.R. Shinde, Z. Ma, B. Varughese, K.V. Ramanujachary, L. Salamanca-Riba, and T. Venkatesan. *Nat. Mater.*, 3, 709, 2004.
70. J. Blinowski and P. Kacman, *Phys. Rev. B.*, 46, 12298, 1992.
71. J.R. Neal, A.M. Fox, A. Behan, H.J. Blythe, R.M. Ibrahim, and G.A. Gehring, *Phys. Rev. Lett.*, 96, 197208, 2006.

2 Synthesis and Characterization of Wide Band-Gap Semiconductor Spintronic Materials

Rong Zhang and Xiangqian Xiu

CONTENTS

Recently, semiconductor-based spintronic materials have obtained much attention because they make it possible to develop novel spintronic devices, such as novel transistors, lasers, integrated magnetic sensors; ultralow power, high-speed memories; and logic and photonic devices by utilizing the spin of charge carriers. The utility of these devices depends on the availability of materials with practical magnetic ordering temperatures. Dietl's work predicted that the Curie temperature will be strongly related to the band gap of semiconductors, and wide band-gap semiconductors, such as GaN and ZnO, may be suitable candidates to realize carrier-induced ferromagnetism at room temperature or higher. A number of research studies have been conducted to exploit high-temperature ferromagnetism in wide band-gap

semiconductor materials. In this chapter, we will review the growth and characterization of wide band-gap dilute magnetic semiconductors, such as GaN and ZnO,
and discuss the origins of the magnetism in these materials.

2.1 INTRODUCTION

In conventional electronics, the charge of an electron is used to achieve functionalities, such as is found in diodes, transistors, detectors, lasers and so on. But the aim
of spintronics is to manipulate electron spin (or the resulting magnetism) to achieve
new or improved functionalities, such as spin controlled valves, switches, modula-
tors, transistors, memories, tunable detectors, lasers, bits (Q-bits) for quantum computing, other novel devices and so forth. Spintronic devices have the advantages of
higher speed and lower power than electronic devices because they are based on the
spin direction and spin coupling.

Electron spin (magnetism) has nonetheless always been important for information storage. For instance, computer hard drives use magnetoresistance — a change
in electrical resistance caused by a magnetic field — to read data stored in magnetic
domains. Although metallic spin devices are now better understood, progress in spin
injection from metals has slowed down. Recently, there have been reports of successful and efficient spin injection from a metal to a semiconductor, even at room
temperature, by ballistic transport,[1] but the realization of functional spintronic
devices requires materials with ferromagnetic ordering at operational temperatures
compatible with existing semiconductor materials.

Semiconductor-based spintronics may offer a greater wealth of possibilities.
Semiconductor-based spintronics could combine storage, detection, logical, and
communication capabilities on a single chip to produce a multifunctional device that
could replace several components. The optical properties of the semiconductors are
also of particular interest to transform magnetic information into an optical signal
(magnetic-opto-electronic devices). These ideas have boosted research on diluted
magnetic semiconductors (DMSs)[2] that can be combined more easily with nonmagnetic semiconductors for spin injection.

Therefore, DMSs — also known as magnetic alloy semiconductors (MASs) or
semimagnetic semiconductors — have gained the attention of the scientific and
industrial communities. DMSs are semiconductors with a portion of their constituent
ions replaced by transition metal ions, such as Mn and Fe. In the presence of an
external magnetic field, these materials behave differently from nonmagnetic semiconductors. Their characteristics make it possible to realize the charge degree of
freedom of electrons to process information and the spin degree of freedom to store
information in the same kind of electronic or optoelectronic materials,[3] which is
very attractive for modern information technology. Another important potential of
DMSs is obtaining high spin injection efficiencies.

DMSs may be considered as containing two interacting subsystems. One is the
system of delocalized conduction and valence band electrons. The other is the
random, diluted system of localized magnetic moments associated with the magnetic
atoms. A good knowledge of both the structure and the electronic properties of the
host crystals is required for studying the basic mechanisms of the magnetic interactions

coupling the spins of the band carriers and the localized spins of magnetic ions. The coupling between the localized moments results in the existence of different magnetic phases (such as ferromagnets, paramagnets, spin glasses, and antiferromagnets).

2.2 SEMICONDUCTOR SPINTRONIC MATERIALS

The most common DMS are II-VI compounds (e.g., CdTe, ZnSe, CdSe, CdS, etc.) with transition metal ions (e.g., Mn, Fe, or Co) substituting their original cations. There are also materials based on IV-VI (e.g., PbTe, SnTe) and recently III-V (e.g., GaAs, InSb) crystals (Figure 2.1). In addition, rare earth elements (e.g., Eu, Gd, Er) can also be used as magnetic atoms in DMS.

After the realization of (In,Mn)As[4] in 1989 and (Ga,Mn)As[5,6] in 1996, studies on the III-V DMSs have become one of the fascinating subjects in the research fields of semiconductors and magnetic materials. However, the highest Curie temperatures reported for these materials ~110 K for (Ga,Mn)As[7] and ~35 K for (In,Mn)As[8] are too low for most practical applications.

There are at least two important reasons to study the preparation of wide band-gap DMSs. First, a theoretical work has shown that the wide band-gap semiconductors, GaN and ZnO, may be suitable candidates to realize carrier-induced ferromagnetism at room temperature or higher.[2] There have been numerous recent reviews on developments in transition metal-doped GaN and other wide band-gap semiconductors.[9–12] Many materials for which room temperature ferromagnetism has been reported include (Cd,Mn)GeP$_2$,[13] (Zn,Mn)GeP$_2$,[14,15] ZnSnAs$_2$,[16] (Zn,Co)O,[17] and (Co,Ti)O$_2$,[18,19] and so forth. Second, the development of optoelectronic systems in the visible wavelength region may require magneto-optical components made of wide band-gap semiconductors.

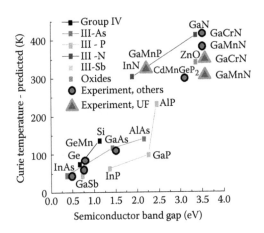

FIGURE 2.1 The most common DMSs. (From S.J. Pearton, C.R. Abernathy, D.P. Norton, A.F. Hebard, Y.D. Park, L.A. Boatner, and J.D. Budai, *Mater. Sci. Eng. R.*, 40, 137, 2003. With permission.)

2.3 SYNTHESIS OF WIDE BAND-GAP SEMICONDUCTOR SPINTRONIC MATERIALS

Since Dietl predicted that wide band-gap semiconductors may exhibit ferromagnetism at room temperature or higher, more research work has been done to seek a breakthrough. A number of different methods have been successfully used for the synthesis of spintronic semiconductor materials, such as ion implantation, metallo-organic chemical vapor deposition (MOCVD), molecular beam epitaxy (MBE), sol-gel, pulsed laser deposition (PLD), hydride vapor phase epitaxy (HVPE), and so on. Clear signatures of room temperature ferromagnetism have been observed.

2.3.1 GaN and GaP

As one of the candidate materials for spintronic semiconductors, GaN has been studied widely by doping with different magnetic ions (Mn, Fe, Cr, V) and using different growth methods. Microcrystalline $Ga_xMn_{1-x}N$ samples grown by the ammonothermal method showed a paramagnetic behavior reported by Zajac et al.[20,21] The Curie temperature in most GaN-based DMSs was found to be in the range of 220 to 400 K,[22,23] and an estimated T_c of 940 K has been reported.[24]

Until now, it has been proposed that the origin of the ferromagnetism may be related to (Ga,Mn)N solid solution or secondary phases (Ga–Mn). Most theories describing DMS ferromagnetism think that the exchange interaction between nearest neighbor transition metal ions (Mn^{2+}) is mediated by the carriers and gives rise to ferromagnetism.[2] Experimentally, the formation of ferromagnetic impurity phases or clustering of the transition metal implant atoms can produce ferromagnetism, but this magnetism need not necessarily be carrier mediated. Ferromagnetic impurity phases, such as MnGa (> 300 K), Mn_4N (745 K), and MnP (291 K),[10] have Curie temperatures equal to or greater than the room temperature. So the impurity phases or clusters may play an important role in most of the III-V wide band-gap semiconductor spintronic materials, such as those hosted by AlN,[25] GaN, GaP,[25] and so on. However, nitrogen-stabilized nanoclusters of Mn atoms are predicted to have large enhanced ferromagnetic moments in (Ga,Mn)N,[26] while some Mn nitrides such as $Mn_6N_{2.58}$ and Mn_3N_2 are antiferromagnetic.

2.3.1.1 MBE

The solubility limit of magnetic elements in III-V semiconductors is very low, but to have ferromagnetism in DMS, a sizable number of magnetic ions are needed. This can only be accomplished by means of nonequilibrium crystal growth techniques, such as low-temperature MBE.

In the initial work, (Ga,Mn)N films grown by MBE with Mn contents of 6 to 9 at.% exhibited magnetization (M) vs. magnetic field (H) curves with clear hysteresis and a coercivity of 52 to 85 Oe at 300 K.[24] Room temperature ferromagnetism in single-phase n-type (Ga,Mn)N grown by MBE has also been reported; the upper concentration limit of magnetic ions is around 10% in MBE GaN.[27] The Curie temperature was found to be in the range of 220 to 370 K, depending on the diffusion conditions.

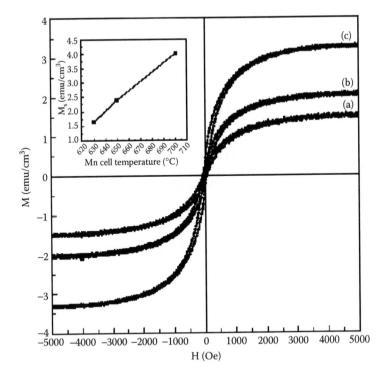

FIGURE 2.2 Hysteresis loops for the Ga(Mg, Mn)N films as a function of Mn effusion temperature (T_{Mn} = (a) 630, (b) 650, and (c) 700°C). The inset clearly shows the linear relation between M_s and Mn effusion cell temperature.

Mg-Mn codoped GaN films grown by plasma-enhanced molecular beam epitaxy (PEMBE) have been reported[28] and room temperature ferromagnetic ordering observed (Figure 2.2). Mg incorporation is not related to ferromagnetic ordering in the films, because the saturation magnetization change was not observed with the variation of Mg effusion cell temperature. Codoping with Mg reduces the Mn incorporation,[29] but increases the conductivity of the GaMnN films. At the same time, the saturation magnetization and coercivity increase. The crucial role of the carriers in carrier-induced ferromagnetism in magnetic semiconductors is observed while the much reduced Mn concentration reached ~0.3%.

In general, no second phases are found for Mn levels below ~10% for growth temperatures of ~750°C in MBE-grown GaMnN within detectable limits. Extended x-ray absorption fine structure (EXAFS) measurements performed on (Ga,Mn)N samples grown by MBE on sapphire at temperatures of 400 to 650°C with Mn concentrations of ~7 × 10^{20} cm^{-3} are shown in Figure 2.3.[30] Figure 2.3 shows that Mn is in fact soluble at these densities. There was also evidence that a fraction of the total Mn concentration could be present as small Mn clusters.[31]

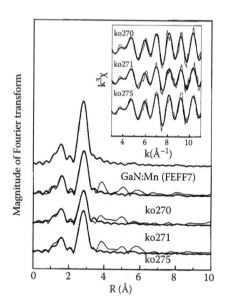

FIGURE 2.3 Fourier transforms of the Mn k-edge EXAFS data from Mn-doped (GaMn)N. The fine lines are the experimental data in these curves, and the coarse lines are the calculated curves assuming 2 at.% of Mn in the GaN. (Adapted from Y.L. Soo, G. Kioseouglou, S. Kim, S. Huang, Y.H. Kaa, S. Kubarawa, S. Owa, T. Kondo, and H. Munekata, *Appl. Phys. Lett.*, 79, 3926, 2001. With permission.)

2.3.1.2 Ion Implantation

Ion implantation is used extensively for integrated circuits in silicon technology because of its reliability, accuracy of ion dose, and reproducibility. In the beginning of research into semiconductor spintronic materials, ion implantation has been paid much attention because it is a useful technique to combine the specific magnetic dopants and host semiconductors for their high-temperature ferromagnetic properties. Transition metal ions such as Mn, Fe, Co, or Ni have been implanted into the host materials including GaN,[32–40] AlN,[25] GaP,[27,41] SiC,[42] ZnO:Sn,[43] ZnGeSiN2,[44] and AlGaP.[45]

Ion implantation has the same advantage as MBE growth to introduce the impurities into III-V compounds beyond solubility limits for obtaining high T_c ferromagnetism, because the strength of the magnetism is proportional to the number of transition metal ions substituted on the group-III cation sites. In our results,[40] the concentration of heavy Mn implantation reached 1.54×10^{21} cm^{-3} and showed room temperature ferromagnetism. However, such a high implantation dose leads to serious lattice damage that must be repaired by thermal annealing. It can also lead to the formation of impurity phases or clusters, thus complicating the interpretation of the origin of the ferromagnetism, especially if the impurity phases are themselves ferromagnetic.

The postimplantation annealing is needed to ensure that most of the implanted dopant ions will migrate to the correct (cation) lattice sites, that is, magnetic impurity

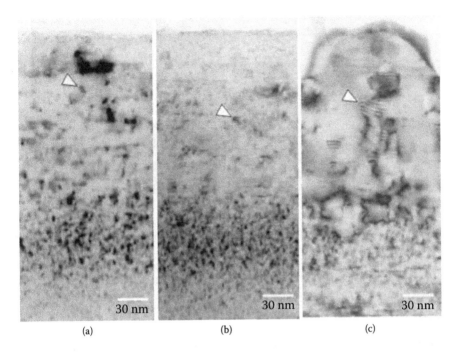

FIGURE 2.4 Cross-sectional bright-field transmission electron microscope (TEM) images of Mn-implanted GaN as a function of annealing temperature: (a) 700°C, (b) 800°C, and (c) 900°C. The cluster size ranged from 3 to 7 nm for the 800°C and 700°C annealed samples. When the annealing temperature increased to 900°C, the cluster size increased to around 30 nm. (From J.M. Baik, H.S. Kim, C.G. Park, and J.-L. Lee, *Appl. Phys. Lett.*, 83(13), 2632, 2003. With permission.)

ions should be activated effectively. The annealing can also remove or reduce the point and line defects induced by ion implantation. However, too high an annealing temperature may cause excessive diffusion of most dopants and even lead to the formation of impurity phases.

Jeong et al. thought that the ferromagnetic property was closely related to microstructural changes of nanoclusters in Mn-implanted p-type GaN. They found that higher temperature annealing (> 900°C) produced antiferromagnetic Mn-nitride nanoclusters, such as $Mn_6N_{2.58}$ and Mn_3N_2, which lead to a poor ferromagnetic signal, which sometimes even disappeared.[31,46] GaN films annealed at a lower temperature (< 800°C) showed weak ferromagnetism, which was attributed to the Mn_3Ga nanoclusters with sizes of 3 to 7 nm (Figure 2.4 and Figure 2.5). Different nanoclusters produced Ga or N vacancies corresponding to the change of net hole concentration, which induced the degradation of the ferromagnetic property as the annealing temperature was increased. It is suggested that an optimum annealing temperature (< 900°C) could be an important parameter in enhancing the ferromagnetism in the Mn-implanted and annealed GaN by suppressing the production of N vacancies. The annealing temperature is usually higher than 800°C; otherwise, the macroscopic

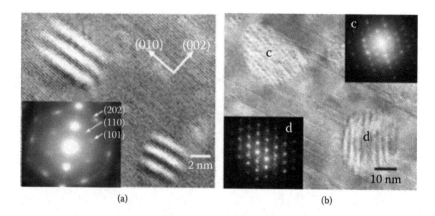

(a)   (b)

FIGURE 2.5 High-resolution TEM image and selected area diffraction patterns (SADPs) for Mn-implanted and annealed GaN samples: (a) annealed at 700°C. The SADPs in the inset support that the phase corresponds to Mn_3Ga. (b) High-resolution image and nanobeam diffraction patterns for the 900°C annealed sample. The nanobeam diffraction patterns of the phase marked as (c) and (d) correspond to $Mn_6N_{2.58}$ and Mn_3N_2, respectively. (From J.M. Baik, H.S. Kim, C.G. Park, and J.-L. Lee, *Appl. Phys. Lett.*, 83(13), 2632, 2003. With permission.)

disorder or defects induced by implantation will be preserved.[47] We presented a similar annealing temperature (800 to 900°C) for Mn-implantation *n*-type GaN.[40]

Baik et al. found that Mn-N coimplanted *p*-type GaMnN has a larger ferromagnetic signal than that of Mn-implanted GaN (Figure 2.6),[48] which originates

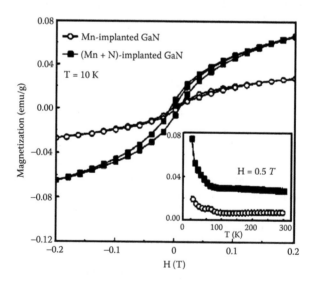

FIGURE 2.6 Magnetization curves for Mn-implanted and Mn-N coimplanted samples annealed at 800°C for 30 s. The temperature dependence of the magnetic moment is plotted in the inset. The EMU/g indicates the magnetization of the entire sample including the substrate. (From J.M. Baik, H.S. Kim, C.G. Park, and J.-L. Lee, *Appl. Phys. Lett.*, 84(7), 1120, 2004. With permission.)

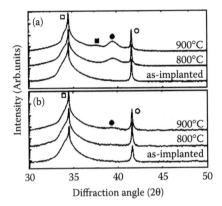

FIGURE 2.7 Change of x-ray diffraction (XRD) scans as a function of annealing temperature: (a) Mn-implanted GaN and (b) Mn-N coimplanted GaN magnetization. (From J.M. Baik, H.S. Kim, C.G. Park, and J.-L. Lee, *Appl. Phys. Lett.,* 84(7), 1120, 2004. With permission.)

from the increase of Ga-Mn magnetic phases. Implantation of N ions increased the Mn-N compounds (such as $Mn_6N_{2.58}$ and Mn_3N_2) (Figure 2.7) and reduced the N vacancies.

2.3.1.3 HVPE

Because of the specific growth chemistry, HVPE can also be used to grow GaN films doped with high Mn concentration. We have prepared Mn-doped GaN films by using an ammonia source HVPE system.[49] Mn-doped GaN films showed clear magnetic signal at room temperature (Figure 2.8). It appears that three different manganese phases MnN, Mn_2N, and $Mn_3GaN_{0.5}$ were formed in the films (Figure 2.9). The lattice parameter of HVPE $Ga_xMn_{1-x}N$ was larger than that of undoped

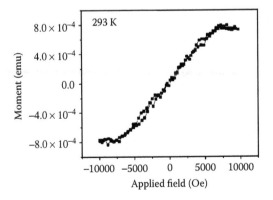

FIGURE 2.8 Magnetization curve for Mn-doped film with Mn concentration of 16%.

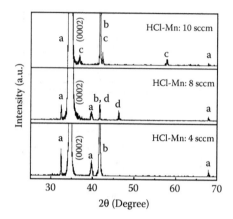

FIGURE 2.9 XRD patterns of three Mn-doped films with HCl-Mn flux of 10, 8, 4 sccm. Reflections of different compounds are marked as (a) $Mn_3GaN_{0.5}$, (b) α-Al_2O_3, (c) Mn_2N, and (d) MnN.

HVPE-grown GaN, which is similar to that of MBE-[50] or MOCVD-grown[51] GaMnN. The films also have metallic characteristics and this mixed structure may be greatly beneficial to spin injection.

The magnetic properties of GaN doped with other transition metal impurities have also been recently reported. For initially p-type samples directly implanted with either Fe or Ni, ferromagnetism has been observed at temperatures of ~250 K,[34] and 185 K,[35] respectively. We have obtained ferromagnetism at room temperature in an Fe-implanted GaN film with a coercivity of 82 Oe (unpublished results). (Ga,Fe)N films grown by MBE showed Curie temperatures $\leq$ 100 K with EXAFS data showing that the majority of the Fe atoms were substitutional on Ga sites.[52] (Ga,Cr)N layers grown in a similar fashion at 700°C on sapphire substrates showed single-phase behavior, clear hysteresis, saturation of magnetization at 300 K and a Curie temperature exceeding 400 K.[53]

GaP is a very attractive host material for spintronics because it is almost lattice matched to Si. Although theories predict relatively low Curie temperatures (< 110 K) for (Ga,Mn)P,[41,54] ferromagnetism of (Ga,Mn)P at room temperature has been reported.[41,54] Carbon-doped p-type GaP samples implanted with ~6 at.% of Mn and then annealed at 700°C show a ferromagnetic Curie temperature of ~236 to 270 K. Hysteresis loops of MBE-grown GaMnP films without secondary phases (such as MnGa or MnP) or clusters were obtained up to 330 K. The magnetic behavior of the (Ga,Mn)P was consistent with mean-field predictions. There are some interesting, different results for GaMnN. The Curie temperature increased at first with increasing Mn concentration and then decreased at higher concentrations.[54] The Curie temperature was also strongly influenced by the carrier density and type, that is, highly p-type samples show much higher values than n-type or undoped samples.

2.3.2 SiC

SiC, as an indirect wide band-gap semiconductor (3.0 eV for the 6H polytype), is at a relatively mature state of development for high-power, high-temperature electronics. It would be a good candidate for spintronic applications because of wide band-gap, excellent transport properties, dopability, and so on. But very little attention has been paid to potential diluted magnetic semiconductor behavior in SiC. As yet no theory has been used to guide the choice of magnetic dopants for SiC.

Recently, 6H-SiC implanted with Ni, Fe, or Mn at different doses has been reported and ferromagnetic ordering temperatures between 50 to 270 K observed.[42] The implanted peak concentrations of these elements was up to ~5 at.% and the samples were annealed for 5 min at 700 to 1000°C under flowing N_2. The direct implantation of Fe, Ni, and Mn into SiC produced significantly different magnetic characteristics. Fairly well-defined ordering temperatures were observed for Mn and Fe between 250 and 270 K, but the Ni led to low values of the ordering temperature (~50 K). The origin of the ferromagnetic contributions in implanted SiC is still to be determined.

2.3.3 ZnO

Most of the initial work in diluted magnetic materials focused on II-VI semiconductors in which a fraction of the group-II sublattice was randomly replaced by Mn atoms.[55] Dietl et al.[2] first predicted a Curie temperature of $\geq$ 300 K for Mn-doped p-type ZnO; whereas, Fe-, Co-, or Ni-doped n-type ZnO was predicted to stabilize high Curie temperature ferromagnetism.[56,57] Carrier-induced ferromagnetism has been addressed theoretically by others in the case of hole doping of ZnO(Mn),[58,59] and methods for improving p-type doping have also been suggested.[60]

Ab initio calculations predict ferromagnetism in n-type ZnO doped with most transition metal ions, including Co and Cr, but predict no ferromagnetism for Mn-doped ZnO.[57] Experimentally, the ferromagnetism has been observed in n-type ZnO crystals. Ueda et al.[17] reported Curie temperatures above 300 K for Co-doped ZnO. J.-H. Kim et al.[61] fabricated $Zn_{0.75}Co_{0.25}O$ thin films on sapphire (0001) substrates at different substrate temperatures (400 to 700°C) using PLD techniques (Figure 2.10).

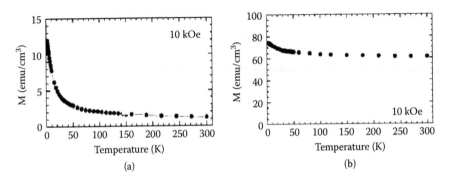

FIGURE 2.10 Magnetization vs. temperature curves for laser-deposited $Zn_{0.75}Co_{0.25}O$ films at substrate temperatures of 600°C (a) and 700°C (b) in an O_2 pressure of 1105 torr.

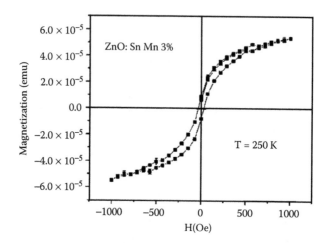

FIGURE 2.11 M vs. H curve for Mn-implanted ZnO:Sn single crystals showing ferromagnetism in 3 at.% Mn implantation doses. (From D.P. Norton, S.J. Pearton, A.F. Hebard, N. Theodoropoulou, L.A. Boatner, and R.G. Wilson, *Appl. Phys. Lett.,* 82, 239, 2003. With permission.)

Their observations of the structural and magnetic properties collectively indicate that the presence of the ferromagnetic hexagonal Co clusters leads to ferromagnetism in Co-doped ZnO films.[61] Prellier et al.[62] showed that homogenous incorporation of Co in $Zn_{1-x}Co_xO$ films produced ferromagnetism with a Curie temperature close to room temperature for $x = 0.08$, and 150 K for $x = 0.05$. Ni-doped ZnO thin films with ferromagnetism at 2 K were reported.[63]

Ion implantation can also be used to screen the magnetic properties of transition metal dopants in oxide semiconductor materials. High-temperature ferromagnetism has been observed in ZnO crystals implanted with transition metal dopants, including (Co,Mn) codoped with Sn (Figure 2.11)[64,65] and (Cr,Fe)-implanted ZnO bulk crystals.[66]

Codoping technology was also used to prepare ZnO-based spintronic materials, such as $Zn_{1-x}Fe_xO$:Cu by solid phase reaction,[67] $Zn_{1-x}(Co_{0.5}Fe_{0.5})_xO$ by reactive magnetron sputtering.[68]

For Mn-doped n-type ZnO, ferromagnetism is not observed. Unfortunately, the realization of p-type ZnO has proven difficult until now. Han et al.[69] have demonstrated that bulk Mn-doped ZnO produced by solid phase reaction without a secondary phase was predominantly paramagnetic. Jung et al.[70] have reported ferromagnetism in Mn-doped ZnO epitaxial films with a Curie temperature of 45 K depending on the film growth conditions. Y.M. Kim et al.[71] have fabricated $Zn_{1-x}Mn_xO$ films with a Curie temperature of ~39 K by the sol-gel method and found a coercive field H_c ~ 2100 Oe for the film with $x = 0.2$. Here, the origin of the ferromagnetic properties of the $Zn_{1-x}Mn_xO$ films may come from precipitates of manganese oxide (Mn_3O_4) at the interface of the films.

Recently, ferromagnetism above room temperature has been observed in Mn-doped ZnO bulk pellets (Figure 2.12A) and transparent films 2 to 3 μm thick (Figure 2.12B).[72] Ferromagnetic resonance studies have been used for detecting ferromagnetic order

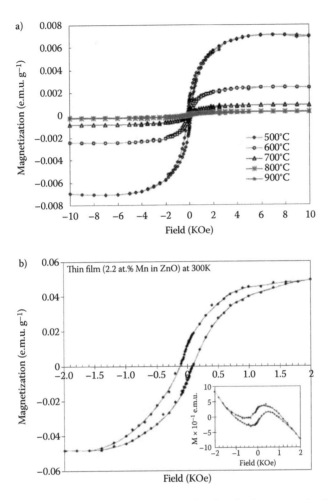

FIGURE 2.12 Room temperature hysteresis loops showing the ferromagnetic phase: (a) for nominal 2 at.% Mn-doped ZnO pellets sintered at various temperatures ranging from 500 to 900°C; (b) for $Zn_{0.978}Mn_{0.022}O$ pulsed-laser-deposited thin film on fused quartz. The curve was obtained after subtracting the diamagnetic contribution from the substrate. Inset: The as-obtained data from the superconducting quantum interference device measurements along with the diamagnetic contribution arising from the fused quartz substrate.

as well as the possible existence of other magnetic species, which clearly show a ferromagnetic spectra with T_c well above 425 K for a nominal 2 at.% Mn-doped ZnO pellet (Figure 2.13).

In the growth of Zn-incorporated Mn oxide using PLD, the ferromagnetic oxygen-vacancy-stabilized $Mn_{2-x}Zn_xO_{3-\delta}$ is also observed in the low-temperature processed Mn-Zn-O system, which is responsible for the high-temperature ferromagnetism up to ~980 K (Figure 2.14).[73] It is thought that the ferromagnetism

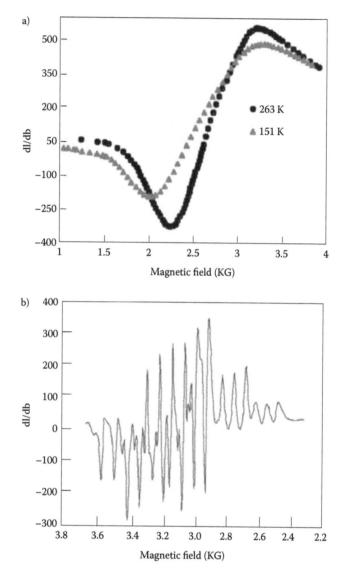

FIGURE 2.13 Ferro- and paramagnetic resonance spectra for a nominal 2 at.% Mn-doped ZnO pellet. (a) Ferromagnetic resonance spectra sintered at 500°C. (b) The room temperature paramagnetic resonance spectra for the same sample sintered at 900°C. (From P. Sharma, A. Gupta, K.V. Rao, F.J. Owens, R. Sharma, R. Ahuja, J.M. Osorio Guillen, B. Johansson, and G.A. Gehring, *Nat. Mater.,* 2, 673, 2003. With permission.)

originates from a metastable phase rather than by carrier-induced interaction between separated Mn atoms in ZnO. The interface diffusion and reaction between thin-film bilayers of Mn and Zn oxides indicate that a uniform solution of Mn in ZnO does not form under low-temperature processing.

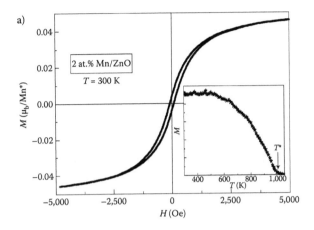

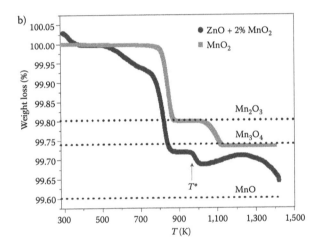

FIGURE 2.14 Magnetization, thermogravimetric analysis (TGA), and XRD data for bulk compounds. (a) Plot of 300 K magnetization (M) vs. applied magnetic field (H) for low-temperature processed 2 at.% MnO_2–ZnO bulk compound. The asterisk on Mn in μB/Mn indicates that the Mn content is taken to be the quantity introduced in making the sample. The inset shows magnetization as a function of temperature above 300 K. (b) TGA data for MnO_2 (gray) and ZnO + 2% MnO_2 powder mixture (black) shown as percentage weight loss. The data for the mixture are slightly offset. The weight loss levels for total conversion of MnO_2 into other stoichiometric forms are indicated by horizontal dashed lines. (c) XRD pattern for ZnO (black), sintered 2 at.% MnO_2–ZnO (gray) mixture, and unsintered 2 at.% MnO_2–ZnO (gray) mixture. The intensity is plotted on a logarithmic scale. (From D.C. Kundaliya, S.B. Ogale, S.E. Lofland, S. Dhar, C.J. Metting, S.R. Shinde, Z. Ma, B. Varughese, K.V. Ramanujachary, L. Salamanca-Riba, and T. Venkatesan, *Nat. Mater.,* 3, 709, 2004. With permission.)

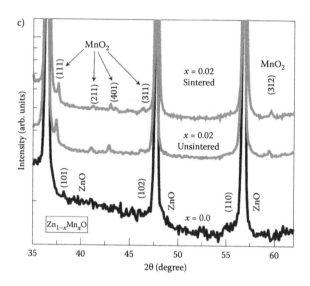

FIGURE 2.14   (Continued).

2.4 MECHANISMS OF FERROMAGNETISM

There are two basic approaches to understanding the origin of ferromagnetism in DMSs. The first one is mean-field theory based on the original Zener model.[74] The theories assume that the diluted magnetic semiconductor is a more or less random alloy, such as (Ga,Mn)N, in which Mn substitutes for one of the lattice constituents. The mean-field model and its variants produce reliable estimates of T_C for materials such as (Ga,Mn)As and (In,Mn)As and predict that (Ga,Mn)N will have a value above room temperature.[2] For the model, a further issue that needs additional exploration in the theories is the role of electrons, rather than holes, in stabilizing the ferromagnetism in DMS materials. This is particularly relevant as ferromagnetism has been observed in samples, such as (Ga,Mn)N, that have very low hole concentrations, in insulating materials and more recently in n-type materials.

The second model suggests that the magnetic atoms form small (a few atoms) clusters that produce the observed ferromagnetism.[75] In fact, it is very difficult to verify the mechanism responsible for the observed experimental magnetic properties. To date, most of the above-mentioned experimental results showing room temperature ferromagnetism in DMS show a trace of nanoclusters of magnetic atoms in precipitates and second phase formation [such as MnGa or Mn_4N in (Ga,Mn)N, and Co clusters in (Zn,Co)O]. Although these clusters are difficult to detect by most characterization techniques, a nanocluster or second phase has been observed by selected area diffraction patterns, transmission electron microscopy, EXAFS, scanning tunneling microscopy, and Z-contrast scanning transmission electron microscopy (Z-contrast STEM).

In spintronics, the application of ferromagnetic wide band-gap semiconductors uses both ferromagnetic and semiconducting properties. If second phases or clusters are responsible for the ferromagnetism and if the charge carriers do not mediate the ferromagnetic interaction, then the application of these materials to spintronics would be questionable. Of course, it is not clear that the presence of ferromagnetic nanoclusters excludes carrier-mediated interaction. Additional further studies need to be done to understand the potential limitations and to confirm the presence of carrier-mediated ferromagnetism, such as anomalous Hall effect measurements for the local magnetization in regions visited by the carriers[76] and magnetic circular dichroism for the interaction between the *sp* carriers and the localized *d* spins.[77]

2.5 SUMMARY

A number of research studies have been undertaken to exploit high-temperature ferromagnetism in wide band-gap semiconductor materials. In this chapter, we review the growth and characterization of wide band-gap diluted magnetic semiconductors, such as GaN and ZnO, and discuss the origins of the magnetism. Complete characterization of many DMSs discussed here has not been achieved, and it is clear that much more work needs to be done to fully exploit the potential of wide band-gap semiconductors as spintronic materials.

ACKNOWLEDGMENTS

This work was supported by Special Funds for Major State Basic Research Project and Natural Science Foundation (Grant Nos. 60390072, 60421003) of China.

REFERENCES

1. I. Malajovich, J.M. Kikkawa, D.D. Awschalom, J.J. Berry, and N. Samarth, *Phys. Rev. Lett.,* 84, 1015, 2000.
2. T. Dietl, H. Ohno, F. Matsukura, J. Cibert, and D. Ferrand, *Science*, 287, 1019, 2000.
3. M. Henini, *III-Vs Rev.*, 13, 32, 2000.
4. H. Munekata, H. Ohno, S. von Molnar, A. Segmuller, L.L. Chang, and L. Esaki, *Phys. Rev. Lett.,* 63, 1849, 1989.
5. H. Ohno, A. Shen, F. Matsukura, A. Oiwa, A. Endo, and Y. Iye, *Appl. Phys. Lett.,* 69, 363, 1996.
6. T. Hayashi, M. Tanaka, T. Nishinaga, H. Shimada, H. Tsuchiya, and Y. Otuka, *J. Cryst. Growth,* 175/176, 1063, 1997.
7. H. Ohno, F. Matsukura, and Y. Ohno, *JSAP Int.,* 5, 4, 2002.
8. H. Ohno, *J. Vac. Sci. Technol. B*, 18, 2039, 2000.
9. S.J. Pearton, C.R. Abernathy, M.E. Overberg, G.T. Thaler, D.P. Norton, N. Theodoropoulou, A.F. Hebard, Y.D. Park, F. Ren, J. Kim, and L.A. Boatner, *J. Appl. Phys.,* 93, 1, 2003.
10. S.J. Pearton, C.R. Abernathy, D.P. Norton, A.F. Hebard, Y.D. Park, L.A. Boatner, and J.D. Budai, *Mater. Sci. Eng. R.,* 40, 137, 2003.

11. T. Graf, S.T.B. Goennenwein, and M.S. Brandt, *Phys. Status Solidi B,* 239, 277, 2003.
12. T. Dietl, *Phys. Status Solidi B,* 243, 344, 2003.
13. G.A. Medvedkin, T. Ishibashi, T. Nishi, and K. Hiyata, *Jpn. J. Appl. Phys.,* 39, L949, 2002.
14. S. Cho, Y. Kim, S. Choi, Y.Z. Zhao, A.J. Freeman, B.J. Kim, B.-C. Choi, Y.C. Kim, G.B. Chao, S.C. Hong, and J.B. Kellerson, *Phys. Rev. Lett.,* 88, 257203, 2002.
15. G.A. Medvedkin, K. Hirose, T. Ishibashi, T. Nishi, V.G. Voevodin, and K. Sato, *J. Cryst. Growth,* 236, 609, 2002.
16. S. Choi, G.-B. Cha, S.C. Hong, S. Cho, Y. Kim, J.B. Kellerson, S.-Y. Jeong, and G.C. Yi, *Solid State Commun.,* 122, 165, 2002.
17. K. Ueda, H. Tahata, and T. Kawai, *Appl. Phys. Lett.,* 79, 988, 2001.
18. S.A. Chambers, *Mater. Today,* (April), 5(4), 34, 2002.
19. Y. Matsumoto, M. Murahami, T. Shono, T. Hasegawa, T. Fukumura, M. Kawasaki, P. Ahmet, T. Chikyow, S. Koshikara, and H. Koinuma, *Science,* 291, 854, 2001.
20. M. Zajac, R. Doradzinski, J. Gosk, J. Szczytko, M. Lefeld- Sosnowska, M. Kaminska, A. Towardowski, M. Palczewska, E. Grzanka, and W. Gebicki, *Appl. Phys. Lett.,* 78, 1276, 2001.
21. M. Zajac, J. Gosk, M. Kaminska, A. Towardowski, T. Szyszko, and S. Podsiadlo, *Appl. Phys. Lett.,* 79, 2432, 2001.
22. M.L. Reed, M.K. Ritums, H.H. Stadelmaier, M.J. Reed, C.A. Parker, S.M. Bedair, and N.A. El-Masry, *Mater. Lett.,* 51, 500, 2001.
23. M.L. Reed, M.K. Ritums, H.H. Stadelmaier, M.J. Reed, C.A. Parker, S.M. Bedair, and N.A. El-Masry, *Appl. Phys. Lett.,* 79, 3473, 2001.
24. S. Sonoda, S. Shimizu, T. Sasaki, Y. Yamamoto, and H. Hori, *J. Cryst. Growth,* 237–239, 1358, 2002.
25. A.F. Hebard, R.P. Rairigh, J.G. Kelly, S.J. Pearton, C.R. Abernathy, S.N.G. Chu, and R.G. Wilson, *J. Phys. D: Appl. Phys.,* 37, 511, 2004.
26. B.K. Rao and P. Jena, *Phys. Rev. Lett.,* 89, 185504, 2002.
27. G.T. Thaler, M.E. Overberg, B. Gila, R. Frazier, C.R. Abernathy, S.J. Pearton, J.S. Lee, S.Y. Lee, Y.D. Park, Z.G. Khim, J. Kim, and F. Ren, *Appl. Phys. Lett.,* 80, 3964, 2002.
28. M.-C. Jeong, M.-H. Ham, J.-M. Myoung, and Sam-Kyu Noh, *Appl. Surf. Sci.,* 222(1–4), 322 (2004).
29. K.H. Kim, K.J. Lee, D.J. Kim, C.S. Kim, H.C. Lee, C.G. Kim, S.H. Yoo, H.J. Kim, and Y.E. Ihm, *J. Appl. Phys.,* 93, 6793, 2003.
30. Y.L. Soo, G. Kioseouglou, S. Kim, S. Huang, Y.H. Kaa, S. Kubarawa, S. Owa, T. Kondo, and H. Munekata, *Appl. Phys. Lett.,* 79, 3926, 2001.
31. J.M. Baik, H.S. Kim, C.G. Park, and J.-L. Lee, *Appl. Phys. Lett.,* 83(13), 2632, 2003.
32. M.E. Overberg, C.R. Abernathy, S.J. Pearton, N.A. Theodoropoulou, K.T. McCarthy, and A.F. Hebard, *Appl. Phys. Lett.,* 79, 1312, 2001.
33. N. Theodoropoulou, A.F. Hebard, M.E. Overberg, C.R. Abernathy, S.J. Pearton, S.N.G. Chu, and R.G. Wilson, *Appl. Phys. Lett.,* 78, 3475, 2001.
34. N. Theodoropoulou, A.F. Hebard, S.N.G. Chu, M.E. Overberg, C.R. Abernathy, S.J. Pearton, R.G. Wilson, and J.M. Zavada, *Appl. Phys. Lett.,* 79, 3452, 2001.
35. S.J. Pearton, M.E. Overberg, G. Thaler, C.R. Abernathy, N. Theodoropoulou, A.F. Hebard, S.N.G. Chu, R.G. Wilson, J.M. Zavada, A.Y. Polyakov, A.V. Osinsky, P.E. Norris, P.P. Chow, A.M. Wowchack, J.M. Van Hove, and Y.D. Park, *J. Vac. Sci. Technol. A (Vacuum, Surfaces, and Films),* 20, 721, 2002.
36. C. Liu, E. Alves, A.R. Ramos, M.F. da Silva, J.C. Soares, T. Matsutani, and M. Muchi, *Nucl. Instrum. Methods Phys. Res., Sec. B (Beam Interactions with Materials and Atoms),* 191, 544, 2002.

37. S.J. Pearton, M.E. Overberg, G.T. Thaler, C.R. Abernathy, J. Kim, F. Ren, N. Theodoropoulou, A.F. Hebard, and Daniel P. Yun, *Phys. Status Solidi A*, 195, 222, 2003.

38. J.M. Baik, H.W. Jang, J.K. Kim, and J.-.W Lee, *Appl. Phys. Lett.*, 82, 583, 2003.

39. M.E. Overberg, B.P. Gila, C.R. Abernathy, S.J. Pearton, N.A. Theodoropoulou, K.T. McCarthy, S.B. Arnason, and A.F. Hebard, *Appl. Phys. Lett.*, 79, 3128, 2001.

40. J. Xu, J. Li, R. Zhang, X.Q. Xiu, D.Q. Lu, H.Q. Yu, S.L. Gu, B. Shen, Y. Shi, Y.D. Ye, and Y.D. Zheng, *Opt. Mater.*, 23(1–2), 163, 2003.

41. N. Theodoropoulou, A.F. Hebard, M.E. Overberg, C.R. Abernathy, and S.J. Pearton, *Phys. Rev. Lett.*, 89, 107203, 2002.

42. N. Theodoropoulou, A.F. Hebard, S.N.G. Chu, M.E. Overberg, C.R. Abernathy, S.J. Pearton, R.G. Wilson, and J.M. Zavada, *Electrochem. Solid State Lett.*, 4, G119, 2001.

43. D.P. Norton, S.J. Pearton, A.F. Hebard, N. Theodoropoulou, L.A. Boatner, and R.G. Wilson, *Appl. Phys. Lett.*, 82, 239, 2003.

44. S.J. Pearton, M.E. Overberg, C.R. Abernathy, N.A. Theodoropoulou, A.F. Hebard, S.N.G. Chu, A. Osinsky, V. Fuflyigin, L.D. Zhu, A.Y. Polyakov, and R.G. Wilson, *J. Appl. Phys.*, 92, 2047, 2002.

45. M.E. Overberg, G.T. Thaler, R.M. Frazier, C.R. Abernathy, S.J. Pearton, R. Rairigh, J. Kelly, N.A. Theodoropoulou, A.F. Hebard, R.G. Wilson, and J.M. Zavada, *J. Appl. Phys.*, 93, 7861, 2003.

46. M.-B. Jeong, U.-K. Sang, M.-K. Yang, W.-K. Tae, and J.-L. Lee, *Electrochem. Solid State Lett.*, 7(12), G313, 2004.

47. M.-B. Jeong, J.-L. Lee, S.Y., and T.-W. Kang, *J. Appl. Phys.*, 93(11), 9024, 2003.

48. J.M. Baik, H.S. Kim, C.G. Park, and J.-L. Lee, *Appl. Phys. Lett.*, 84(7), 1120, 2004.

49. Y.Y. Yu, R. Zhang, X.Q. Xiu, Z.L. Xie, H.Q. Yu, Y. Shi, B. Shen, S.L. Gu, Y.D. Zheng, *J. Cryst. Growth*, 269, 270, 2004.

50. Y. Cui and L. Li, *Appl. Phys. Lett.*, 80, 4139, 2002.

51. M.C. Park, K.S. Huh, J.M. Myoung, J.M. Lee, J.Y. Chang, K.I. Lee, S.H. Han, and W.Y. Lee, *Solid State Commun.*, 124, 11, 2002.

52. H. Akinaga, S. Nemeth, J. De Boeck, L. Nistor, H. Bender, G. Borghs, H. Ofuchi, and M. Oshima, *Appl. Phys. Lett.*, 77, 4377, 2000.

53. M. Hashimoto, Y.Z. Zhou, M. Kanamura, and H. Asahi, *Solid State Commun.*, 122, 37, 2002.

54. M.E. Overberg, B.P. Gila, G.T. Thaler, C.R. Abernathy, S.J. Pearton, N. Theodoropoulou, K.T. McCarthy, S.B. Arnason, A.F. Hebard, S.N.G. Chu, R.G. Wilson, J.M. Zavada, and Y.D. Park, *J. Vac. Sci. Technol. B*, 20, 969, 2002.

55. T. Aoki, D.C. Look, and Y. Hatanaka, *Appl. Phys. Lett.*, 76, 3257, 2000.

56. K. Sato and H. Katayama-Yoshida, *Jpn. J. Appl. Phys.*, 40, L334, 2001.

57. K. Sato and H. Katayama-Yoshida, *Jpn. J. Appl. Phys.*, 39, L555, 2000.

58. K. Sato and H. Katayama-Yoshida, *Physica E*, 10, 251, 2001.

59. K. Sato and H. Katayama-Yoshida, *Mat. Res. Soc. Symp. Proc.*, 666, F4.6.1, 2001.

60. T. Yamamoto and H. Katayama-Yoshida, *Jpn. J. Appl. Phys.*, 38, L166, 1999.

61. J.-H. Kim, H. Kim, D. Kim, Y.E. Ihm, and W. K. Choo, *J. Eur. Ceram. Soc.*, 24, 1847, 2004.

62. W. Prellier, A. Fouchet, Ch. Simon, and B. Mercey, *Mater. Sci. Engineer. B*, 109, 192, 2004.

63. T. Wakano, N. Fujimura, Y. Morinaga, N. Abe, A. Ashida, and T. Ito, *Physica E*, 10, 260, 2001.

64. D.P. Norton, S.J. Pearton, A.F. Hebard, N. Theodoropoulou, L.A. Boatner, and R.G. Wilson, *Appl. Phys. Lett.*, 82, 239, 2003.

65. D.P. Norton, M.E. Overberg, S.J. Pearton, K. Pruessner, J.D. Budai, L.A. Boatner, M.F. Chisholm, J.S. Lee, Z.G. Khim, Y.D. Park, and R.G. Wilson, *Appl. Phys. Lett.*, 83, 5488, 2003.

66. A.Y. Polyakov, A.V. Govorkov, N.B. Smirnov, N.V. Pashkova, S.J. Pearton, K. Ip, R.M. Frazier, C.R. Abernathy, D.P. Norton, J.M. Zavada, and R.G. Wilson, *Mater. Sci. Semicond. Proc.,* 7, 77, 2004.

67. S.J. Han, J.W. Song, C.H. Yang, S.H. Park, J.H. Park, Y.H. Jeong, and K.W. Rhie, *Appl. Phys. Lett.,* 81(22), 4212, 2002.

68. Y.M. Cho, W.K. Choo, H. Kim, D. Kim, and Y.E. Ihm, *Appl. Phys. Lett.,* 80, 3358, 2002.

69. S.J. Han, T.-H. Jang, Y.B. Kim, B.G. Park, J.H. Park, and Y.H. Jeong, *Appl. Phys. Lett.,* 83(5), 920, 2003.

70. S.W. Jung, S.J. An, G.C. Yi, C.U. Jung, S.I. Lee, and S. Cho, *Appl. Phys. Lett.,* 80, 4561, 2002.

71. Y.M. Kim, M. Yoon, I.W. Park, Y.J. Park, J.H. Lyou, *Solid State Commun.,* 129,175, 2004.

72. P. Sharma, A. Gupta, K.V. Rao, F.J. Owens, R. Sharma, R. Ahuja, J.M. Osorio Guillen, B. Johansson, and G.A. Gehring, *Nat. Mater.,* 2, 673, 2003.

73. D.C. Kundaliya, S.B. Ogale, S.E. Lofland, S. Dhar, C.J. Metting, S.R. Shinde, Z. Ma, B. Varughese, K.V. Ramanujachary, L. Salamanca-Riba, and T. Venkatesan, *Nat. Mater.,* 3, 709, 2004.

74. C. Zener, *Phys. Rev. B,* 81, 440, 1951.

75. M. Van Schilfgaarde and O.N. Myrasov, *Phys. Rev. B,* 63, 233205, 2001.

76. T. Jungwirth, Q. Niu, and A.H. MacDonald, *Phys. Rev. Lett.,* 88, 207208, 2002.

77. K. Ando, T. Hayashi, M. Tanaka, and A. Twardowski, *J. Appl. Phys.,* 83, 6548, 1998.

3 Magnetic Properties of (Ga,Mn)As-Based Magnetic Semiconductors

Maciej Sawicki

CONTENTS

3.1 INTRODUCTION

The discovery of carrier-mediated ferromagnetism in (III,Mn)V and (II,Mn)VI dilute magnetic semiconductors (DMSs) grown by molecular beam epitaxy makes it possible to examine the interplay between the physical properties of semiconductor quantum structures and ferromagnetic materials.[1,2] At the same time, the complementary properties of these systems open doors for novel functionalities and devices. A considerable effort in this field is focused on identifying methods for the mutual manipulations of semiconductor and magnetic properties as well as on developing DMSs in which the ferromagnetism can persist to above room temperature. In this context, (Ga,Mn)As serves as a valuable testing ground for ferro-DMSs, due to the relatively high T_C and its compatibility with the well-characterized GaAs system. The Mn dopant in this III-V host matrix is expected to substitute for the Ga site and fulfill two roles: to supply a local spin 5/2 magnetic moment and to act as an acceptor, providing itinerant holes that mediate the ferromagnetic order. The theoretical understanding of this phenomenon[3] is built on Zener's model of ferromagnetism, the Ginzburg–Landau approach to the phase transitions, and the Kohn–Luttinger kp theory of semiconductors. Within this model and its variants, the magnitude of the Curie temperature T_C in Mn-doped GaAs, InAs, GaSb, InSb,[4–6] as well as in p-CdTe,

p-ZnTe, and Ge,[7] is understood assuming that the long-range ferromagnetic inter-actions between the localized spins are mediated by delocalized holes in the weakly perturbed valence band.[8] The assumption that the relevant carriers reside in the p-like valence band makes it possible to describe various magneto-optical[4,9] and magnetotransport properties of (Ga,Mn)As, including the anomalous Hall effect and anisotropic magnetoresistance,[9,10] as well the negative magnetoresistance caused by the orbital weak-localization effect.[11] From this point of view, (Ga,Mn)As, and related compounds, emerge as the best understood ferromagnets, providing a basis for the development of novel methods enabling magnetization manipulation and switching.[12–14]

Here, a brief review of micromagnetic properties of (Ga,Mn)As is given. Inter-estingly, despite much lower spin and carrier concentrations compared to ferromag-netic metals, (III,Mn)V exhibit excellent micromagnetic characteristics, including well-defined magnetic anisotropy and large ferromagnetic domains separated by usually straight-line domain walls. It turns out that the above-mentioned p-d Zener model explains the influence of strain on magnetic anisotropy as well as describing the magnitudes of the anisotropy field and domain width. Importantly, the experi-mentally observed various magnetic easy axis reorientation transitions as functions of temperature and hole concentration are readily accounted for. This applies also to a weak in-plane magnetic anisotropy that has been detected in these systems. Its presence points to a symmetry breaking, but the origin has not yet been identified.

3.2 CURIE TEMPERATURE

Ferromagnetic ordering of the relatively widely spaced Mn dopants in the semiconduc-tor host arises from antiferromagnetic exchange interactions between Mn $3d$ magnetic moments and the delocalized charge carriers. The wide-ranging experimental studies of (Ga,Mn)As of the past few years have revealed several curiosities that have triggered intensive theoretical debate, namely: (i) the hole density p is often much smaller than the Mn density x;[15] (ii) the saturation magnetization M_{sat} may be smaller than the expected 4 to 5 μ_B per Mn atom;[16,17] (iii) the Curie temperature T_C saturates or even tails off as x is increased above around 5%.[15,17] These peculiarities proved to be inher-ently related to the growth of (Ga,Mn)As with the Mn and hole concentrations surpass-ing thermal equilibrium limits by use of low-temperature epitaxy. This leads to a high density of defects such as As antisites, As_{Ga}, and at the early stages of the investigation all the above-mentioned features were assigned to a disordered nature of these materials. It turns out however, that the quality of the material can be improved substantially. It has been demonstrated that (iv) p, T_C, and M_{sat} can all be increased by annealing at temperatures comparable to[17,18] or even lower than[19] the growth temperature.

Recently, simultaneous channeling Rutherford backscattering and particle-induced x-ray emission experiments[20] revealed that low-temperature growth results in a high concentration of Mn interstitials, Mn_I, which are removed from their sites during postgrowth annealing. Because Mn_I is a double donor in (Ga,Mn)As, it compensates holes provided by two substitutional Mn_{Ga}. According to experimental studies on low-temperature annealing effects and *ab initio* calculations summarized in Reference 21, it has been established that upon annealing, Mn_I out-diffuses to the surface where it gets passivated — the low binding energy of Mn_I in (Ga,Mn)As

(0.7 to 0.8 eV) makes these defects mobile at temperatures as low as ~150°C. This reduces the electrical compensation, increases the hole density, and, hence, the T_C. Further, tight-binding[22] and density-functional[21] calculations indicate that Mn_I couples antiferromagnetically to neighboring Mn_{Ga}, which combined with the former effect suppresses ferromagnetism even further. So at least two steps have been identified to minimize defect densities: (i) careful control of growth to minimize As_{Ga} density followed by (ii) postgrowth annealing to reduce Mn_I.

Figure 3.1 shows the results of magnetic investigations performed on such a carefully prepared[23] set of (Ga,Mn)As layers. The annealing clearly has a pronounced effect on T_C, especially at high Mn concentrations (see the left panel). This is a clear manifestation of the removal of compensating interstitial Mn from the bulk of the layers to the free surface.[19–20,24] The resistivity and hole density show similar trends.[25] Figure 3.1b exemplifies the highest T_C reached so far in (Ga,Mn)As layers. $T_C = 173$ K has been obtained in an annealed 25 nm thick layer with a nominal 8% of Mn.[26] The remnant magnetization and inverse paramagnetic susceptibility vs. temperature for this sample is shown in Figure 3.1b. To further confirm this finding, a clear ferromagnetic hysteresis at 172 K for this sample is shown in the inset. Direct magnetic measurements performed on the same set of layers also allow us to address the so-called magnetization deficit in (Ga,Mn)As.

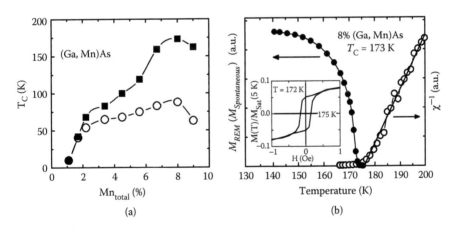

(a)   (b)

FIGURE 3.1 (a) Ferromagnetic transition temperatures vs. total Mn concentration for as-grown (open) and annealed (filled) (Ga,Mn)As thin films. The total Mn concentration is determined from the Mn/Ga flux ratio, which was calibrated by secondary ion mass spectroscopy measurements on 1 mm thick samples grown under otherwise the same conditions. (b) Temperature dependence of remnant magnetization and inverse paramagnetic susceptibility for nominally 8% sample Inset: Hysteresis loops for the same sample just below and above $T_C = 173$ K. (After K.Y. Wang, K.W. Edmonds, R.P. Campion, B.L. Gallagher, N.R.L. Farley, C.T. Foxon, M. Sawicki, P. Bogusawski, and T. Dietl, *J. Appl. Phys.*, 95, 6512, 2004; K.Y. Wang, R.P. Campion, K.W. Edmonds, M. Sawicki, T. Dietl, C.T. Foxon, and B.L. Gallagher, in *Proceedings of the 27th International Conference on Physics of Semiconductors*, Flagstaff, AZ, July 2004, J. Menendez, and Ch. Van de Walle, Eds., New York, 2005, p. 333; cond-mat/0411475. With permission.)

Figure 3.2a shows the average magnetic moment per each Mn ion, μ_{total}, as a function of the total Mn concentration. From first principles calculations, a value close to 4 μ_B per Mn is expected in the absence of compensation, because a single Mn ion moment of 5 μ_B should be partially compensated by ~1 μ_B per each hole due to antiferromagnetic p-d coupling. The measured moment is close to this value at low Mn concentrations. However, as has been found elsewhere, the magnetic moment per Mn ion decreases with increasing x_{total} in the as-grown samples. This is not surprising if the total Mn concentration includes an increasing proportion of Mn_I. The latter can be estimated from hole concentration measurements before and after annealing, because for every interstitial defect reaching the free surface, two holes are recovered in the system. Having estimated substitutional and interstitial Mn concentrations, denoted as x_{sub} and x_I, respectively, the magnetic moment per magnetically active substitutional Mn can be determined, according to $\mu_{sub} = (x_{total} \cdot \mu_{total} + p)/x_{eff}$. The contribution from the charge carriers has been accounted for, so that an ideal value of 5 μ_B per Mn_{Ga} is expected. Two scenarios can be considered then: (i) that the Mn_I is paramagnetic and so makes a negligible contribution to the measured magnetic moment ($x_{eff} = x_{sub}$); and (ii) that Mn_I is antiferromagnetically coupled to Mn_{Ga} ($x_{eff} = x_{sub} - x_I$). The corrected magnetic moments under these two assumptions are shown in Figure 3.3b and Figure 3.3c, respectively. It can be seen that the presence of interstitial Mn accounts for the apparent magnetization deficit

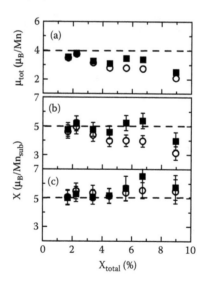

FIGURE 3.2 (a) Measured magnetic moment per Mn in as-grown (circles) and annealed (squares) (Ga,Mn)As. (b) Estimated magnetic moment per substitutional Mn assuming zero contribution to the magnetization from interstitial Mn. (c) Estimated magnetic moment per uncoupled substitutional Mn assuming interstitial and substitutional Mn from antiferromagnetically coupled pairs. Dashed lines depict the ideal-case expected values. (After K.Y. Wang, K.W. Edmonds, R.P. Campion, B.L. Gallagher, N.R.L. Farley, C.T. Foxon, M. Sawicki, P. Bogusawski, and T. Dietl, *J. Appl. Phys.*, 95, 6512, 2004. With permission.)

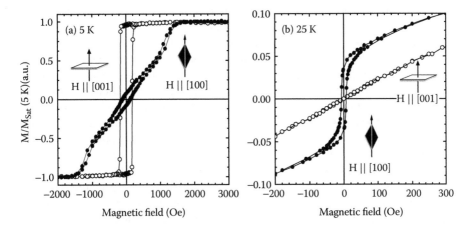

FIGURE 3.3 Magnetic field loops for $Ga_{0.977}Mn_{0.023}As$ film on GaAs substrate measured for two orientations with respect to the magnetizing field: full points — parallel, open ones — perpendicular. At low temperature (a), a perfect square hysteresis is obtained when the magnetic field is perpendicular to the film surface, and an elongated loop is seen when the in-plane orientation is probed. This clearly evidences that the easy axis is perpendicular. An opposite behavior is observed at higher temperatures (b). The reversed character of the hysteresis loops indicates the flip of the easy axis direction between these two temperatures. (After M. Sawicki, F. Matsukura, A. Idziaszek, T. Dietl, G.M. Schott, C. Ruester, G. Karczewski, G. Schmidt, and L.W. Molenkamp, cond-mat/0212511, unpublished.)

in these samples. The magnetic moment per Mn_{Ga} in the annealed samples under assumption (i), and in both sets of samples under assumption (ii), is close to the expected value of 5 μ_B across the whole range of Mn concentrations.

3.3 ORIGIN OF MAGNETIC ANISOTROPY IN FERROMAGNETIC ZINC-BLENDE DMSs

The magnetic dipolar anisotropy, or shape anisotropy, is mediated by dipolar interactions. Because it is long-range, its contribution depends on the shape of the sample, and in thin films, the shape anisotropy often results in the in-plane alignment of the moments. One of the most intriguing findings concerning magnetic properties of (III,Mn)V ferro-DMSs was the discovery of a perpendicular ferromagnetic order [perpendicular magnetic anisotropy (PMA)] in p-(In,Mn)As.[27] Similar studies of the ferromagnetic phase in (Ga,Mn)As[28–30] have also demonstrated the existence of PMA in some special cases. As all the layers are very thin (typically a fraction of micrometer thick) and of macroscopic lateral dimensions, the observation of PMA in these layers points to the existence of a strong, microscopic mechanism that counteracts the shape-imposed in-plane arrangement of the magnetization. The low-temperature growth of (III,V)Mn precludes misfit dislocation formations and the resulting layers are pseudomorphic with respect to the GaAs substrate. This leads to large epitaxial (lattice-mismatch-driven) strain persisting well beyond the critical thickness.

The main outcome of the above-mentioned studies was that, generally, for compressive biaxial strain [as in canonical (Ga,Mn)As on a GaAs substrate], an in-plane magnetic easy axis (in-plane magnetic anisotropy [IMA]) develops. By contrast, for layers under tensile biaxial strain [like (Ga,Mn)As on an (In,Ga)As buffer], a perpendicular-to-plane magnetic easy axis (PMA) is observed. The matter turns out, however, to be more complex. For example, perpendicular orientation of spontaneous magnetization can also be found in *some* of the (Ga,Mn)As/GaAs layers,[31–33] which is when IMA would be expected due to the presence of the compressive biaxial strain. Moreover, it has simultaneously become clear that other factors, such as hole concentration and temperature, play a role here, and that the anisotropy is determined by a combination of all of these factors. Figure 3.3 gives a clear example of such a (Ga,Mn)As/GaAs layer that despite the strain-related expectations exhibits PMA (the left panel). This, however, takes place only at low temperatures. As the right panel indicates, the *same* layer exhibits IMA at elevated temperatures. In fact, it has been established recently that such a rich sensitivity to epitaxial strain, hole density, and temperature is a ubiquitous property of carrier-mediated ferromagnetism. Also, it is solely due to the anisotropy of the carrier-mediated exchange interaction reflecting the *anisotropic properties* of the top of the valence band. This should not be too surprising given that we are dealing with magnetically diluted systems and that the shape anisotropy field that must be overcome is not particularly strong. In the considered case, as for thin films, the shape anisotropy energy per unit volume is given by $E = \frac{1}{2} \mu_0 M_S^2 \cos^2\theta$ (M_S is the saturation magnetization and θ is the angle that M_S subtends to the plane normal), which gives the shape anisotropy field $\mu_0 H_A = \mu_0 M_S$ only of about 0.06 T for 5% (Ga,Mn)As, as compared to 2.2 T for iron.

Knowing the *p-d* exchange energy and *kp* parameters of the valence band, it is possible to compute the magnetic anisotropy energy in the studied compounds.[3,4,34] In fact, the published results agree with the experimental data with remarkably good accuracy.[30,35–37] Nevertheless, it is instructive to consider a simplified case of the model, that is, the nearly empty top of the valence band in biaxial strained zinc-blende compounds.[38] When the strain is present, the valence band splits and the energetic distance between the heavy-hole $j_z = 3/2$ and light-hole $j_z = 1/2$ subbands depends on the strain, see Figure 3.4. For the biaxial compressive strain, the ground-state subband assumes a heavy-hole character. Then, if only the ground-state subband is occupied, the hole spins are oriented along the growth direction. Now, because the *p-d* exchange interaction has a scalar form, $H_{pd} \sim s \cdot S$, the in-plane Mn spin magnetization M will not affect the heavy-hole subband. This means that perpendicular magnetic anisotropy is expected, because only for such magnetization orientation can the holes lower their energy by the coupling to the Mn spins. In the opposite, tensile strain case, the in-plane component of the hole spin is greater than the perpendicular component, so a stronger exchange splitting occurs for the in-plane orientation of M. Hence, if only the light-hole subband is occupied — the in-plane anisotropy is expected.

It is worth remarking here that PMA is not a unique property of (III,Mn)V ferro-DMSs. PMA is typically realized in a compressive strained (Cd,Mn)Te quantum well[39,35] and recent work has showed that in (II,Mn)VI/II,VI structures the magnetic

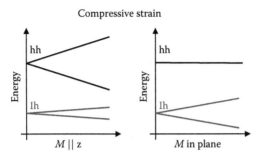

FIGURE 3.4 Illustration of valence band splitting of tetrahedrally coordinated semiconductors for compressive strain and for two orientations of magnetization M in respect to sample plane. (After P. Kossacki, W. Pacuski, W. Malana, J.A. Gaj, M. Bertolini, D. Ferrand, S. Tatarenko, and J. Cibert, *Physica E,* 21, 943, 2004. With permission.)

anisotropy can be tailored by adequate strain engineering.[39] For the compressive strain, a ferromagnetic state related to splitting of the luminescence line is found only for the perpendicular orientation. However, when large enough tensile strain is built into that system, an in-plane direction of the easy axis was observed.

By nature, (III,Mn)V DMS systems are heavily populated with holes. But even when the Fermi energy is comparable or larger than the heavy-hole–light-hole splitting, when strong mixing takes place, we can still talk about either heavy- or light-hole-like characteristics of the holes. Therefore, the lines of reasoning sketched above remain valid to a large extent, and such a simplified approach can still serve as a helpful guideline. Therefore, it will be used again to sketch how in (III,Mn)V ferro-DMSs, the direction of the magnetic easy axis can be set or altered by changing the hole density or temperature. The model description will be exemplified solely by direct magnetization measurements. These are performed in superconducting quantum interference device (SQUID) magnetometers because the typically very small volume of the investigated films calls for the most sensitive setups. But SQUID magnetometry is by no means the only method suited for magnetic anisotropy studies. Among others, the ferromagnetic resonance (FMR) technique should be mentioned as it complements the conventional studies by its ability for precise mapping of the angular dependence of the resonance conditions; therefore, relevant phenomenological description of the magnetic anisotropy can be established. A recent broad summary of the most recent FMR studies for (Ga,Mn)As and other (III,Mn)V structures can be found in Reference 40.

3.4 REORIENTATIONS OF MAGNETIC EASY AXIS

The Zener model dictates that Mn moments (specifically, their collective macroscopic magnetization) adjust their orientation to minimize the total energy of the carriers required to support ferromagnetic ordering of the Mn ions. In particular, depending on the Fermi level position within the valence band or the value of the exchange splitting (that is depending on magnetization and thus also on temperature) different orientations of magnetization can be required to drive the system to its energy minimum. Therefore, by changing hole concentration or temperature, the corresponding changes of the overall orbital momentum of the hole liquid may force a *spontaneous* reorientation of magnetization. It is relatively easy to trace such an effect when we consider hole-concentration-induced reorientation. Referring again to Figure 3.4, we note that by introducing more holes into the system, we populate the second, light-hole subband for which the ferromagnetic state can only be realized with in-plane magnetization orientation. This effect gets even stronger because the heavy holes acquire a light-hole character on increasing Fermi energy E_F. It is therefore expected that at some critical value of the hole concentration, it will be more favorable for the entire system if holes experience the in-plane Mn magnetization. So, an isothermal change of the hole density[41] can lead to the out-of-plane to in-plane reorientation of the magnetic anisotropy, providing the population of the *light* holes gets large enough. A similar mechanism operates when the temperature is used as a lever for easy axis switching. The only difference is that this time, the heavy-hole or light-hole population is changed via temperature-induced changes of the valence subbands exchange splitting. Now, because the spin splitting is proportional to Mn magnetization $M_S(T)$ that varies according to the Brillouin-like function, the character of magnetic anisotropy depends on the temperature. Accordingly, in compressively strained structures, PMA occurs at both low temperatures and hole concentrations, while otherwise IMA will be realized. This implies that there exists a class of samples for which the material parameters are such that within the experimental temperature range, the reorientation of the easy axis from easy z axis to easy plane should occur on increasing temperature. It must, however, be strongly underlined that despite the general understanding of the physical mechanisms responsible for the effects, the strong mixing of the valence band states results in such large anisotropic and nonparabolic valence subband dispersions that it is not possible to specify a single set of heavy-hole:light-hole ratio or hole density, magnetization, or temperature values necessary to trigger the magnetization easy axis switch. This has to be computed up to the fullest possible extent of the mean-field model for each particular sample individually.

Postgrowth low-temperature annealing reduces electrical compensation in (Ga,Mn)As. When performed in stages, the annealing allows researchers to gradually increase the hole density and so to trace the magnetic easy axis dependence on both the hole density and temperature in technically the same material. Figure 3.5 depicts the pertinent experimental results for a 5.3% (Ga,Mn)As/GaAs family of samples originating from the same wafer with increasing hole density prepared by low-temperature annealing.[33] The top panel reconfirms that the parent, the as-grown layer, similarly to the sample presented in Figure 3.3 exhibits PMA at low temperatures (Figure 3.5A), and that it switches to IMA when the temperature rises (Figure 3.5B). In order to trace the magnetization reorientation more closely, we have examined the temperature dependence of the remnant magnetization M_{REM}. For this measurement, the sample is oriented along the investigated crystallographic direction and cooled down through T_C in an external magnetic field that must be at least a few times greater than the maximum coercive field in the studied material. Then the field is removed at low temperature and the M_{REM} component along this direction

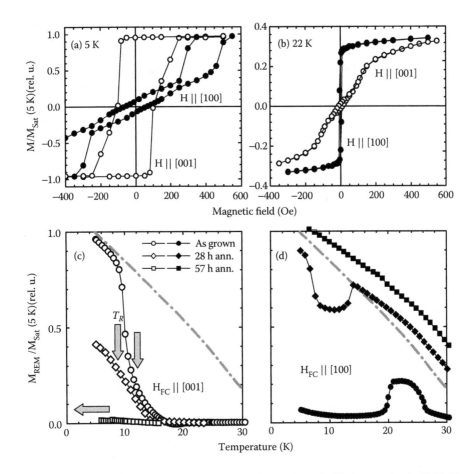

FIGURE 3.5 Magnetic reorientation studies in $Ga_{0.947}Mn_{0.053}As/GaAs$ sample. (a, b) Field dependence of magnetization normalized to its saturation value at 5 K for two orientations of the film in respect to the magnetic field measured at two temperatures for the as-grown layer. Full and empty circles denote the data taken for the magnetic field along the [100] and [001] crystal direction, respectively. Note the flip of the easy axis direction from the perpendicular to the in-plane orientation on increasing temperature. (c, d) Temperature dependence of the remnant magnetization measured in perpendicular [001] (c) and in plane [100] (d) configurations to annealing (circles) and after annealing (diamonds and squares). Note that upon annealing, the development of the in-plane component of M is accompanied by an equivalent quench of the perpendicular one. Bulk arrows mark the reorientation temperature T_R when the crossover to in-plane magnetic anisotropy takes place. The thick dashed gray lines mark the results of the field-cooled measurements at 1 kOe for the as-grown layer and thus mimic the temperature dependence of the saturation (spontaneous) magnetization in this sample. (After M. Sawicki, F. Matsukura, A. Idziaszek, T. Dietl, G.M. Schott, C. Ruester, G. Karczewski, G. Schmidt, and L.W. Molenkamp, *Phys. Rev. B*, 70, 245325, 2004. With permission.)

measured as a function of increasing temperature. It should be noted that to perform an experiment like this, a dedicated (SQUID) magnetometer has to be used. In particular, special care must be devoted to screening the sample from external fields and to keep the parasitic remnant fields generated by the magnet, H_r, at the lowest possible level (preferably a small fraction of an Oersted for investigation of III-V ferro-DMSs). Such a low value of H_r allows M to rotate exactly to the nearest easy direction set only by the torques exerted on M by internal anisotropic field(s). In general, this procedure, if repeated for the main crystallographic orientations, allows unambiguous determination of the orientation of M across the whole temperature range up to T_C.

The bottom panel of Figure 3.5 collects the in-plane (c) and perpendicular (d) M_{REM} components measured as a function of temperature. Starting with the sample having the lowest hole density (as-grown [circles]), we find M_{REM} at low temperatures to be tenfold larger for the perpendicular experimental arrangement as compared to the parallel one. This further confirms the appearance of PMA at low temperatures in this film despite the presence of a sizable compressive strain. This however reverses at elevated temperatures: both components swap their relative intensities above the reorientation temperature T_R, which is defined as the temperature above which perpendicular M_{REM} gets considerably smaller with respect to M_S. Further, on annealing, as the hole density increases, the intensities of the whole $M_{REM}(T)$ traces are found to exchange with each other: the in-plane M_{REM} clearly grows at the expense of perpendicular M_{REM}. The latter finally disappears if the sample has been annealed for a sufficiently long time. At the same time, T_R gets pushed below the experimental temperature range, and so the last sample exhibits only IMA.

Figure 3.6 presents a computed phase diagram for the $x = 5.3\%$ sample, with the theoretical values of T_R plotted as a function of the hole concentration p (the thickest line). The theoretical model and material parameters adopted to obtain this diagram are described in detail in Reference 4. Shape anisotropy is taken into account, and it shifts the PMA $\Rightarrow$ IMA phase boundary by about 20% toward lower p values. We see that the theoretical model confirms the appearance of PMA and correctly describes the temperature-driven phase transformation PMA $\Rightarrow$ IMA in this sample for all stages of the annealing. Considering that the theory has been developed with no adjustable parameters, the agreement between experimental and computed p and T corresponding to the reorientation transition is very good. The same temperature-induced crossover from PMA to IMA described above has been found in other samples of (Al,Ga,Mn)As and (Ga,Mn)As.[31,42] Remarkably, the opposite behavior occurs in tensile strained (In,Mn)As/GaAs, where the easy axis switches from in-plane to out-of-plane on warming,[43] further endorsing the model.

At this point, it is worth emphasizing that although the reorientation transition is a general feature of heavily doped ferro-DMSs, it is only a sample-specific property. In particular, for a given strain, if the hole concentration is either too small

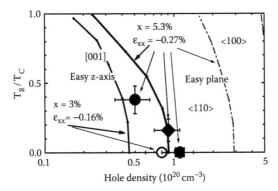

FIGURE 3.6 Computed values of the ratio of the reorientation to Curie temperature for the perpendicular to in-plane magnetic anisotropy transition (thick lines) for $x = 5.3\%$ and 3% samples. Experimental values for $x = 5.3\%$ sample are taken from Figure 3.5. For $x = 3\%$ sample (the open circle), this transition is not detected above 5 K, which is in agreement with the presented calculations. The dashed line marks expected temperatures for the reorientation of the easy axis between <100> and <110> in-plane directions. (After M. Sawicki, F. Matsukura, A. Idziaszek, T. Dietl, G.M. Schott, C. Ruester, G. Karczewski, G. Schmidt, and L.W. Molenkamp, *Phys. Rev. B,* 70, 245325, 2004. With permission.)

or too large, no reorientation transition is expected for any value of magnetization (temperature). In fact, as Figure 3.6 demonstrates, the range of hole densities for which the reorientation can occur is quite narrow. On the other hand, for an appropriate combination of strain and hole concentration, even a minute change of magnetization (temperature) switches the easy axis between the two directions. This feature, characteristic of a second-order phase transition, is confirmed by a comparison of our $M_{REM}(T)$ and quasi-M_S(T) data for the as-grown sample. As seen in Figure 3.5C and Figure 3.5D, M_S is undoubtedly a smooth and slowly varying function of temperature, and despite this, the reorientation still takes place.

It is worth noting that although a change of the hole density by annealing as an irreversible process cannot be employed in any working device, p can be changed by applying an appropriate gate voltage;[41] therefore, a local control of the magnetic easy axis in ferro-DMSs seems feasible. Such an experiment has not been performed so far, but as is demonstrated in Figure 3.7, instead of modulating p, a narrow-range temperature cycling ($\Delta T = 2$ K) can produce reversible, almost on-off type, changes of the in-plane component of the spontaneous magnetization. Because the projection of M on a given direction can be detected by various resistive methods, this phenomenon can serve as a prototype of a spintronic device with functionality based on the direction of the magnetic anisotropy.

Finally, the experimentally found absolute values of biaxial strain-induced uniaxial anisotropy fields[30,35–37] also remain in a very good agreement with the theoretical calculations if large hole densities are assumed.

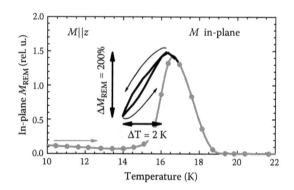

FIGURE 3.7 Temperature cycling of the remnant in-plane magnetization in $x = 0.01$ (Ga,Mn)As at the border of switching of the perpendicular magnetic anisotropy to the in-plane one — black curve. The procedure is fully reproducible and allows production even bigger than indicated changes of the in-plane magnetization, but on expenditure of a much stronger temperature hysteresis. The background gray curve represents the results of "normal" remnant magnetization measurement on warming. The peak structure originates from the conversion of the perpendicular to the planar magnetic anisotropy on the left-hand side of it and to the destructive role of the thermal fluctuation on magnetization on the other side. (After M. Sawicki, F. Matsukura, A. Idziaszek, T. Dietl, G.M. Schott, C. Ruester, G. Karczewski, G. Schmidt, and L.W. Molenkamp, cond-mat/0212511, unpublished.)

3.5   IN-PLANE MAGNETIC ANISOTROPY

According to the discussion above, the easy axis assumes the in-plane orientation for typical carrier concentrations in (Ga,Mn)As/GaAs. In this case, according to the theoretical predictions presented in Figure 3.6 (as well in Figure 9 of Reference 4 and in Figure 6 of Reference 34), fourfold magnetic symmetry is expected with the easy axis predicted to switch between the <100> and <110> in-plane cubic directions as a function of p or T. This biaxial magnetic symmetry is indeed observed at low temperatures, however, with the easy axis assuming exclusively [100] in-plane orientations.[30,33,36,44–49] To the author's knowledge, no <100> ⇔ <110> reorientation transition has been detected in (Ga,Mn)As to date. It is possible that anisotropy of the hole magnetic moment, neglected in the theoretical calculations,[4,34] stabilizes the <100> orientation of the easy axis. It is also possible that the stabilization energy comes from broken magnetic bonds at the film surfaces, an effect observed in (Fe,Co)/GaAs films.[50] However, whether such a model will simultaneously explain the recently reported <110> biaxial symmetry in (In,Mn)As/(In,Al)As films,[51] remains to be seen. Nevertheless, the corresponding in-plane anisotropy field assumes the expected magnitude, of the order of 0.2 T at low temperatures, which is typically 2 to 3 times smaller than that corresponding to the strain-induced in-plane of magnetic anisotropy.[30,35–37]

In addition to the cubic in-plane anisotropy, the accumulated data for both (Ga,Mn)As/GaAs[30,33,36,44–49] and (In,Mn)As/(In,Al)As[51] point to a nonequivalence of [110] and [−110] directions, which leads to the in-plane uniaxial magnetic anisotropy.

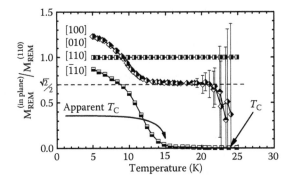

FIGURE 3.8 Experimental evidence for the uniaxial anisotropy along [110] direction in $Ga_{0.97}Mn_{0.03}As$ film. The remnant magnetization is measured for four major in-plane directions and its magnitude is normalized by the data of the [110] case. Note, that the sudden drop of M along [$\bar{1}$10] at $T < T_C$ may wrongly indicate too low a value of T_C, if only this orientation is probed. (After M. Sawicki, F. Matsukura, A. Idziaszek, T. Dietl, G.M. Schott, C. Ruester, G. Karczewski, G. Schmidt, and L.W. Molenkamp, *Phys. Rev. B*, 70, 245325, 2004. With permission.)

As shown in Figure 3.8, M_{REM} measured along the [$\bar{1}$10] direction vanishes completely above 15 K indicating that this is the hard direction in this film. We also note that when $M_{[\bar{1}10]}$ vanishes, the $M_{[100]}/M_{[110]}$ ratio drops to $1/\sqrt{2}$, as expected for the easy axis along [110]. Such a remarkably good correspondence of a simple geometrical picture and experimental findings strongly suggests that the macroscopically sized layers (typically 5×5 mm²) behave as single magnetic domains. This is a rather unexpected property for such a magnetically diluted system. Phenomenologically, it can be said that it originates from the very long-range nature of the Zener mechanism. On the microscopic level, it has been shown[52] that the enhancement of the spatial uniformity of the spin ordering stems from p-like character of the valence band wave functions — a characteristic feature of zinc-blende (ferro)-DMSs. When calculated within the framework of the model, the magnetic stiffness largely exceeds that expected for a simple spin degenerate band. Remarkably, these calculations correctly reproduce the stripe domain's width of 1.5 μm, observed in demagnetized PMA (Ga,Mn)As.[53]

The single domain character of the IMA (Ga,Mn)As has been confirmed by direct domain mapping using magneto-optical techniques,[47] and by full compliance to the Stoner–Wohlfarth model of coherent rotation driven by the application of the external magnetic field, which has been reported on many occasions.[30,33,36,44–49,54] In particular, it has been found that the total energy of (Ga,Mn)As with *in-plane* magnetization can be remarkably well described taking only the lowest order contributions of biaxial (in-plane cubic) and uniaxial terms. The former, as stated above, has always been found to have <100> easy axes, for the latter, the easy axis is found to be tilted by 45°, along one of the <110> directions. Because the cubic-like anisotropy energy is proportional to M^4, whereas the uniaxial one to M^2, the latter, though initially weaker, is dominating at high temperatures where M is small.

Such a uniaxial anisotropy is not expected for D_{2d} symmetry of a T_d crystal under epitaxial strain. This shows the existence of a symmetry-breaking mechanism, whose microscopic origin seems still to be unidentified. The accumulated data so far indicate that the magnitude of the corresponding anisotropy field appears to be independent of the film thickness, both for films as thick[49] as 7 μm and as thin as 50 nm.[48] Also, a surface etching experiment[48] confirms that even further thickness reduction down to 25 nm has a marginal influence on the observed anisotropic properties. The last experiment rules out the effect of Mn oxide accumulated at the free surface,[21,55] and all together strongly suggest that the mechanism responsible for the uniaxial anisotropy may be a *bulk* property of the layered material. Recently, an effect connected with surface reconstruction and induced preferential Mn incorporation repeated at every step of layer-by-layer growth has been advocated.[47,49] At the same time, the present author with coworkers[32,33] argued that a unidirectional character of the growth process or differences between (Ga,Mn)As/GaAs and (Ga,Mn)As/vacuum interfaces may lower the D_{2d} symmetry compared to the C_{2v}. There, the three principal directions [001], [110], and [$\bar{1}$10] are not equivalent. In C_{2v} the [110] ⇔ [$\bar{1}$10] symmetry gets broken, while the [100] ⇔ [010] one is maintained, conforming with the presented results. Another interesting approach points out that if not the character of the growth itself, the growth direction directed gradient of either the composition of structural defects or Mn concentration, or even both, can also effectively lead to the D_{2d} to C_{2v} symmetry lowering.[56] Such a Mn composition gradient is indeed observed in as-grown (Ga,Mn)As.[57] However, the same experimental method points out that upon annealing, the layers become largely homogenized. Because the magnetic uniaxiality is observed in both as-grown and annealed films,[48] the latter concept needs further refinement.

Perhaps a valuable clue comes from recent work,[48] where detailed studies of the *x, p,* and *T* dependencies shed new light on this issue. In particular, the authors show that the uniaxial easy axis of (Ga,Mn)As films is associated with *particular* crystallographic axes and that it can rotate 90° from the [$\bar{1}$10] to the [110] direction on annealing. Figure 3.9 gives an example of such an annealing-induced rotation. Moreover, for a specific combination of *p* and *x,* the easy axis becomes temperature dependent too, see Figure 3.10. These rotations, however, take place only if the annealing-driven increase of the hole density exceeds some well-defined value, found to be approximately $p \cong 6 \times 10^{20}$ cm^{-3} for the series of samples studied. Such a universal dependence on *p,* together with the aforementioned lack of both thickness dependence and surface effects, decisively points to a symmetry lowering mechanism existing inside the body of the film. Remarkably, the magnitude of the required symmetry lowering perturbation that explains the reported effects can be evaluated through the incorporation into the *p-d* Zener theory of ferromagnetism[3,4,34] of a trigonal distortion described by the deformation tensor component $\varepsilon_{xy} \neq 0$.[48] Even though such a distortion has not yet been seen in other experiments, it is thought to be either associated with magnetostriction, or result from a nonisotropic Mn distribution, caused for instance by the presence of surface dimers oriented along the [$\bar{1}$10] direction during the epitaxy.

The anisotropy field computed in this way corresponding to the in-plane uniaxial anisotropy $\mu_0 H_{un}$ is shown in Figure 3.11 as a function of the hole concentration

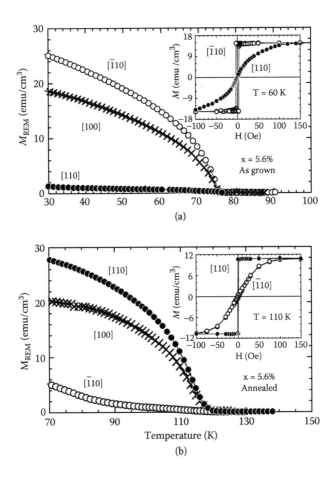

FIGURE 3.9 High temperature dependence of the three major in-plane components of the remnant magnetization in $Ga_{0.944}Mn_{0.056}As$ before (a) and after annealing (b). Insets: Magnetization curves at temperatures close to T_C measured for [110] and [$\bar{1}$10] orientations. Note the switch of the magnetic easy axis from [$\bar{1}$10] to [110] upon annealing. (After M. Sawicki, K.-Y. Wang, K.W. Edmonds, R.P. Campion, C.R. Staddon, N.R.S. Farley, C.T. Foxon, E. Papis, E. Kamiska, A. Piotrowska, T. Dietl, and B.L. Gallagher, *Phys. Rev. B*, 71, 121302(R), 2005. With permission.)

and the valence band spin-splitting parameter $B_G = A_F\beta M(T)/6g\mu_B$, where $g = 2.0$, $A_F = 1.2$ is the Fermi liquid parameter, and $\beta = -0.054$ eVnm³ is the *p-d* exchange integral. In the relevant region of hole concentration and for $\varepsilon_{xy} = 0.05\%$, $\mu_0 H_{un}$ has the experimental value of 0.1 T. For a given x, Figure 3.11 shows that a switch of the easy axis from [$\bar{1}$10] to [110] will occur on increasing hole density. As M and so B_G are decreasing functions of T, the experimentally observed uniaxial easy axis reorientation transition [$\bar{1}$10] $\Rightarrow$ [110] on increasing T is also well explained. Therefore, the model qualitatively reproduces the observed change in the easy axis direction as a function of both hole concentration and temperature. Such a correspondence

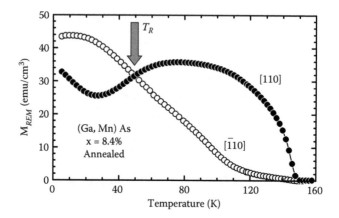

FIGURE 3.10 Temperature dependence of the [110] and [$\bar{1}$10] in-plane components of the remnant magnetization in $Ga_{0.916}Mn_{0.084}As$ annealed layer. Note the switch of the uniaxial easy direction at temperature $T_R \sim 50$ K. (After M. Sawicki, K.-Y. Wang, K.W. Edmonds, R.P. Campion, C.R. Staddon, N.R.S. Farley, C.T. Foxon, E. Papis, E. Kamiska, A. Piotrowska, T. Dietl, and B.L. Gallagher, *Phys. Rev. B,* 71, 121302(R), 2005. With permission.)

between the experimental findings and the model computations strongly underline that the origin of the uniaxial anisotropy discussed here and its rotations stems from the same, anisotropic properties of the hole liquid (valence band) subjected this time to an arbitrary direction strain (that is additional to the epitaxial one).

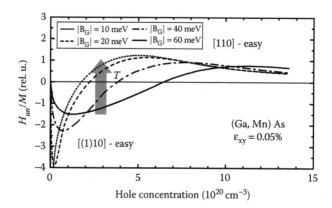

FIGURE 3.11 In-plane uniaxial anisotropy field (in International System of Units units and normalized to sample saturation magnetization) vs. hole density computed for various valence band spin splittings in (Ga,Mn)As. The thick arrow illustrates the possibility of [110] $\Rightarrow$ [$\bar{1}$10] easy axis rotation on increasing temperature as observed experimentally and depicted on Figure 3.10. (After M. Sawicki, K.-Y. Wang, K.W. Edmonds, R.P. Campion, C.R. Staddon, N.R.S. Farley, C.T. Foxon, E. Papis, E. Kamiska, A. Piotrowska, T. Dietl, and B.L. Gallagher, *Phys. Rev. B,* 71, 121302(R), 2005. With permission.)

Selecting only the xy component, the fourfold symmetry in the xy plane is broken, and the uniaxial property along <110> directions has been created. Because the valence band holes mediate the magnetic order, it is no surprise that strong x, p, and T dependencies are seen. Conversely, the effectiveness of this strain-to-magnetization transfer of information means that any foreseeable source of residual strain, such as sample bending, thermal contraction or expansion (sample holders), patterning, metallization, and so on, will result in a corresponding valence band splitting and selection of a new broken symmetry plane, and so introducing new x-, p-, and T-dependent uniaxial properties. It may be argued that this is the somehow nonsymmetrical strain imposed on (Ga,Mn)As during the whole lithographic process used to fabricate the metal–insulator–ferro-DMS tunneling device that resulted not only in uniaxial properties along [100] direction, but also in the temperature-driven change of sign of tunneling anisotropic magnetoresistance observed in the device.[14]

One more important aspect is that as the uniaxial properties dominate at high temperatures, harnessing of the strain may be either the main challenge or a source of new functionalities in future room temperature devices based on p-type ferro-DMSs. This is still to come. At present, it has to be concluded, that even taking for granted the correctness of the $\varepsilon_{xy} \neq 0$ model, the underlying microscopic mechanism that breaks D_{2d} symmetry in unperturbed (III,Mn)V epitaxial films is still to be elucidated.

3.6 CONCLUSIONS

In summary, the experimental and theoretical work presented here demonstrates how rich characteristics of magnetic properties are obtained if the magnetic subsystem is almost exclusively controlled by the spin anisotropy of the valence band subbands. Remarkably, the model calculations,[4,34] taking into account the influence of strain on the valence subbands shape and the influence of hole density and temperature on the splitting and population, are found not only to correctly describe the experimentally emerging picture, but also to quantitatively reproduce the relevant experimental findings with no adjustable parameters. From one point of view, such a great sensitivity to minute details of the growth procedure or external parameters may spell some problems in any possible utilization for hole-mediated ferromagnets; alternatively, it opens a sea of new functionalities and makes ferro-DMS an extremely attractive subject for investigations.

ACKNOWLEDGMENTS

I would like to acknowledge the constant support and encouragement of Tomasz Dietl, experimental effort of A. Idziaszek and K.Y. Wang, and fruitful collaboration with groups of Laurence Molenkamp at Wuerzburg University and of Bryan Gallagher at University of Nottingham. I am also indebted to T. Dietl and K. Edmonds for critical reading of the manuscript. The work was supported by FENIKS (EC: G5RD-CT-2001–00535), Exploratory Research for Advanced Technology Office's Semiconductor Spintronics Project of Japanese Science and Technology Corporation, and KBN Grant No. PBZ-KBN-044/P03/2001.

REFERENCES

1. F. Matsukura, H. Ohno, and T. Dietl, III-V ferromagnetic semiconductors, in *Handbook of Magnetic Materials*, Vol. 14, K.H.J. Buschow, Ed. Elsevier, Amsterdam, 2002, pp. 1–87.
2. T. Dietl, *Semicond. Sci. Technol.,* 17, 377, 2002.
3. T. Dietl, H. Ohno, F. Matsukura, F. Cibert, and D. Ferrand, *Science,* 287, 1019, 2000.
4. T. Dietl, H. Ohno, and F. Matsukura, *Phys. Rev. B,* 63, 195205, 2001.
5. T. Jungwirth, J. König, J. Sinova, J. Kucera, and A.H. MacDonald, *Phys. Rev. B,* 66, 012402, 2002.
6. I. Vurgaftman and J.R. Meyer, *Phys. Rev. B,* 67, 125209, 2003.
7. T. Dietl, III-V and II-VI Mn-based ferromagnetic semiconductors, in *Advances in Solid State Physics*, B. Kramer, Ed., Springer, Berlin, 2003, pp. 413–426.
8. H. Kpa, L. Van Khoi, C.M. Brown, M. Sawicki, J.K. Furdyna, T.M. Giebutowicz, and T. Dietl, *Phys. Rev. Lett.,* 91, 087205, 2003.
9. J. Sinova, T. Jungwirth, and J. Černe, *Int. J. Mod. Phys. B,* 18, 1083, 2004.
10. T. Jungwirth, J. Sinova, K.Y. Wang, K.W. Edmonds, R.P. Campion, B.L. Gallagher, C.T. Foxon, Q. Niu, and A.H. MacDonald, *Appl. Phys. Lett.,* 83, 320, 2003.
11. F. Matsukura, M. Sawicki, T. Dietl, D. Chiba, and H. Ohno, *Physica E,* 21, 1032, 2004.
12. H. Ohno, *J. Phys.: Condens. Matter,* 16, S5693, 2004.
13. C. Rüster, T. Borzenko, C. Gould, G. Schmidt, L.W. Molenkamp, X. Liu, T.J. Wojtowicz, J.K. Furdyna, Z.G. Yu, and M.E. Flatté, *Phys. Rev. Lett.,* 91, 216602, 2003.
14. C. Gould, C. Rüster, T. Jungwirth, E. Girgis, G.M. Schott, R. Giraud, K. Brunner, G. Schmidt, and L.W. Molenkamp, *Phys. Rev. Lett.,* 93, 117203, 2004.
15. F. Matsukura, H. Ohno, A. Shen, and Y. Sugawara, *Phys. Rev. B,* 57, 2037, 1998.
16. P.A. Korzhavyi, I.A. Abrikosov, E.A. Smirnova, L. Bergqvist, P. Mohn, R. Mathieu, P. Svedlindh, J. Sadowski, E.I. Isaev, Yu. Kh. Vekilov, and O. Eriksson, *Phys. Rev. Lett.,* 88, 187202, 2002.
17. S.J. Potashnik, K.C. Ku, S.H. Chun, J.J. Berry, N. Samarth, and P. Schiffrer, *Phys. Rev. B,* 66, 012408, 2002.
18. T. Hayashi, Y. Hashimoto, S. Katsumoto, and Y. Iye, *Appl. Phys. Lett.,* 78, 1691, 2001; S.J. Potashnik et al., *Appl. Phys. Lett.,* 79, 1495, 2001.
19. K.W. Edmonds, K.Y. Wang, R.P. Campion, A.C. Neumann, N.R.S. Farley, B.L. Gallagher, and C.T. Foxon, *Appl. Phys. Lett.,* 81, 4991, 2002.
20. K.M. Yu, W. Walukiewicz, T. Wojtowicz, I. Kuryliszyn, X. Liu, Y. Sasaki, and J.K. Furdyna, *Phys. Rev. B,* 65, 201303(R), 2002.
21. K.W. Edmonds, P. Bogusawski, B.L. Gallagher, R.P. Campion, K.Y. Wang, N.R.S. Farley, C.T. Foxon, M. Sawicki, T. Dietl, M.B. Nardelli, and J. Bernholc, *Phys. Rev. Lett.,* 92, 037201, 2004.
22. J. Blinowski and P. Kacman, *Phys. Rev. B,* 67, 121204, 2003.
23. R.P. Campion, K.W. Edmonds, L.X. Zhao, K.Y. Wang, C.T. Foxon, B.L. Gallagher, and C.R. Staddon, *J. Cryst. Growth,* 247, 42, 2003.
24. K.C. Ku, S.J. Potashnik, R.F. Wang, M.J. Seong, E. Johnson-Halperin, R.C. Meyers, S.H. Chun, A. Mascarenhas, A.C. Gossard, D.D. Awshalom, P. Schiffer, and N. Samarth, *Appl. Phys. Lett.,* 82, 2302, 2003.
25. K.Y. Wang, K.W. Edmonds, R.P. Campion, B.L. Gallagher, N.R.L. Farley, C.T. Foxon, M. Sawicki, P. Bogusawski, and T. Dietl, *J. Appl. Phys.,* 95, 6512, 2004.

26. K.Y. Wang, R.P. Campion, K.W. Edmonds, M. Sawicki, T. Dietl, C.T. Foxon, and B.L. Gallagher, in *Proceedings of the 27th International Conference on Physics of Semiconductors*, Flagstaff, AZ, July 2004, J. Menendez, and Ch. Van de Walle, Eds., New York, 2005, p. 333; cond-mat/0411475.

27. H. Munekata, A. Zaslavsky, P. Fumagalli, and R.J. Gambino, *Appl. Phys. Lett.*, 63, 2929, 1993.

28. A. Shen, H. Ohno, F. Matsukura, Y. Sugawara, N. Akiba, T. Kuroiwa, A. Oiwa, A. Endo, S. Katsumoto, and Y. Iye, *J. Cryst. Growth*, 175/176, 1069, 1997.

29. H. Ohno, F. Matsukura, A. Shen, Y. Sugawara, A. Oiwa, A. Endo, S. Katsumoto, and Y. Iye, in *Proceedings of the 23rd International Conference on Physics of Semiconductors*, Berlin 1996, M. Scheffler and R. Zimmermann, Eds., World Scientific, Singapore, 1996, p. 405.

30. X. Liu, Y. Sasaki, and J.K. Furdyna, *Phys. Rev. B*, 67, 205204, 2003.

31. M. Sawicki, F. Matsukura, T. Dietl, G.M. Schott, C. Ruester, G. Schmidt, L.W. Molenkamp, and G. Karczewski, *J. Supercond./Novel Magn.*, 16, 7, 2003.

32. M. Sawicki, F. Matsukura, A. Idziaszek, T. Dietl, G.M. Schott, C. Ruester, G. Karczewski, G. Schmidt, and L.W. Molenkamp, cond-mat/0212511, unpublished.

33. M. Sawicki, F. Matsukura, A. Idziaszek, T. Dietl, G.M. Schott, C. Ruester, G. Karczewski, G. Schmidt, and L.W. Molenkamp, *Phys. Rev. B*, 70, 245325, 2004.

34. M. Abolfath, T. Jungwirth, J. Brum, and A.H. MacDonald, *Phys. Rev. B*, 63, 054418, 2001.

35. H. Boukari, P. Kossacki, M. Bertolini, D. Ferrand, J. Cibert, S. Tatarenko, A. Wasiela, J.A. Gaj, and T. Dietl, *Phys. Rev. Lett.*, 88, 207204, 2002.

36. G.P. Moore, J. Ferré, A. Mougin, A. Moreno, and L. Däweritz, *J. Appl. Phys.*, 94, 4530, 2003.

37. P. Van Dorpe, Z. Liu, W. Van Roy, V.F. Motsnyi, M. Sawicki, G. Borghs, and J. De Boeck, *Appl. Phys. Lett.*, 84, 3495, 2004.

38. T. Dietl, *J. Phys.: Condens. Matter*, 16, S5471, 2004.

39. P. Kossacki, W. Pacuski, W. Malana, J.A. Gaj, M. Bertolini, D. Ferrand, S. Tatarenko, and J. Cibert, *Physica E*, 21, 943, 2004.

40. J.K. Furdyna, X. Liu, T. Wojtowicz, W.L. Lim, U. Welp, and V.K. Vlasko-Vlasow, in *Advances in Solid State Physics*, B. Kramer, Ed. Springer, Berlin, 2004, pp. 515–530.

41. H. Ohno, D. Chiba, F. Matsukura, T. Omiya, E. Abe, T. Dietl, Y. Ohno, and K. Ohtani, *Nature*, 408, 944, 2000.

42. K. Takamura, F. Matsukura, D. Chiba, and H. Ohno, *Appl. Phys. Lett.*, 81, 2590, 2002.

43. T. Endo, T. Supiski, S. Yanagi, A. Oiwa, and H. Munekata, unpublished.

44. S. Katsumoto, A. Oiwa, Y. Iye, H. Ohno, F. Matsukura, A. Shen, and Y. Sugawara, *Phys. Status Solidi B*, 205, 115, 1998.

45. D. Hrabovsky, E. Vanelle, A.R. Fert, D.S. Yee, and J.P. Redoules, J. Sadowski, J. Kanski, and L. Ilver, *Appl. Phys. Lett.*, 81, 2806, 2002.

46. H.X. Tang, R.K. Kawakami, D.D. Awschalom, and M.L. Roukes, *Phys. Rev. Lett.*, 90, 107201, 2003.

47. U. Welp, V.K. Vlasko-Vlasov, X. Liu, J.K. Furdyna, and T. Wojtowicz, *Phys. Rev. Lett.*, 90, 167206, 2003.

48. M. Sawicki, K.-Y. Wang, K.W. Edmonds, R.P. Campion, C.R. Staddon, N.R.S. Farley, C.T. Foxon, E. Papis, E. Kamiska, A. Piotrowska, T. Dietl, and B.L. Gallagher, *Phys. Rev. B*, 71, 121302(R), 2005.

49. U. Welp, V.K. Vlasko-Vlasov, X. Liu, J.K. Furdyna, and T. Wojtowicz, *Appl. Phys. Lett.*, 85, 260, 2004.

50. M. Dumm, B. Uhl, M. Zölfl, W. Kipferl, and G. Bayreuther, *J. Appl. Phys.*, 91, 8763, 2002.
51. X. Liu, W.L. Lim, Z. Ge, S. Shen, T. Wojtowicz, K.M. Yu, W. Walukiewicz, M. Dobrowolska, and J.K. Furdyna, in *Proceedings of the 27th International Conference on Physics of Semiconductors,* Flagstaff, AZ, AIP, Melville, NY, in press; *Appl. Phys. Lett.,* in press.
52. T. Dietl, J. Koenig, and A.H. MacDonald, *Phys. Rev. B,* 64, R241201, 2001.
53. T. Shono, T. Hasegawa, T. Fukumura, F. Matsukura, and H. Ohno, *Appl. Phys. Lett.,* 77, 1363, 2000.
54. K.-Y. Wang, M. Sawicki, K.W. Edmonds, R.P. Campion, C.R. Staddon, N.R.S. Farley, C.T. Foxon, T. Dietl, and B.L. Gallagher, *Phys. Rev Lett.,* 95, 217204, 2005.
55. J.K. Furdyna, T. Wojtowicz, X. Liu, K.M. Yu, and W. Walukiewicz, *J. Phys.: Condens. Matter,* in press.
56. Z. Wilamowski et al., in preparation.
57. B.J. Kirby, J.A. Borchers, J.J. Rhyne, S.G.E. te Velthuis, A. Hoffmann, K.V. O'Donovan, T. Wojtowich, X. Liu, W.L. Lim, and J.K. Furdyna, *Phys. Rev. B,* 69, R081307, 2004.

4 Soft X-Ray Resonant Magnetic Scattering from Magnetic Nanostructures

Gerrit van der Laan

CONTENTS

4.1 INTRODUCTION

Confinement of the dimensions of a magnetic structure to a microscopic scale can lead to intriguing phenomena that are not encountered in bulk materials. Controlled thin film deposition and lithographic patterning have been established as the technology promoters in magnetism research ever since the engineering of multilayers of magnetic and nonmagnetic metals led to the discovery of new phenomena such as interlayer exchange coupling, giant magnetoresistance (GMR), and nonvolatile magnetic random access memory (MRAM). Several different methods are available to characterize the magnetic properties of thin films and confined magnetic structures. Among these, magneto-optical methods are most prominent and have the advantage that they can be performed under applied magnetic and electric fields. However, magneto-optics in the visible region lack element specificity and, moreover, the ultimate spatial resolution is limited by the wavelength. The discovery of strong magneto-optical effects in the core level x-ray absorption edges of magnetic elements[1] has opened up new possibilities, such as the separation of the spin and orbital part of the magnetic moments[2] and the determination of the magnetic anisotropy.[3] Such techniques strongly benefit from the advent of synchrotron radiation undulator devices offering variable linear and circular polarization with tunable photon energy.

X-ray magnetic scattering experiments were pioneered by de Bergevin and Brunel on NiO using a tube source.[4] Early calculations by Blume[5] showed that resonance effects occur near an absorption edge and a very small effect was indeed observed by Namikawa et al.[6] in ferromagnetic Ni by tuning the incident x-ray energy to the K edge. Large resonant enhancements of the magnetic scattering were discovered at the $L_{2,3}$ edges of the rare earths[7] and the $M_{4,5}$ edges of the actinides,[8] which could be explained in terms of atomic physics.[9] Strong magnetic dichroism was predicted[10] and, soon after, observed[11] in x-ray absorption (i.e., forward scattering) proportional to the element-specific magnetic moment. The selection rules in x-ray magnetic circular dichroism (XMCD) and x-ray magnetic linear dichroism (XMLD)[12,13] give a difference in the transition probabilities for left- and right-circular polarized x-rays into the unoccupied polarized valence band. Thus, x-ray resonant magnetic scattering combines the advantages of x-ray scattering with those of the magnetic x-ray dichroism techniques.

In the past, x-ray scattering experiments have been mainly restricted to the region of the hard x-rays, offering high spatial resolution and large penetration. However, in the *soft* x-ray range, the resonant magnetic scattering cross sections are much stronger. Although soft x-rays usually have prohibitively long wavelengths for Bragg diffraction from crystal lattices, in $3d$ transition metals, the wavelength of the $L_{2,3}$ absorption edges[14] matches perfectly onto the nanometric length scale of artificial multilayers[15] and periodic domain structures.[16,17] Soft x-ray resonant magnetic scattering (SXRMS) has the ability to cover the length scale from 1 to 1000 nm and the measurement is relatively easy to perform on *ex situ* prepared samples.[18,19] Indeed, it has grown into a workhorse technique within the time span of a few years after pioneering experiments by Dürr et al.[17] who measured the circular dichroism of the in-plane diffracted intensity from periodic stripe domains in FePd thin films in reflection geometry. Applications of SXRMS also include the characterization of structural and magnetic properties of layered and domain systems, especially their interface roughness, induced magnetic order in nonmagnetic spacer layer, and layer-resolved magnetic moments.[20–26]

Element specificity is obtained by tuning the x-ray energy to the appropriate absorption edge. In $3d$ transition metals, the excitation of $2p$ electrons into unoccupied $3d$ states leads to a large enhancement of the scattering cross section in the soft x-ray region.[27] Large effects are also observed in rare earths for the $3d$ to $4f$ transitions.[28–30] Considerable enhancements occur also at the $2p$ absorption edges of $4d$ and $5d$ transition metals[31] and at the $3d$ and $4d$ absorption edges of actinides.[32,33] Even absorption edges as low as the $3p$ to $3d$ transitions (located at around 50 eV) in $3d$ transition metals have been successfully explored.[34] The tunability of the synchrotron radiation allows us to scan the photon energy across the resonance for each of the accessible Bragg peaks. X-ray anomalous diffraction permits us to investigate the energy dependence of the anomalous part of the atomic scattering factor using energies of the incident photon in the neighborhood of an atomic absorption edge.[35,36] This technique consists of recording the intensity of a Bragg reflection as a function of the energy of the scattered photons crossing an atomic absorption edge. Due to the fact that the energies of the absorption edges are specific to an atom and an electronic level, x-ray anomalous diffraction has all the capabilities of diffraction and x-ray absorption combined in a single technique. It provides short-range order information about the set of long-range ordered atoms selected by the diffraction condition and it is chemical-, valence-, and site-specific. X-ray resonant scattering is an x-ray anomalous diffraction technique, where the reflections investigated are either Thomson forbidden, or have a very low intensity. The main contribution to the x-ray structure factor for these reflections arises from the anomalous part of the atomic scattering factor of a specific atom (selected by tuning the photon energy to its absorption edge). As the structure factor for forbidden (or almost forbidden) reflections is given by the difference of atomic scattering factors of particular atoms in the lattice, the intensity of this reflection are only visible at selected photon energies (i.e., around the resonance energies) where the contrast of the anomalous part of the scattering factor is enhanced. The atomic anomalous scattering factor is intimately correlated with the x-ray absorption coefficient and is a rank-two tensor for electric-dipole transitions. Thus resonant effects can be observed as arising from differences among the terms of the scattering tensor or from a different orientation of this tensor in the lattice, which gives rise to anisotropic tensor scattering (ATS) reflections. In the first case, the absorption edges (or the anomalous scattering factors) of atoms with different valence states are shifted in energy (chemical shift). Consequently, a strong resonance at the absorption edge must be observed for charge densities with the periodicity of the reflection studied in the crystal. The appearance of a resonance at the absorption edge does not guarantee the occurrence of charge ordering, as ATS reflections can also give rise to the same kind of resonance. In this second case, the intensity of the ATS reflections depends on the x-ray polarization and azimuthal angle of the diffraction plane. Thus the azimuthal and polarization dependence of the x-ray resonant scattering allows us to discriminate between ATS and charge-ordering reflections. We can summarize the above-mentioned extensions to x-ray scattering as follows:

Extension to X-Ray Scattering:	Allows Probing of:
Tunable x-rays at resonance	Element, site, and valence specificity
Polarized x-rays	Magnetic orbital and spin structure
Soft x-rays	Nanoscale sensitivity
Coherent radiation	Local configuration
Pulsed radiation	Dynamics

All this is available with synchrotron radiation.

We will first present two examples exploiting the polarization properties of the x-rays at the (fixed) resonance energy: (i) the study of magnetic stripe domains (Section 4.2 to Section 4.5), (ii) the scattering with coherent radiation (Section 4.6 and Section 4.7). The two final examples take specific advantage of the energy dependence of the Bragg reflections: (iii) the analysis of the magnetic profile in artificially grown multilayers (Section 4.8 to Section 4.9) and (iv) the study of the orbital ordering in manganites (Section 4.12).

The synopsis of the remainder of this chapter is as follows.

Example (i) — The magnetic stripe domains in FePd thin films, which show varying degrees of perpendicular magnetic anisotropy (PMA), will serve as the first example. The films are grown by codepositing Fe and Pd at elevated temperatures and depending on the precise growth conditions, the films are more or less chemically ordered, with Fe and Pd occupying alternating layers in a tetragonally distorted face-centered-cubic phase. This chemical ordering leads to the PMA, so that the anisotropy increases with the degree of order. The competition between the PMA and the thin-film shape anisotropy leads to well-ordered stripe domains with up and down magnetization perpendicular to the film plane. The magnetic stripes with a period of typically 100 nm can be imaged by magnetic force microscopy (MFM) and modeled by micromagnetic calculations, and they give rise to magnetic peaks in the diffraction pattern measured by SXRMS. Formation of magnetic closure domains will reduce the energy of the magnetic flux lines outside the sample. Dürr et al.[17] demonstrated the presence of these closure domains in FePd thin films with SXRMS using circularly polarized x-rays. Closure domains with in-plane magnetization become energetically unfavorable in samples with strong PMA because of the preference for the easy axis to be perpendicular to the surface. Dudzik et al.[37] studied the influence of PMA on closure domains in FePd films using SXRMS at the Fe L_3 absorption edge by fitting the circular dichroism in the magnetic scattering and by fitting the structure in the magnetic rod scans.

Section 4.2 discusses stripe domain formation in FePd thin film alloys, showing some of the variations that can be observed with MFM on samples with different magnetic anisotropies. Section 4.3 shows the results of micromagnetic calculations for the samples. In Section 4.4, we discuss the principles of resonant scattering and give the selection rules of the magnetization components for linearly and circularly polarized light. In Section 4.5, we show some experimental SXRMS results for the FePd thin films and discuss the observation of the magnetic closure domains in circular dichroism.[38]

Example (ii) — With the advent of x-ray free electron lasers, the future for coherent x-rays will be bright. In anticipation, we present the magnetic speckle pattern from an 8 μm FePd wire measured using coherent soft x-rays with the scattering detected by a charge-coupled device (CCD) camera.[39] A high degree of coherence is obtained by using a pinhole in front of the sample, as evidenced from the observed intensity fluctuations in the scattering image. The speckle pattern can provide information about the local (dis)order of the magnetic stripe domains in the wire sample. Thus using coherent radiation, it is possible to study the local magnetic configuration and SXRMS has great potential in the time domain to study dynamical processes, even in circumstances where the global magnetic configuration stays the same.

Section 4.6 introduces the coherence of the x-ray source. Section 4.7 shows some speckle patterns and discusses the intensity fluctuations in scattering image using coherent x-rays.[40]

Example (iii) — X-ray resonant magnetic scattering (XRMS) across the W L_3 absorption edge, in combination with x-ray magnetic circular dichroism and anomalous x-ray reflectivity, was used to obtain the magnetic polarization profile in a Fe(30 Å)/W(11 Å) multilayer. Analysis of the experimental results shows that there is a strong oscillatory behavior of the magnetization in the W layer. This tendency is in good agreement with recent theoretical predictions.

In Section 4.8, we present the study on Fe/W multilayers.[31] The experimental results are given in Section 4.9, and the analysis method is discussed in Section 4.10. In Section 4.11, we show that the energy dependence of the Bragg reflections across the W L_3 edge provides evidence for the oscillatory behavior of the W $5d$ magnetic moment.

Example (iv) — In order to demonstrate the power of the soft x-ray resonant diffraction technique, we briefly give some results for the orbital ordering in the layered perovskite $La_{0.5}Sr_{1.5}MnO_4$. The lattice parameter in this crystal is sufficiently large to do resonant scattering in the soft x-ray region. The energy dependence of the orbital ordering diffraction peak over the $L_{2,3}$ edges is compared to ligand-field calculations allowing a distinction between the influences of Jahn–Teller distortions and spin correlations.

In Section 4.12, we study the orbital ordering in manganites.[41] Finally, Section 4.13 gives some conclusions and an outlook.

4.2 MAGNETIC STRIPE DOMAIN FORMATION IN FePd ALLOYS

4.2.1 SAMPLE PREPARATION

The systems of interest here are self-organizing magnetic patterns as formed in thin films of FePd alloys. The magnetization in these systems breaks up in domains of different directions to reduce the stray field energy. The formation of these domains

TABLE 4.1
Description and Magnetic Characteristics of FePd Thin Films with Different Magnetic Anisotropies as Quantified by the Quality Factor Q

	Sample M1	Sample M2	Sample M3
Anisotropy	Weak	Medium	Strong
Composition	MgO/Cr/Pd/FePd/Pd	MgO/FePd/Pd	MgO/Cr/Pd/FePd/Pd
FePd thickness	44 nm	40 nm	34 nm
Growth	Layer-by-layer (RT)	Codeposit (220°C)	Codeposit (350°C)
Q (VSM)	0.35	0.6	1.5
Stripes (MFM)	Strongly aligned	Aligned	Badly aligned
Stripe period	110 nm	100 nm	~100 nm

Note: The corresponding MFM images of the three samples are shown in Figure 4.1.

coincides with the creation of domain walls where the magnetization rotates from one direction to the other at the expense of the exchange and anisotropy energies.

The thin films were grown at the Commissariat à l'Energie Atomique (CEA) in Grenoble, France by depositing ~40 nm FePd with molecular beam epitaxy (MBE) onto a MgO(001) substrate. The layer was capped with a 20 Å thick Pd layer to prevent oxidation and contamination.[42,43] The codeposition of Fe and Pd at equiatomic composition was performed at an elevated temperature of the substrate to control the partial chemical ordering.[44] Growth conditions were varied to obtain a range of different anisotropies. Samples ranging from low, medium, to high magnetic anisotropy are listed in Table 4.1 together with their properties.

4.2.2 MAGNETIC ANISOTROPY AND MFM IMAGES

The perpendicular magnetic anisotropy can be quantified by the quality factor $Q = K_u/2\pi M_s^2$, where K_u is the perpendicular anisotropy constant and M_s is the saturation magnetization per volume unit. The magnetic configuration depends both on the layer thickness and on the quality factor Q, which strongly depend on the disorder as controlled by the deposition temperature. Figure 4.1 shows the MFM images for three rather different FePd layers of ~40 nm with the anisotropies listed in Table 4.1. Sample M1, obtained using layer-by-layer growth at room temperature (RT), shows a weak anisotropy ($Q = 0.35$) as estimated by vibrating sample magnetometry (VSM). The MFM image, shown in Figure 4.1a, displays well-aligned stripes. Sample M2, obtained by codeposition at 220°C, has a medium anisotropy ($Q = 0.8$) and the MFM image in Figure 4.1b shows aligned stripes but with the presence of many forks and bifurcations. Finally, Sample M3, obtained by codeposition at 350°C, has a strong anisotropy ($Q = 1.5$) and the MFM image in Figure 4.1c shows highly disordered stripes.

The FePd thin films that were studied, and whose MFM images are presented in Figure 4.1, exhibit magnetic stripes in the natural as-deposited state. Under certain

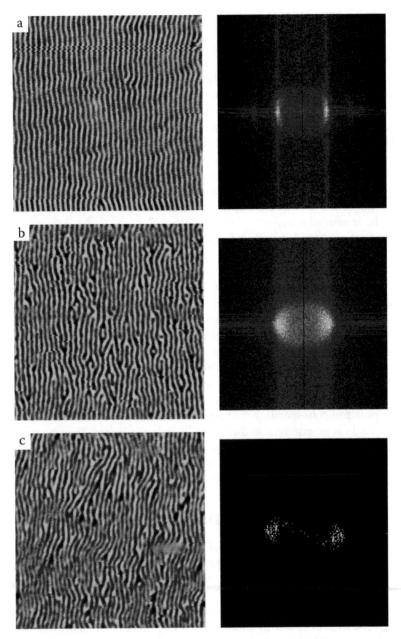

FIGURE 4.1 MFM images of the magnetic stripe domains in thin-film FePd alloys. (a) M1 with weak anisotropy; (b) M2 with medium anisotropy; (c) M3 with strong anisotropy. The sample details are given in Table 4.1. The domains show up in dark and light shading corresponding to up and down magnetization. The size of the MFM images is 4 μm. At the right-hand side, the images are represented by their Fourier transforms.

conditions (thickness and growth temperature), the stripes are naturally ordered and there is no need for in-plane demagnetizing or field cycling the sample to form them. It appears that in the case of low anisotropy, the stripes are spontaneously aligned, probably because of energy minimization (in the case of low anisotropy, the in-plane component is very strong, and therefore the energy cost of the domain walls is lower when they are aligned parallel instead of disordered). In other words, all MFM images that are presented here were obtained in a natural as-deposited state and were not obtained after any specific magnetic field history or different response to the same field history.

The evolution observed in the stripe disorder in Figure 4.1 is due to the difference in the processes that create the magnetic walls. When the anisotropy is weak ($Q < 1$) closure domains will lead to a gradual rotation of the magnetization over the period of the stripe; whereas for strong anisotropy ($Q > 1$), the energy of the domain walls is much larger and their creation occurs by nucleation processes in a way that gives rise to the interlaced domains.

Although we have indirectly inferred the presence of closure domains from the stripe ordering observed in the MFM image, this technique is not able to resolve the details of the magnetic closure domains. MFM is probing the magnetic stray fields above the sample surface, so that the images reveal only the magnetization in the up and down directions. SXRMS, however, gives access to complementary information, such as the in-plane magnetization, depth profile and correlation length. SXRMS is also insensitive to an applied magnetic field, so that it is a suitable technique to perform hysteresis studies.

The right-hand side of Figure 4.1 shows the Fourier transforms of the MFM images, which provide already a first impression of the periodicity distribution of the stripes. At both sides, symmetrically around the middle of the transformed image, there is peak that corresponds to the periodicity of the magnetic structure. The width of these peaks is proportional to the inverse correlation length. Thus the correlation length of the stripes is much larger for the well-ordered Sample M1, as is indeed expected. To obtain any information from within the sample as opposed to above the sample, we will have to do SXRMS.

4.3 MICROMAGNETIC MODELING

The microscopic magnetic structure of the layers can be calculated by solving numerically the position-dependent magnetization vectors using the micromagnetic equations and minimizing the total energy. Using the Gilbert–Landau–Lifshitz (GLL) formulation,[45,46] the magnetization is split up in elementary cells. The total energy contains the Zeeman energy due to the external field, the exchange energy, the magnetic anisotropy energy connected with K_u, and the demagnetization energy, which depends on the form of the volume magnetization and the saturation magnetization M_s. The energy is minimized under the constraints that the total length of the magnetization vector for each cell is normalized and that the derivative of the magnetization vanishes perpendicular to the surface.

Figure 4.2 shows the result of the GLL calculation for a 40 nm FePd layer with weak and strong magnetic anisotropy (comparable to Sample M1 and Sample M3, respectively, in Table 4.1). The figure shows that for constant thickness, the magnetic

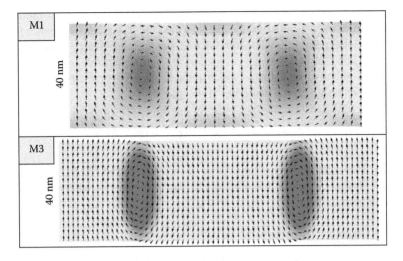

FIGURE 4.2 Micromagnetic simulation for FePd layers M1 with Q = 0.35 and M3 with Q = 1.5 corresponding to a weak and strong magnetic anisotropy, respectively. The layer thickness is 40 nm. The magnetic periods obtained from the calculation are 110 and 124 nm, respectively. Each cell (~1 nm containing ~10 Fe atoms) is represented by a normalized magnetization vector, shown in the m_z and m_y (transverse) direction and the third component along the m_x (longitudinal) direction is represented by the change in gray scale gradually increasing from light gray to dark gray.

period increases when the anisotropy increases. Near the surface, we observe a pattern ↑→↓← . The effective width of the in-plane → and ← magnetization of the closure domains is shrinking with increasing anisotropy.

4.4 PRINCIPLES OF X-RAY RESONANT SCATTERING

4.4.1 SCATTERING AMPLITUDE

The kinetic energy operator $(\mathbf{p} - e\mathbf{A}/c)^2$ in the photon-matter interaction contains a first-order term in the squared vector potential $\mathbf{A}^2$, which gives rise to the (nonresonant) Thomson scattering, and a second-order term in $\mathbf{p} \cdot \mathbf{A}$, describing the resonant scattering. The $\mathbf{p} \cdot \mathbf{A}$ term will become dominant at the resonance energy for excitation of a $2p$ core electron into an unoccupied $3d$ valence state. Using the electric-dipole approximation, $\mathbf{A} = \varepsilon \exp(i\mathbf{k} \cdot \mathbf{r}) \approx \varepsilon$, the matrix element for the coherent amplitude of resonant photon scattering can then be written as[9,47]

$$\sum_i \langle g | f_i | g \rangle \equiv \sum_i \frac{\langle g | \varepsilon'^* \cdot \mathbf{p} | i \rangle \langle i | \mathbf{p} \cdot \varepsilon | g \rangle}{E_i - E_g - \omega - i\Gamma_i / 2} \tag{4.1}$$

where ε (ε') is the polarization vector of the incident (scattered) x-ray beam. The intermediate state $|i\rangle$ makes the scattering selective to the environment (element, site, magnetism, and charge anisotropy).

The photon energy ω selects a set of intermediate states $|i\rangle$ with lifetime Γ_i, assessable from the ground state $|g\rangle$. The intermediate state in the resonance process allows additional paths otherwise forbidden in the direct transition, such as spin-flip transitions. The matrix element in Equation 4.1 can be written as the expectation value of an effective operator f_i, which after recoupling gives the amplitude for coherent resonant magnetic scattering at the site i[9,48]

$$f_i(\omega) = (\hat{\varepsilon}'^* \cdot \hat{\varepsilon})F^0 - i(\hat{\varepsilon}'^* \times \hat{\varepsilon}) \cdot \hat{\mathbf{M}}_i F^1 + (\hat{\varepsilon}'^* \cdot \hat{\mathbf{M}}_i)(\hat{\varepsilon} \cdot \hat{\mathbf{M}}_i)F^2 \qquad (4.2)$$

where $\hat{\mathbf{M}}_i$ is the magnetization unit vector at the ith site, and F^0, F^1, and F^2 are the monopole, magnetic dipole, and quadrupole part of the energy dependent resonance amplitude, respectively.

4.4.2 SCATTERING INTENSITY

The magnetic diffraction intensity is a function of both ω and the momentum transfer, given by the scattering vector $\mathbf{q} = \mathbf{k}' - \mathbf{k}$, where $\mathbf{k}$ ($\mathbf{k}'$) is the wave vector of the incident (scattered) x-ray beam. Within the first Born approximation, one obtains

$$I(\omega, \mathbf{q}) = \left| \sum_i f_i \, e^{i\mathbf{q}\cdot\mathbf{r}} \right|^2 \qquad (4.3)$$

where the summation is over the sites at $\mathbf{r}_i$.

For a simple collinear structure, with all the magnetization vectors parallel, the polarization dependence of the scattering is the same as for the single site in Equation 4.2. This gives an overlap of magnetic and charge peaks, and to study ferromagnetic materials, one has to use the interference between the two amplitudes that results in a change in the scattering signal upon reversal of the sample magnetization. Antiferromagnetically coupled layers or domains offer, like spiral structures, the advantage that the magnetic diffraction peaks are separated from the structural diffraction peaks. Doubling of the periodicity in real space results in magnetic peaks with half the structural period in reciprocal space. Distinction between magnetic scattering and Thomson scattering is achieved by tuning on and off the resonance energy.

4.4.3 LINEAR POLARIZATION

The direction of the magnetization vector at site i can be assessed using the polarization dependence given in Equation 4.2. We are using the convention to denote linear polarization perpendicular and parallel to the scattering plane as σ and π polarization, respectively. Thus in Figure 4.4, the unit vectors $\hat{\sigma}$ and $\hat{\sigma}'$ are along the y direction (transverse direction) and $\hat{\pi}$ and $\hat{\pi}'$ are in the xz plane. Because $\hat{\sigma}' \times \hat{\sigma} = 0$, $\hat{\pi}' \times \hat{\sigma} = -\hat{\mathbf{k}}'$, $\hat{\sigma}' \times \hat{\pi} = \hat{\mathbf{k}}$, and $\hat{\pi}' \times \hat{\pi} = (\hat{\mathbf{k}} \times \hat{\mathbf{k}})$, where $\hat{\mathbf{k}}$ ($\hat{\mathbf{k}}'$) is the unit vector along the direction of the ingoing (outgoing) photon direction, the in-product $(\hat{\varepsilon}'^{*} \cdot \hat{\varepsilon})$ and the vector product $(\hat{\varepsilon}'^{*} \times \hat{\varepsilon}) \cdot \mathbf{M}_i$ in Equation 4.2 result in the following "selection rules" for linear polarization:

$$\sigma \rightarrow \sigma' : \text{only charge scattering}$$

$$\sigma \rightarrow \pi' : -\hat{\mathbf{k}}' \cdot \hat{\mathbf{M}}$$

$$\pi \rightarrow \sigma' : \hat{\mathbf{k}} \cdot \hat{\mathbf{M}}$$

$$\pi \rightarrow \pi' : (\hat{\mathbf{k}}' \times \hat{\mathbf{k}}) \cdot \hat{\mathbf{M}} + \text{charge scattering}$$

Thus the two rotated polarization channels $\sigma \rightarrow \pi'$ and $\pi \rightarrow \sigma'$ measure $\mathbf{M}_{//}$ the magnetization component parallel to the scattering plane, which is spanned by $\hat{\mathbf{k}}$ and $\hat{\mathbf{k}}'$. The $\pi \rightarrow \pi'$ polarization channel measures $\mathbf{M}_{\perp}$, the magnetization component perpendicular to the scattering plane. Note that even without polarization analysis of the outgoing x-rays, one can obtain information about the magnetization direction, because σ polarized light probes only the in-plane component and not to the perpendicular component of the magnetization vector. However, π polarized light probes both the in-plane and perpendicular component of the magnetization, except at very grazing incidence angles $(\hat{\mathbf{k}}' \approx \hat{\mathbf{k}})$, where the perpendicular component vanishes.

4.4.4 Circular Polarization

Left- and right-circularly polarized light are described by the two polarization vectors $\hat{\varepsilon}_{\pm} = (\hat{\sigma} \pm i\hat{\pi}) / \sqrt{2}$. For coherent scattering, the polarization of the final state must be the same. Because only π' polarization can give a nonzero product with both σ and π, the only nonzero vector product combination is $(\hat{\pi}' \times \hat{\sigma}) \cdot \mathbf{M}$ with $i(\hat{\pi}' \times \hat{\pi}) \cdot \mathbf{M}$. The cross section for circular polarized light can be written as:

$$\left(\frac{d\sigma}{d\omega}\right)_{\pm} \propto \pm \text{Im}\{[(\hat{\pi}' \times \hat{\sigma}) \cdot \mathbf{M}]^{*}[(\hat{\pi}' \times \hat{\pi}) \cdot \mathbf{M}]\} = \pm \text{Im}\{(-\hat{\mathbf{k}}' \cdot \hat{\mathbf{M}})^{*}[(\hat{\mathbf{k}}' \times \hat{\mathbf{k}}) \cdot \hat{\mathbf{M}}]\}$$

(4.4)

This requires both a real and imaginary part for the magnetization, which means that the components in the two perpendicular planes must be out-of-phase, that is, $\mathbf{M}_{\pm} = \mathbf{M}_{//} \pm i\mathbf{M}_{\perp}$. This is accomplished in a sinusoidal, helical, or chiral magnetization pattern, in which case circular dichroism, that is, the difference between left- and right-circularly polarized light, gives first-order magnetic peak intensities of opposite sign.[17,49] Analysis of the polarization of the scattered light would show that the dichroism vanishes for σ' polarization.

Magnetic peaks can be observed for both σ and π polarization of the incident beam, but it is only the out-of-phase interference between the scattering amplitudes from different domains that can lead to π' polarization of the outgoing beam. Thus the phase relation, which is lost for linear polarization, is retrieved with circular polarization. The pattern ↑→↓← (as encountered near the surface of the FePd film, cf. Figure 4.2) is invariant for an operation where the unit magnetization vectors are translated by $\pm\pi/2\tau$ (where π stands for 3.14) and simultaneously rotated by $\pm\pi/2 = \pm 90°$. The rotation is clockwise (anticlockwise) for a translation in the positive (negative) direction. Therefore, right- and left-circularly polarized light give magnetic scattering peaks at $-\tau$ and $+\tau$, respectively.

4.4.5 MICROMAGNETIC MODELING OF THE SCATTERING

It is perhaps interesting to mention that using Equation 4.2 and Equation 4.3, the SXRMS can be calculated starting from micromagnetic model obtained using the GLL equations (cf. Section 4.3) that give the position-dependent magnetic moments in the domain pattern. Reference 50 and Reference 51 have reported on the numerical simulation of the SXRMS measurements, including both the structural and magnetic scattering by the Fourier transform over all cells from the micromagnetic simulation. The magnetic profile $\hat{\mathbf{M}}_i$ couples with the x-ray polarization vectors to the scalar products $(\hat{\varepsilon}'^* \times \hat{\varepsilon}) \cdot \mathbf{M}_i$ and $(\hat{\varepsilon}'^* \cdot \mathbf{M}_i)(\hat{\varepsilon} \cdot \mathbf{M}_i)$, where the index i runs over the magnetic cells. These calculations take into account roughness, diffusion, and the x-ray absorption attenuation at the resonance energy.

4.5 EXPERIMENTAL SXRMS RESULTS FOR FEPD FILMS

The magnetic scattering experiments discussed here were carried out in the Daresbury two-circle diffractometer using circularly polarized x-rays from both beam line 1.1 of the Synchrotron Radiation Source (SRS) at Daresbury Laboratory, United Kingdom and beam line ID12B (currently ID08) at the European Synchrotron Radiation Facility (ESRF) in Grenoble. The $\theta - 2\theta$ diffractometer is under high vacuum to prevent x-ray absorption by air. The diffractometer has been described elsewhere.[18,19,52]

Figure 4.3 shows a two-dimensional (2D) diffraction scan from a FePd alloy thin film measured at the Fe L_3 resonance energy. Symmetrically around the specular peak, one observes first-order magnetic peaks with a modulation vector, τ, corresponding to a domain periodicity $2\pi/\tau$. Strong Kiessig fringes are present along both the specular rod and the magnetic rods due to interference of the scattering amplitudes from the top and bottom interface of the FePd layer. The period of these oscillations are different because the magnetic interfaces do not coincide with the structural interfaces.

Figure 4.4 shows the SXRMS at the Fe L_3 energy from a FePd alloy sample of type M2 with striped magnetic domain structures (MFM period of 90 nm). The sample was measured with left- (solid line) and right- (dotted line) circularly polarized x-rays in the geometry with the scattering plane parallel to the magnetic stripes.[17,49] This experimental geometry is shown in the inset. The detector is scanned horizontally across the diffraction pattern. The magnetic peaks are observed around the central, specularly reflected x-ray beam, at $q_y/2\pi = \pm 0.011$ Å^{-1} confirming the

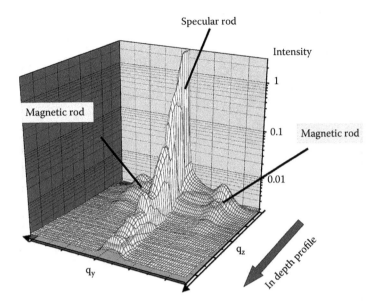

FIGURE 4.3 2D soft x-ray resonant magnetic scattering at the Fe L_3 resonance from a FePd trilayer with different anisotropies for each layer. Shown is the $q_y q_z$ plane, where q_y (q_z) is parallel (perpendicular) to the sample plane. Sample: MgO(001)/Pd(buffer, 60 nm)/FePd (10 nm)/Pd (2 nm)/FePd (10 nm)/Pd (2 nm)/FePd (10 nm).

real space period of $\tau = 2\pi/q_y = 90$ nm (y is transverse direction). The magnetic peaks are only observed at photon energies within a few electron volts of the absorption edge, which confirms their magnetic origin. The magnetic peaks exhibit a strong circular dichroism. As explained in Section 4.4, this requires a periodic magnetic ordering near the surface with a pattern $\uparrow \rightarrow \downarrow \leftarrow$, hence evidencing the presence of closure domains. A magnetic pattern $\uparrow \downarrow \uparrow \downarrow$ would not give any circular dichroism.

Depth profiling of the closure domains can be performed by measuring the scattering intensity as a function of Bragg angle. Modeling of the asymmetry ratio of the first- and second-order magnetic peaks as a function of the Bragg angle gave a value of ~8.5 nm for the effective depth of the closure domains for ~40 nm FePd films with low to medium anisotropy.[37] This effective depth is in good agreement with the modeled micromagnetic structure in Figure 4.2 and also with the results obtained from a fit of the measured magnetic rod scans.[37]

It is also possible to determine directly the magnetic correlation length. The inset of Figure 4.4 shows the peak width in q_y for the magnetic reflection measured in the geometry with the scattering plane perpendicular to the magnetic stripes, where the sample was rocked while the detector was kept at constant angle. The inset compares Sample I (of type M1) with weak anisotropy ($Q = 0.4$) and Sample II (of type M2) with medium anisotropy ($Q = 0.8$). The correlation length obtained from the peak width is 1.6 μm (~18 periods) for Sample I and 0.63 μm (~7 periods) for Sample II. This confirms that the stripe domains are much better ordered in the low-PMA Sample I.

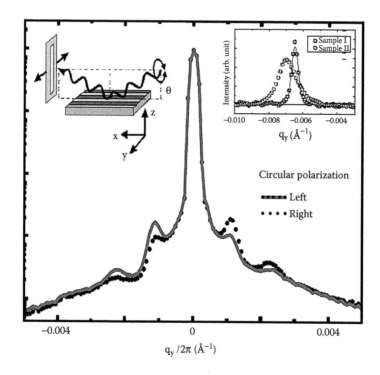

FIGURE 4.4 X-ray resonant magnetic scattering (on a logarithmic scale) at the Fe L_3 edge of a FePd alloy thin film with striped magnetic domain structures (period of ~90 nm) measured with left- (solid line) and right- (dotted line) circularly polarized x-rays in a geometry with the scattering plane parallel to the magnetic stripes for a sample of type M2 ($Q = 0.8$). See left inset for the experimental geometry. The magnetic peaks around the central, specularly reflected x-ray beam show strong magnetic circular dichroism. The right inset shows the peak width of the magnetic peaks measured in the geometry with the scattering plane perpendicular to the magnetic stripes for Sample I (of type M1) with weak anisotropy ($Q = 0.4$) and Sample II (of type M2) with medium anisotropy ($Q = 0.8$). The diffuse background has been subtracted and Gaussian functions fitted for both data sets to obtain the peak width. The difference in peak width shows that the low anisotropy Sample I has a longer magnetic correlation length.

4.6 COHERENCE OF X-RAYS

Synchrotron radiation, like any other finite source, can be made coherent at the expense of flux, so that normally a third-generation synchrotron source is required to produce a sufficient coherent flux. The conditions for both longitudinal coherence and transverse coherence have to be fulfilled.[53]

The longitudinal (or temporal) coherence length $\xi_l = \lambda^2/(2\Delta\lambda)$ is determined by the resolving power $\lambda/\Delta\lambda$ of the monochromator at given wavelength λ. In the soft x-ray region, $\lambda/\Delta\lambda$ is typical between 1,000 and 10,000, so that ξ_l is a few μm. Given that the absorption length μ^{-1} in Fe metal at the L_3 edge is ~30 nm, the longitudinal coherence condition is largely fulfilled at this resonance.

The transverse (or spatial) coherence is obtained from the diffraction limit as $\xi_l = \lambda z/(2\pi d_s)$, for a distance z from a source with diameter d_s. The required transverse coherence can be achieved by spatial filtering, that is, by putting a pinhole with a diameter of typically a few tens of µm at close range in front of the sample. It is clear that the beam cannot be truly coherent due to both the finite spectral width and the finite physical extent. However, using a pinhole it is possible to achieve 80 to 90% coherence.

4.6.1 SPECKLE PATTERN

Speckle can be described as the random intensity variation in the scattering of coherent radiation from a random sample. The speckle is caused by the random path differences in the paths of radiation scattered from different locations in the sample, which interfere at the detector in the far field to produce a random interference pattern. Speckle patterns are sensitive to the particular configuration of the random sample, whereas incoherent light scattering is only sensitive to statistical averages of the structure. By measuring the time dependence of the speckle pattern, one can study the dynamics (fluctuation, deformation, etc.) of the system. This would mean that one can study the dynamics of the local magnetic moments, even when the average sample magnetization is constant. Although the use of coherent visible radiation is well established, only the shorter wavelengths offered by synchrotron radiation will allow us to probe shorter length scales than is possible with visible light. Because for hard x-rays the scattering is small, it is beneficial — using the advantages of the continuous synchrotron radiation spectrum — to tune into a specific absorption edge, thereby enhancing the scattering and becoming element specific. Polarized radiation gives sensitivity to the magnetization, and to benefit from strong dichroic effects, one needs to go to the soft x-ray region.

A simple physical understanding of speckle formation can be obtained as follows.[40] First consider the coherent illumination of a pair of slits, which leads to a diffraction pattern containing Young's fringes. The angular separation of the slits can be determined from the spatial frequency of the fringes. If the slits scatter equally, the fringes will have unit contrast (zero intensity in their minima), otherwise the contrast will be lower and its value will yield the ratio of the scattering intensities. Now consider a configuration of illuminated objects with random locations and with random scattering amplitudes. From Huygens' principle, it is obvious that this will produce a complicated interference pattern — a speckle pattern. Changing one of the random objects will give a change in the entire speckle pattern, which demonstrates the sensitivity to the local configuration. However, this does not immediately imply that the speckle pattern allows a reconstruction of the local configuration of the illuminated objects. To reconstruct the local configuration of the illuminated objects from the speckle pattern would require the measurement of the signal $F(\mathbf{u}) = |F(\mathbf{u})|\, e^{i\phi(\mathbf{u})}$, when only the intensity (i.e., the squared amplitude $|F(\mathbf{u})|^2$) can be detected and the phase $\phi(\mathbf{u})$ is lost. Only under very favorable circumstances, and for a small amount of objects, the phase problem can be solved and the illuminated sample can be reconstructed directly from its coherent scattering pattern.[54]

The presence of magnetic speckles, displaying a very high magnetic contrast, opens up the possibility to use SXRMS for the study of static and dynamic magnetic disorder on length scales relevant for nanomagnetism. Combined with the time structure of synchrotron radiation, speckle measurements would enable us to study rapid fluctuations in the local magnetic order, giving access to dynamical processes, and ultimately leading to improvements in the read-write speed of magnetic storage devices. This photon in–photon out technique allows us to study reversal processes as a function of magnetic field. For example, it allows us to determine the degree to which the microscopic domain pattern is retained after magnetization cycling.[55,56]

4.6.2 X-Ray Scattering Using a CCD Camera

A CCD camera is an essential requisite to record speckle patterns. The CCD is a silicon-based semiconductor chip containing a 2D array of photosensors (pixels) laid down via photolithography. Commercial chips are available and have typically up to 2048×2048 pixels where the pixel size is usually either 25 μm or 13.5 μm. Compared to photographic media, CCD detectors provide a high quantum efficiency, high linearity, low noise, large dynamical range, and digital output signals. The camera can be fully controlled by software that allows us to change parameters including pixel readout rate, binning patterns (i.e., combining charge from two or more pixels prior to readout), exposure time, and operating temperature. Figure 4.5 shows an example of a CCD image of the SXRMS from FePd at the Fe L_3 resonance.

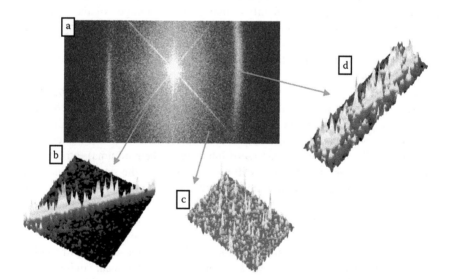

FIGURE 4.5 Soft x-ray resonant magnetic scattering image from FePd at the Fe L_3 resonance using circularly polarized coherent x-rays obtained in 1000 s. (a) Complete CCD image showing the specular peak flanked by the magnetic peaks; (b) three-dimensional (3D) view of the fluctuations along the specular ring; (c) 3D view of the diffuse background showing individual x-rays well that are above the noise; (d) 3D view of the intensity fluctuations in the magnetic peak.

4.7   COHERENT X-RAY SCATTERING FROM FePd NANOWIRE

4.7.1   2D Image of the Magnetic Scattering

We will demonstrate the measurement of "magnetic speckle" patterns using coherent x-rays. The global shape of the images provides information about the structural and magnetic order, such as periodicities and correlation lengths. In addition, when the sample is illuminated with coherent x-rays, there are static intensity fluctuations, which give information on the local disorder instead of the periodic structure as in the case of incoherent x-rays. It is not our purpose here to analyze the speckle pattern in detail, we would only like to demonstrate that it is possible to measure such patterns with great accuracy.[39,55]

To obtain a high degree of coherence, we not only used a pinhole (of 22 μm diameter) but also a "pin-hole sample," consisting of an 8 μm wide FePd wire sample prepared using ultraviolet (UV) lithography and Ar$^+$ ion bombardment. The geometry is schematically shown in the lower part of Figure 4.6a. The SXRMS measurements were performed on ESRF beam line ID08 using the Daresbury in-vacuum diffractometer.[18,19] The sample was placed in reflection geometry at an incidence angle of θ = 22.5° and the CCD camera was connected to an exit port of the in-vacuum diffractometer at a fixed angle of 2θ = 45° and at a distance of 0.46 m from the sample. The scattering plane was along the magnetic stripes. To enhance the magnetic contrast the sample was illuminated with 100% circularly polarized soft x-rays with the energy tuned to the Fe L_3 resonance.

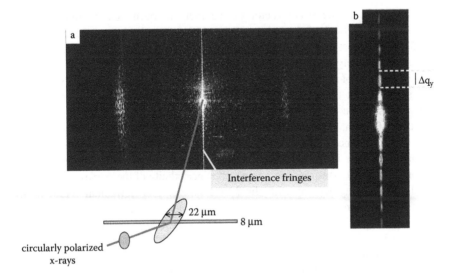

FIGURE 4.6 Soft x-ray resonant magnetic scattering from a 8 μm wide FePd wire at the Fe L_3 resonance using circularly polarized coherent x-rays. (a) CCD image of 1237 × 594 pixels corresponding to an accumulation of 200 exposures of 10 sec. each; (b) enlarged image showing the oscillating streak along the vertical rod, which period Δq_y in reciprocal space corresponds to the width of the wire (8 μm).

Figure 4.6 shows an image obtained from an accumulation of 200 exposures of 10 sec, which amounts to a total of 33 min. After each exposure, the fixed pattern noise was removed using a dark exposure and then the dark noise below 20 ADU/photon was removed in order to extract the true x-rays. The resulting image in Figure 4.6 exhibits a very intense central streak due to the specular reflection. The magnetic peaks on each site of the specular peak are several orders weaker in intensity and only visible when the energy is tuned to the Fe resonance. The specular maximum contains 10^5 photons/pixel and there are ~100 and ~20 photons/pixel in the maximum of the left and right magnetic peaks, respectively. The presence of some individual photons in the low-intensity regions demonstrates the large dynamical range of the detection method.

In the reflection geometry, the three components of the scattering vector $\mathbf{q}$, along the transverse direction (q_y) and the longitudinal direction (q_x) are given by

$$q_x = k[\cos\theta - \cos(\theta + \gamma)] \approx (k\sin\theta\,)\gamma \qquad (4.5a)$$

$$q_y = k\sin\alpha \approx k\alpha \qquad (4.5b)$$

$$q_z = k[\sin\theta + \sin(\theta + \gamma)] \approx 2k\sin\theta \qquad (4.5c)$$

where $k = 2\pi/\lambda$ is the wave vector of the incident beam and α and γ are the horizontal and vertical angles with respect to the specular beam. Because α and $\gamma < 1.5°$, q_z is almost constant (z is perpendicular to sample surface), whereas q_y and q_x are approximately proportional to α and γ, respectively.

The central spot in Figure 4.6 consists of a vertical rod with a modulation arising from diffraction at the wire edges. The sharp oscillations in this rod confirm the abrupt edges of the wire. The oscillation period (Δq_y) corresponds to a real-space width of $8 \pm 0.1\,\mu m$, which confirms the width of the wire as prepared. The elongated shape of the magnetic peaks arises mainly from the reflection geometry, which stretches the image in the vertical direction (cf. Equation 4.5). The magnetic peak position in k space corresponds to a transversal periodicity of the stripes of 95 ± 2 nm in real space, in agreement with the MFM image. The horizontal width of the magnetic peak can be used to estimate the periodicity fluctuation and gives a transversal magnetic correlation length, $l_{c,y} = 2\pi/\delta q_y \cong 2\mu m \cong 20$ periods. The vertical extension, δq_x, of the magnetic peaks provides an estimate of the longitudinal correlation length along the stripes. This gives a longitudinal correlation length $l_{c,x} = 2\pi/\delta q_x \cong 700$ nm, which is the average distance between defects, forks, and meandering in the stripes.

Figure 4.6A shows that the left magnetic peak is more intense than the right one. The peak intensities reverse with the light helicity. As explained in Section 4.4, the dichroism in the transverse geometry is due to the presence of closure domains

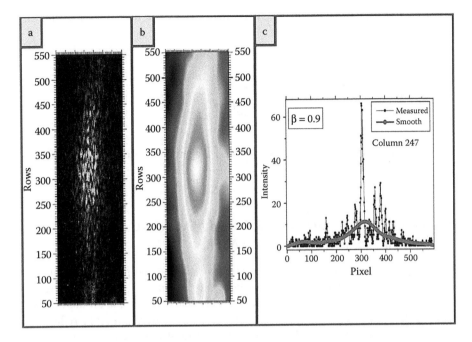

FIGURE 4.7 Smoothing of the magnetic peak and estimation of the degree of coherence. (A) $I(\mathbf{q})$, enlarged image of the magnetic peak from Figure 4.6a; (b) $I_{inc}(\mathbf{q})$, result of an anisotropic smoothing after 1000 iterations; (c): comparison of $I_{inc}(\mathbf{q})$ and $I(\mathbf{q})$ along a vertical cut across the image. The smoothed curve represents the calculated incoherent scattering. The degree of coherence is estimated to be $\beta = 0.9 \pm 0.1$.

at the surface, which leads to interference between the scattering amplitudes from the perpendicular and in-plane magnetized domains.

4.7.2 INTENSITY FLUCTUATION

The intensity fluctuations that can be observed in the magnetic peaks contain information about the magnetic disorder. Figure 4.7A shows an enlarged image of the intensity fluctuations in the left magnetic peak. After anisotropic smoothing of this pattern,[57] we obtain the intensity profile shown in Figure 4.7B. The obtained intensity pattern, $I_{inc}(\mathbf{q})$, corresponds to the intensity distribution that would be observed using an *incoherent beam*. Figure 4.7C shows the measured intensity profile $I(\mathbf{q})$ (thin line) along a vertical line across the image of the magnetic peak on the detector, which is compared with $I_{inc}(\mathbf{q})$ (thick curve).

Despite the low counting statistics, the image displays a high speckle contrast, as expected from the coherence conditions. Due to the wire shape, the speckle size is larger in the vertical direction. The overall degree of coherence β is given by the main square difference between $I(\mathbf{q})$ and $I_{inc}(\mathbf{q})$. The quantity β is evaluated by averaging this squared difference over a chosen surface S, which in this case corresponds to the area of the magnetic satellite. By taking the Poisson noise into account, this corresponds to[58]

$$\beta = \frac{\left\langle [I(\mathbf{q}) - I_{inc}(\mathbf{q})]^2 - I_{inc}(\mathbf{q}) \right\rangle_s}{\left\langle [I_{inc}(\mathbf{q})]^2 \right\rangle_s} \tag{4.6}$$

From the analysis of the image in Figure 4.7, a degree of coherence of $\beta = 0.85 \pm 0.05$ is obtained.

4.8 OSCILLATORY BEHAVIOR OF 5d MAGNETIC MOMENTS IN Fe/W MULTILAYERS

Driven by the unique properties of synchrotron radiation and progress in materials science in supplying materials with ever increasing complexity, scientific interest in the use of x-ray scattering to determine the magnetic structure and orbital ordering has significantly grown over the last few years. Traditionally, x-ray reflectivity measurements have been used to determine structural information, such as layer thickness, effective electron density, and interfacial roughness. However, conventional x-ray reflectivity is limited when it comes to separating structural parameters for layers of different elements in the periodic table. One way to enhance the chemical contrast is to take advantage of the tunability of the incident photon energy using synchrotron radiation. Close to an absorption edge of an atom, the scattering factor is resonantly enhanced. Moreover, by using polarized x-rays, the resonant scattering becomes sensitive to the magnetization profile in the material. In the following sections, we report the magnetic behavior of W in a Fe/W multilayer. Electron hybridization at interfaces can change the spin magnetic moments of the magnetic layers and induce magnetization in adjacent "non-magnetic" layers.[59] In giant magnetoresistive (GMR) materials, such as Fe/Cr multilayers, the magnetic layers are (anti)ferromagnetically coupled by the Ruderman–Kittel–Kasuya–Yosida (RKKY) interaction, which induces an oscillatory spin magnetic moment in the nonmagnetic layer. Furthermore, such systems that combine a large $3d$ spin moment with a strong $5d$ spin-orbit interaction are of special interest because of their enhanced magneto-optical responses.

Improved understanding of the mechanism, giving rise to magnetic interaction in multilayered systems, can be gained by studying directly the induced magnetic properties of the nonmagnetic constituent near the interface. Recent x-ray magnetic circular dichroism (XMCD) studies at the W $L_{2,3}$ absorption edge, probing specifically the $5d$ valence states in Fe/W multilayers, evidenced a significant induced magnetic moment in an ultrathin W spacer layer.[60] However, XMCD only gives the magnetic moments averaged over all W layers (properly corrected for the sampling depth). XRMS would give the layer-resolved magnetic moments if we are able to detect the magnetic contribution to the different order Bragg peaks using circular dichroism.[29,61,62] In the particular case of Fe/W, the line shape of the W L_3 XMCD gives a characteristic signature of the magnetic polarization,[60] so that the energy dependence of the asymmetry ratio for the different Bragg peaks can be used to provide a probe for the profile of the magnetic polarization. Furthermore, measurement of the intensity at the Bragg peaks, which originate from the interference process between periodically stacked layers, strongly reduces the influence of the capping layer, buffer layer, and substrate.

In Section 4.9, we describe the sample preparation and the structural characterization of the Fe/W multilayer. We present experimental results using x-ray diffraction (XRD), XMCD, and XRMS. In Section 4.10, we briefly explain the data analysis of the XRMS results, which can give the layer-resolved induced magnetization within the W layer. The obtained results are discussed in Section 4.11 and compared to those from recent band-structure calculations and conclusions are drawn.

4.9   EXPERIMENTAL RESULTS ON Fe/W MULTILAYERS

The Fe/W multilayer was grown using ion-beam sputtering onto a sapphire ($1\bar{1}20$) substrate. Prior to the multilayer deposition, a 4 nm Mo buffer layer was deposited at 1020 K to favor epitaxial growth. The anticipated thicknesses for the Fe and W layers were 30 and 10 Å, respectively. Twenty-five alternating layers were deposited giving a nominal multilayer thickness of 100 nm.

The structural characterization of the multilayer was performed with high-angle XRD using a conventional Cu rotating anode source and proportional counter. To extract the lattice parameter required for the XRMS analysis, the measured spectra were fitted by using a modified version of the superlattice refinement code SUPREX.[63] The position of the main diffraction peak gives directly the average d spacing and the multilayer period. The fitting algorithm is used to reproduce the entire XRD pattern — the peaks positions, relative intensities, and lines profiles. The results of the XRD analysis for the Fe(30 Å)/W(11 Å) multilayer are the following: the texture is (110) with d spacings (normal to the layers) of $d(\text{Fe}) = 2.06$ Å (bulk like) and $d(\text{W}) = 2.235$ Å. In order to account for the important Fe and W lattice mismatch of 9.4%, two atomic planes were allowed to expand or contract at the bottom and the top of each layer assuming an exponential profile. The result of the fit indicates a relative expansion of 0.12 Å for $d(\text{W})$ and a contraction of 0.07 Å for $d(\text{Fe})$ following this exponential law. Simultaneously, to compensate for this behavior, according to the elastic theory, the best fit of the measured XRD profile was obtained with a 5% increase of the in-plane density for the W layers (lattice contraction) and a mandatory 5% reduction of the in-plane density of the two first Fe planes at the interface (lattice expansion). In agreement with the thicknesses of the W and Fe layers, it was found that the whole W layer was constrained.

Anomalous reflectivity, XRMS and XMCD experiments were carried out using 95% circularly polarized x-rays from beam line ID12 of ESRF at Grenoble.[64] For the XMCD measurements, the sample was mounted at grazing incidence in the bore of a 7 T superconducting magnet. The anomalous reflectivity and XRMS were measured using an ultrahigh violet two-circle reflectometer at low grazing incidence angles θ of the x-rays from beam line ID12. The sample was magnetically saturated

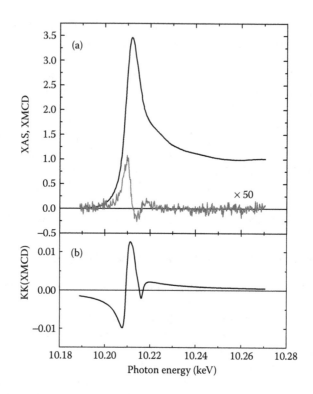

FIGURE 4.8 (a) XAS and XMCD (jagged line) recorded at the W L_3 edge; (b) Real part (m') of the magnetic scattering factor calculated as the Kramers–Kronig transform of the XMCD.

using a rotatable array of permanent magnets producing a field of 0.5 T, which was applied along the intercept of the sample surface and the diffraction plane. A field of few hundred gausses is sufficient to fully saturate the sample. To avoid any experimental artifacts, the XMCD and XRMS were recorded by reversing the photon helicity vector as well as by reversing the direction of applied magnetic field. All data have been corrected for the incomplete degree of circular polarization.

Figure 4.8 shows the x-ray absorption spectroscopy (XAS) and XMCD recorded at the W L_3 edges of the Fe/W multilayer at $T = 300$ K and $H = 4$ T. Despite its very small amplitude (multiplied by 50 for clarity in Figure 4.8), a clear XMCD signal was detected ensuring the presence of a nonvanishing induced magnetization in the W $5d$ states. The XMCD spectrum agrees well with the shape measured previously on a similar sample[60] but was smaller in magnitude due to the thicker W layer. Applying the magneto-optical sum rules[2] to the measured W $L_{2,3}$ edges gives an average total $5d$ magnetic moment $\mu_{av} = -0.038$ μ_B/atom and an orbital-to-spin magnetic moment ratio of 0.085.

The layer-resolved magnetization can be obtained by analyzing the XRMS of the four distinct Bragg peaks. Figure 4.9 shows the energy dependence of the asymmetry ratio $R = (I^+ - I^-)/(I^+ + I^-)$ across the W L_3 edge measured on top of the Bragg peaks, where I^+ and I^- are the diffracted intensities for opposite align-ments between magnetic field and photon helicity. The strongly different energy dependence of R for each Bragg peak is a direct indication that the magnetic polarization is not constant throughout the W layer. These variations could only be observed because of the sufficiently small energy step in the measurement (0.25 eV) associated with the high stability of the x-ray beam. If the asymmetry ratio would have displayed the same shape for each diffraction order, then both the charge and the magnetic structure factors have the same q dependence.

4.10 ANALYSIS METHOD

To explain how the magnetization profile can be extracted from the XRMS, the theoretical analysis of the spectra is briefly outlined. To first order in the magneti-zation, the atomic scattering factor can be written as Equation 4.2 with the charge- and magnetization-dependent scattering amplitudes

$$F^0(E) = f_0 + f'(E) - if''(E) \tag{4.7}$$

$$F^1(E) = m'(E) - im''(E) \tag{4.8}$$

respectively, where E is the energy of the incident x-rays. f_0 is the regular charge scattering factor, and f' and f'' are the energy-dependent resonant contributions associated with the absorption edge. m'' is the imaginary part of the forward scat-tering, which is directly obtained from the XMCD. The energy dependence of m' is derived from that of m' by using the Kramers–Kronig relation and the result is displayed in Figure 4.9B.

Away from the critical angle, the first Born approximation remains valid, and we can define the complex charge and magnetic structure factors

$$F(\mathbf{q}, E) = F' - iF'' = \sum_j C_j \sigma_j F^0(E) \, e^{i\mathbf{q}\cdot\mathbf{r}_j} \tag{4.9}$$

$$M(\mathbf{q}, E) = M' - iM'' = \sum_j \frac{C_j \mu_j}{\mu_{av}} \sigma_j F^1(E) \, e^{i\mathbf{q}\cdot\mathbf{r}_j} \tag{4.10}$$

respectively, where j runs over the different layers, whose composition, especially in the interface region, may be altered by changing the concentration C_j of the different atomic species. σ_j corresponds to the planar atomic density divided by the layer thickness, μ_j is the magnetic moment per atom. The average magnetic moment $\mu_{av} = \sum_j C_j \mu_j / \sum_j C_j$ is obtained from the XMCD. In the complex structure

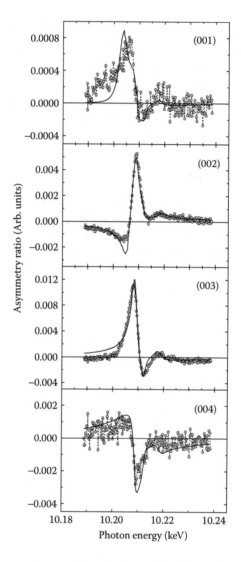

FIGURE 4.9 X-ray resonant magnetic scattering results. Energy dependence of the asymmetry ratio R measured for the first four low-order Bragg peaks across the W L_3 edge. Experimental (open circles) and calculated curves (drawn lines) leading to the magnetic profile given in Table 4.2.

factor, $\mathbf{r}_j$ gives the position of the layers and $\mathbf{q}$ the scattering vector perpendicular to the surface. Because the XRMS is sensitive to $C_j\mu_j$/atom, x-ray reflectivity measurements are required to find C_j. Therefore, the real structure characterization of the sample is an important prerequisite for the analysis of the XRMS. The element-specific magnetic profile is obtained using a refinement procedure[29,61] that calculates the energy dependence of the asymmetry ratio R, which for a longitudinal geometry and circular polarization can be written as

$$R = \frac{-2\cos^3\theta \; (F'M' - F''M'')}{1 - \frac{1}{2}\sin^2 2\theta \; \left|F^2\right| \; + \; \cos^2\theta \; \left|M^2\right|} \qquad (4.11)$$

Details about the refinement procedure can be found in several papers.[29,61,65,66] We only like to mention here that the number of atomic planes, their interplanar distances, and the concentrations C_j are directly obtained from the analysis of the XRD and reflectivity data. To restrict as much as possible the number of free parameters, the structural parameters mentioned above were kept fixed during the refinement of the XRMS spectra. Furthermore, an additional constraint was added imposing that the average magnetic moment $\mu_{av} = \Sigma_j C_j \mu_j / \Sigma_j C_j$ in the W layer is equal to the one obtained from XMCD.

4.11 DISCUSSION OF THE Fe/W MULTILAYER RESULTS

The results of the simultaneous refinement of the asymmetry ratios, measured across the W L_3, are shown by the drawn curves in Figure 4.9, and yield the W $5d$-induced magnetic profile presented in Table 4.2 for the different layers. It was found that the W magnetic moment in the first layer at both interfaces is antiparallel to the Fe magnetic moment, which is ~$2\mu_B$. Interestingly, the W moment of the second layer is antiparallel to that of the first layer. The magnetization profile therefore shows a damped oscillation similar to that expected from the RKKY interaction. This tendency is in agreement with results from recent band-structure calculations,[59] which are reproduced in Table 4.2 for comparison. The theoretical results are for ideal interfaces without any roughness and give therefore a profile that is symmetric around the middle of the stack. This is clearly not the case for the experimental results (cf. Table 4.2), where the interdiffusion at the two interfaces is different because W on Fe has a different growth than Fe on W. Furthermore, the measured W moments for the interface layers are much larger than the theoretical values. This increase in moment can be ascribed to the higher Fe coordination of the W atoms. This is supported by calculations, which find increased W moments for a 5 ML (monolayer) Fe/1 ML W multilayer $(-0.311\mu_B)$[59] and for W impurities in Fe $(-0.45\ \mu_B)$.[67] It is important to note that despite the unavoidable interface roughness and disorder, the oscillatory behavior does not disappear. However, the damping is stronger and the period is shorter than in the ideal theoretical case, which does not include interdiffusion.

4.12 ORBITAL ORDERING IN MANGANITES

We present an example of orbital ordering in the layered perovskite $La_{0.5}Sr_{1.5}MnO_4$ that shows an enhanced sensitivity of soft x-ray resonant diffraction at the Mn L edges. The energy dependence of the orbital ordering diffraction peak over the $L_{2,3}$ edges can be compared to ligand-field calculations allowing a distinction between the influences of Jahn–Teller distortions and spin correlations. [41]

TABLE 4.2

Layer-Dependent W Atomic Concentrations, C_j, and 5d Total Magnetic Moments μ_j/Atom μ_B) for the Fe/W Multilayer Obtained from the Simultaneous Refinement of the Experimental X-Ray Reflectivity and W L_3 XRMS Results. Comparison with Theoretical Results for μ_j/Atom

| Layer | Experiment | | μ_j/Atom (Theory) | |
j	C_j	μ_φ/Atom	5Fe/7W[a]	5Fe/5W[b]
1	0.4	-0.377 ± 0.017	-0.127	-0.136
2	0.8	0.063 ± 0.015	0.025	0.019
3	1	-0.025 ± 0.012	0.032	0.049
4	1	0.041 ± 0.041	0	0.019
5	1	-0.017 ± 0.014	0.032	-0.136
6	0.6	0.094 ± 0.013	0.025	—
7	0.3	-0.427 ± 0.014	-0.127	—

Note: Dashes indicate layer is not present.

[a]5 ML Fe/7 ML W(001) multilayer with $C_j = 1$ (see Reference 59).

[b]5 ML Fe/5 ML W(001) multilayer with $C_j = 1$ (see Reference 59).

Orbital ordering, which manifests itself in the spatial distribution of the outermost valence electrons, is an important topic in current research of transition metal oxides, as the magnetic and transport properties are closely related to the orbital and charge degrees of freedom. Transition metal oxides are examples of strongly correlated electron systems, displaying new, unusual, and unexpected behavior due to nanoscale features in the quantum realm. Advanced properties such as magnetism, colossal magnetoresistance, and superconductivity underpin the development of advanced materials. We can now control the electronic and magnetic phases of correlated electron materials in unconventional ways, in some cases with ultrafast response times. Such control offers the prospect that correlated electron systems may provide a basis for novel future electronics.

XRD is mainly sensitive to the rather isotropic electron distribution, but tuning the photon energy to an absorption edge gives enhanced sensitivity to valence states. Resonant XRD, involving virtual excitations from core to valence states, can then probe the anisotropic valence charge density allowing forbidden diffraction peaks to appear. Mn K edge resonant diffraction from $La_{0.5}Sr_{1.5}MnO_4$ implied that orbital order is established at the same temperature as charge order (217 K) and that it is a precursor to the complex antiferromagnetic spin ordering observed at lower temperatures. However, the Mn K edge resonance involves virtual excitations to the unoccupied 4p levels, so that sensitivity to the 3d orbital order is indirect and considered to be highly controversial. The influence of Jahn–Teller distortions, 4p band-structure effects and on-site and intersite 3d-4p Coulomb interactions have all been proposed to explain the sensitivity of the Mn K edge diffraction to orbital ordering. Soft x-ray resonant diffraction from single crystals allows us to probe directly the Mn $3d$ states during orbital ordering using the enhanced sensitivity at the Mn $L_{2,3}$ edges.[41,68,69] Figure 4.10 shows energy spectrum of the (1/4,1/4,1) soft XRD peak across the Mn $L_{2,3}$ resonance in $La_{0.5}Sr_{1.5}MnO_4$. These results directly demonstrate the wealth of information available from soft XRD. This technique therefore provides a direct probe of the orbital ordering and gives information on the electronic configuration and the underlying mechanism of the orbital order. In $La_{0.5}Sr_{1.5}MnO_4$, both orbital and charge order causes a configurational ordering of the Mn^{3+} and Mn^{4+} ions into a checkerboard pattern. Jahn–Teller distortions of the Mn^{3+} ions causes considerable distortions of the MnO_6 octahedra and the individual Mn-O bond lengths and angles, which simultaneously causes cooperative displacements of the two Mn sublattices.

4.13 CONCLUSIONS

We have compared a series of FePd alloy thin-film samples with low, medium, and high PMA using a combined study of MFM, micromagnetic modeling, and SXRMS. This allowed us to investigate the influence of the PMA on the formation of closure domains and to construct a detailed map of the magnetic structure within the thin film. Closure domains are observed in samples with low and medium magnetic anisotropy. However, they are absent in samples with strong PMA because the in-plane magnetization becomes energetically unfavorable. Analysis of the SXRMS shows that samples of a film thickness of ~40 nm with low to medium PMA have closure domains with in-plane magnetization to a depth of ~85 Å. This is in good agreement with the magnetization distribution obtained from micromagnetic modeling, whereas also the increased disorder observed in MFM images of the stripe pattern suggests that the closure domains disappear for strong PMA. Soft x-ray magnetic speckle patterns in reflection geometry have been measured using 2D scattering. Because soft x-ray absorption spectra show a huge magnetic dichroism compared to their hard x-ray counterparts, the magnetic peak intensities are more straightforward to model. This new technique clearly holds large promises for the future of magnetic scattering, with imaging of dynamic disorder among the possibilities.

In the case of Fe/W multilayers, we have used a combination of complementary x-ray-based techniques to resolve experimentally the in-depth W $5d$ magnetization. X-ray anomalous reflectivity was used to obtain the multilayer period, the W and Fe layer thicknesses, and the interface roughness. The crystallographic parameters have been extracted from XRD at high angles. XMCD was used to obtain the mean value of the induced W $5d$ polarization averaged over the W layer. It was shown that, in addition to the detailed structural analyses combined with magnetic dichroism, the x-ray magnetic scattering in resonance with the W L_3 absorption can be utilized to resolve the W $5d$ magnetization profile. The usefulness of this technique was demonstrated by verifying the existence of the recently predicted W $5d$ spin oscillations in Fe/W multilayers. It was found that the W magnetic moment in the first layer at both interfaces is antiparallel to the Fe magnetic moment. Further, the

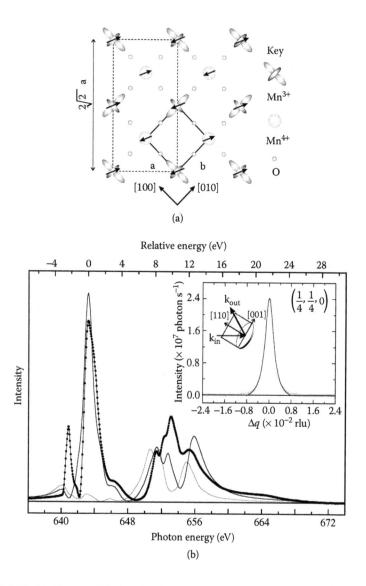

FIGURE 4.10 (a) Charge, orbital, and spin ordering in the MnO_2 planes of $La_{0.5}Sr_{1.5}MnO_4$; the arrows represent the magnetic moments. The primitive I4/*mmm* unit cell is shown by the solid line ($a = b = 3.864$ Å) and the orbital-ordered unit cell by the dashed line. (b) The energy dependence (solid circles with line) of the (1/4,1/4,1) reflection recorded over the Mn $L_{2,3}$ edges at 134 K and the calculated spectrum in D_{2h} symmetry for large (medium gray line) and small (light gray line) Jahn–Teller distortion. The fit is much better for the medium gray curve indicating that these distortions are significant. However, other contributions are also important as can be seen in Reference 41. The inset shows a (1/4 + Δq,1/4 + Δq,1) scan of the forbidden reflection arising from orbital ordering at a photon energy of 643.8 eV at 134 K (open circles). The solid line is a Lorentzian fit to the peak. The experimental geometry is also shown.

moment of the second W layer is antiparallel to that of the first W layer. The magnetization profile shows a damped oscillation in agreement with recent results from band-structure calculations.[59] Despite the unavoidable interface roughness and disorder, the oscillatory behavior does not disappear; however, the damping is stronger and the period is shorter than in the ideal theoretical case that does not include interdiffusion. The presented results open the way for further studies of periodic magnetic nanostructures containing $5d$ metals, of high potential interest for GMR, where XRMS measures the weakly induced $5d$ magnetization, independently from the large $3d$ transition metal moments. XRMS offers a large sampling depth ideal for buried interfaces and can be performed in the presence of externally applied magnetic and electric fields. The use of the kinematical approximation, which proved to work in the specific case of multilayers, allows us to use a rather simple refinement procedure to obtain the magnetization of the probed atom in each atomic plane without having to assume a specific model. We would like to point out that such an approach using diffraction peaks provided by the chemical modulation of the multilayer can be extended to the soft x-rays range, even though the accessible reciprocal space is restricted, however, at the expense of having to use dynamical diffraction theory.

In the case of the layered perovskite $La_{0.5}Sr_{1.5}MnO_4$, the energy dependence of the intense diffraction peak, which arises from $3d$ orbital ordering, was compared with theoretical calculations, implying that Jahn–Teller distortions together with spin correlations drive the orbital ordering process. It is expected that the complex relationship between the charge, orbital, and spin degrees of freedom can be explored in great detail using the increased sensitivity of SXRD. These results suggest that soft XRD will become a major technique for the study of charge, spin, and orbital ordering in a wide range of materials. It is likely that in the near future such soft x-ray techniques will become more important at many synchrotrons worldwide.

ACKNOWLEDGMENTS

The author gratefully acknowledges the contributions of his collaborators, in particular, J.P. Attané, M. Belakhovsky, P. Bencok, G. Beutier, N.B. Brookes, K. Chesnel, S.P. Collins, S.S. Dhesi, E. Dudzik, H.A. Dürr, A. Haznar, N. Jaouen, T.K. Johal, I.W. Kirkman, K. Larsson, F. Livet, A. Marty, A. Mirone, S. Mylonas, L. Ortega, A. Rogalev, M.D. Roper, A. Tagliaferri, N.D. Telling, and F. Wilhelm.

REFERENCES

1. G. van der Laan, B.T. Thole, G.A. Sawatzky, J.B. Goedkoop, J.C. Fuggle, J.M. Esteva, R.C. Karnatak, J.P. Remeika, and H.A. Dabkowska, *Phys. Rev. B,* 34, 6529, 1986.
2. B.T. Thole, P. Carra, F. Sette, and G. van der Laan, *Phys. Rev. Lett.,* 68, 1943, 1992.
3. H.A. Dürr, G.Y. Guo, G. van der Laan, J. Lee, G. Lauhoff, and J.A.C. Bland, *Science,* 277, 213, 1997.

4. F. de Bergevin and M. Brunel, *Phys. Lett. A*, 39, 141, 1972.

5. M. Blume, *J. Appl. Phys.*, 57, 3615, 1985.

6. K. Namikawa, M. Ando, T. Nakajima, and H. Kawata, *J. Phys. Soc. Jpn.*, 54, 4099, 1985.

7. D. Gibbs, D.R. Harshman, E.D. Isaacs, D.B. McWhan, D. Mills, and C. Vettier, *Phys. Rev. Lett.*, 61, 1241 1988.

8. E.D. Isaacs, D.B. McWhan, C. Peters, G.E. Ice, D.P. Siddons, J.B. Hastings, C. Vettier, and O. Vogt, *Phys. Rev. Lett.*, 62, 1671, 1989.

9. J.P. Hannon, G.T. Trammell, M. Blume, and D. Gibbs, *Phys. Rev. Lett.*, 61, 1245, 1988.

10. B.T. Thole, G. van der Laan, and G.A. Sawatzky, *Phys. Rev. Lett.*, 55, 2086, 1985.

11. G. van der Laan, B.T. Thole, G.A. Sawatzky, J.B. Goedkoop, J.C. Fuggle, J.M. Esteva, R. Karnatak, J.P. Remeika, and H.A. Dabkowska, *Phys. Rev. B*, 34, 6529, 1986.

12. G. van der Laan, *Phys. Rev. Lett.*, 82, 640, 1999.

13. S.S. Dhesi, G. van der Laan, E. Dudzik, and A.B. Shick, *Phys. Rev. Lett.*, 87, 067201, 2001.

14. C. Kao, J.B. Hastings, E.D. Johnson, D.P. Siddons, G.C. Smith, and G.A. Prinz, *Phys. Rev. Lett.*, 65, 373, 1990.

15. J.M. Tonnerre, L. Sève, D. Raoux, G. Soullié, B. Rodmacq, and P. Wolfers, *Phys. Rev. Lett.*, 75, 740, 1995.

16. A. Haznar, G. van der Laan, S.P. Collins, C.A.F. Vaz, J.A.C. Bland, and S.S. Dhesi, *J. Synchr. Rad.*, 11, 254, 2004.

17. H.A. Dürr, E. Dudzik, S.S. Dhesi, J.B. Goedkoop, G. van der Laan, M. Belakhovsky, C. Mocuta, A. Marty, and Y. Samson, *Science*, 284, 2166, 1999.

18. G. van der Laan, H.A. Dürr, E. Dudzik, M.D. Roper, S.P. Collins, T.P.A. Hase, and I. Pape, *Synchr. Rad. News*, 12(3), 5, 1999.

19. G. van der Laan, *Synchr. Rad. News*, 14(5), 32, 2001.

20. J.-M. Tonnerre, *X-Ray Magnetic Scattering* in *Magnetism and Synchrotron Radiation*, E. Beaurepaire, B. Carrière, and J.-P. Kappler, Eds., Les Editions de Physique, Les Ulis, 1996, p. 245.

21. T.P.A. Hase, I. Pape, B.K. Tanner, H.A. Dürr, E. Dudzik, G. van der Laan, C.H. Marrows, and B.J. Hickey, *Phys. Rev. B*, 61, R3792, 2000.

22. J.F. MacKay, C. Teichert, D.E. Savage, and M.G. Lagally, *Phys. Rev. Lett.*, 77, 3925, 1996.

23. J.F. MacKay, C. Teichert, and M.G. Lagally, *J. Appl. Phys.*, 81, 4353, 1997.

24. R.M. Osgood, S.K. Sinha, J.W. Freeland, Y.U. Idzerda, and S.D. Bader, *J. Appl. Phys.*, 85, 4619, 1999.

25. J.B. Kortright and S.K. Kim, *Phys. Rev. B*, 62, 12216, 2000.

26. J.B. Kortright, S.K. Kim, G.P. Denbeaux, G. Zeltzer, K. Takano, and E.E. Fullerton, *Phys. Rev. B*, 64, 092401, 2001.

27. G. van der Laan and B.T. Thole, *Phys. Rev. B*, 43, 13401, 1991.

28. B.T. Thole, G. van der Laan, J.C. Fuggle, G.A. Sawatzky, R.C. Karnatak, and J.M. Esteva, *Phys. Rev. B*, 32, 5107, 1985.

29. N. Jaouen, J.-M. Tonnerre, D. Raoux, E. Bontempi, L. Ortega, M. Muenzenberg, W. Felsch, A. Rogalev, H.A. Dürr, E. Dudzik, G. van der Laan, M. Suzuki, and H. Maruyama, *Phys. Rev. B*, 66, 134420, 2002.

30. J.F. Peters, J. Miguel, M.A. de Vries, O.M. Toulemonde, J.B. Goedkoop, S.S. Dhesi, and N.B. Brookes, *Phys. Rev. B*, 70, 224417, 2004.

31. N. Jaouen, G. van der Laan, T.K. Johal, F. Wilhelm, A. Rogalev, S. Mylonas, and L. Ortega, *Phys. Rev. B*, 70, 094417, 2004.

32. G. van der Laan and B.T. Thole, *Phys. Rev. B*, 53, 14458, 1996.
33. G. van der Laan, K.T. Moore, J.G. Tobin, B.W. Chung, M.A. Wall, and A.J. Schwartz, *Phys. Rev. Lett.*, 93, 097401, 2004.
34. N.D. Telling, A. Haznar, G. van der Laan, M.D. Roper, F. Schedin, and G. Thornton, *Physica B*, 345, 157, 2004.
35. G. Materlik, C.J. Sparks, and K. Fisher, Eds., *Resonant Anomalous X-Ray Scattering, Theory and Applications*, Elsevier Science, Amsterdam, 1994.
36. Garcia and G. Subias, *J. Phys.: Condens. Matter*, 16, R145, 2004.
37. E. Dudzik, S.S. Dhesi, H.A. Dürr, S.P. Collins, M.D. Roper, G. van der Laan, K. Chesnel, M. Belakhovsky, A. Marty, and Y. Samson, *Phys. Rev. B*, 62, 5779, 2000.
38. G. van der Laan, K. Chesnel, M. Belakhovsky, A. Marty, F. Livet, S.P. Collins, E. Dudzik, A. Haznar, and J.P. Attané, *Superlattices and Microstructures*, 34, 107, 2003.
39. K. Chesnel, M. Belakhovsky, F. Livet, S.P. Collins, G. van der Laan, S.S. Dhesi, J.P. Attané, and A. Marty, *Phys. Rev. B*, 66, 172404, 2002.
40. G. van der Laan, *Physica B*, 345, 137, 2004.
41. S.S. Dhesi, A. Mirone, C. De Nadai, P. Ohresser, P. Bencok, N.B. Brookes, P. Reutler, A. Revcolevschi, A. Tagliaferri, O. Toulemonde, and G. van der Laan, *Phys. Rev. Lett.*, 92, 056403, 2004.
42. V. Gehanno, A. Marty, and B. Gilles, *Phys. Rev. B*, 55, 12552, 1997.
43. Y. Samson, A. Marty, R. Hoffmann, V. Gehanno, and B. Gilles, *J. Appl. Phys.*, 85, 4604, 1999.
44. P. Kamp, A. Marty, B. Gilles, R. Hoffmann, S. Marchesini, M. Belakhovsky, C. Boeglin, H.A. Dürr, S.S. Dhesi, G. van der Laan, and A. Rogalev, *Phys. Rev. B*, 59, 1105, 1999.
45. A. Hubert and R. Schaeffer, *Magnetic Domains: The Analysis of Magnetic Microstructures*, Springer-Verlag, Berlin, 1998.
46. T. Schrefl, J. Fidler, and J. Chapman, *J. Phys. D*, 29, 2352, 1996.
47. P. Carra and B.T. Thole, *Rev. Mod. Phys.*, 66, 1509, 1994.
48. M. Blume and D. Gibbs, *Phys. Rev. B*, 37, 1779, 1988.
49. G. van der Laan, E. Dudzik, S.P. Collins, S.S. Dhesi, H.A. Dürr, M. Belakhovsky, K. Chesnel, A. Marty, Y. Samson, and B. Gilles, *Physica B*, 283, 171, 2000.
50. G. Beutier, A. Marty, K. Chesnel, M. Belakhovsky, J.C. Toussaint, B. Gilles, G. van der Laan, S.P. Collins, and E. Dudzik, *Physica B*, 345, 143, 2004.
51. G. Beutier, G. van der Laan, K. Chesnel, A. Marty, M. Belakhovsky, S.P. Collins, E. Dudzik, J.-C. Toussaint, and B. Gilles, *Phys. Rev. B*, 71, 184436, 2005.
52. M.D. Roper, G. van der Laan, H.A. Dürr, E. Dudzik, S.P. Collins, M.C. Miller, and S.P. Thompson, *Nucl. Instrum. Meth. A*, 467–468, 1101, 2001.
53. M. Born and E. Wolf, *Principles of Optics*, 6th ed., Cambridge University Press, Cambridge, 1997, chap.10.
54. S. Eisebitt, M. Lorgen, W. Eberhardt, J. Luning, J. Stohr, C.T. Rettner, O. Hellwig, E.E. Fullerton, and G. Denbeaux, *Phys. Rev. B*, 68, 104419, 2003.
55. K. Chesnel, M. Belakhovsky, G. van der Laan, F. Livet, A. Marty, G. Beutier, S.P. Collins, and A. Haznar, *Phys. Rev. B*, 70, 180402, 2004.
56. M.S. Pierce, R.G. Moore, L.B. Sorensen, S.D. Kevan, O. Hellwig, E.E. Fullerton, and J.B. Kortright, *Phys. Rev. Lett.*, 90, 175502, 2003.
57. F. Livet, F. Bley, J. Mainville, M. Sutton, S.G.J. Mochrie, E. Geissler, G. Dolino, D.L. Abernathy, and G. Grubel, *Nucl. Instr. Meth. A*, 451, 596, 2000.
58. K. Chesnel, G. van der Laan, F. Livet, G. Beutier, A. Marty, M. Belakhovsky, A. Haznar, and S.P. Collins, *J. Synchr. Rad.*, 11, 469, 2004.
59. R. Tyer, G. van der Laan, W.M. Temmerman, Z. Szotek, and H. Ebert, *Phys. Rev. B*, 67, 104409, 2003.

60. F. Wilhelm, P. Poulopoulos, H. Wende, A. Scherz, K. Baberschke, M. Angelakeris, N.K. Flevaris, and A. Rogalev, *Phys. Rev. Lett.,* 87, 207202, 2001.
61. L. Sève, N. Jaouen, J.M. Tonnerre, D. Raoux, F. Bartolomé, M. Arend, W. Felsch, A. Rogalev, J. Goulon, C. Gautier, and J.F. Bérar, *Phys. Rev. B*, 60, 9662, 1999.
62. N. Ishimatsu, H. Hashizume, S. Hamada, N. Hosoito, C.S. Nelson, C.T. Venkataraman, G. Srajer, and J.C. Lang, *Phys. Rev. B*, 60, 9596, 1999.
63. E.E. Fullerton, J. Pearson, C.H. Sowers, S.D. Bader, X.Z. Wu, and S.K. Sinha, *Phys. Rev. B*, 48, 17432, 1993.
64. J. Goulon, A. Rogalev, C. Gautier, C. Goulon-Ginet, S. Paste, R. Signorato, C. Neumann, L. Varga, and C. Malgrange, *J. Synchr. Rad.,* 5, 232, 1998.
65. N. Jaouen, J.M. Tonnerre, D. Raoux, L. Ortega, E. Bontempi, M. Muenzenberg, W. Felsch, M. Suzuki, H. Maruyama, H.A. Dürr, E. Dudzik, and G. van der Laan, *Appl. Phys. A*, 73, 711, 2001.
66. N. Jaouen, J.M. Tonnerre, D. Raoux, M. Muenzenberg, W. Felsch, A. Rogalev, N.B. Brookes, H.A. Dürr, and G. van der Laan, *Acta Phys. Pol. B*, 34, 1403, 2003.
67. R. Tyer, G. van der Laan, W.M. Temmerman, and Z. Szotek, *Phys. Rev. Lett.*, 90, 129701, 2003.
68. S.B. Wilkins, P.D. Hatton, M.D. Roper, D. Phabhakaran, and A.T. Boothroyd, *Phys. Rev. Lett.*, 90, 187201, 2003.
69. K.J. Thomas, J.P. Hill, Y.J. Kim, S. Grenier, P. Abbamonte, L. Venema, A. Rusydi, Y. Tomioko, Y. Tokura, D.F. McMorrow, G.A. Sawatzky, and M. van Veenendaal, *Phys. Rev. Lett.,* 92, 237204, 2004.

5 The Effect of Ru on Magnetization Switching and CPP-GMR Enhancement

N. Tezuka, S. Abe, Y. Jiang, and K. Inomata

CONTENTS

5.1 INTRODUCTION

There has been considerable interest in ferromagnetic structures with nanometer scale lateral dimensions, mainly because these systems have the potential to provide spintronic devices such as magnetic random access memories (MRAMs) and magnetoresistive read heads.[1–3]

Current efforts to manufacture high-density MRAMs using magnetic tunnel junctions entail the production of both large area arrays and cells scaled down to 100 nm dimensions. One difficulty in such scale devices is how to control the switching of the magnetization direction and attain a low switching field H_{sw} because the demagnetization field arising from the poles at the edge of the nanometer scale elements is significantly large, and gives rise to complex magnetic domain structures that depend on the aspect ratio of the elements. A high-aspect ratio results in a large switching field,[4] whereas a low-aspect ratio produces multidomain structures, which are not suitable for MRAM devices and hamper the development of high-density MRAM. We have proposed an element structure for reducing the switching field of nanometer scale devices and permitting lower aspect ratio for MRAM devices, which is a synthetic antiferromagnet (SyAF) consisting of two magnetic layers antiferromagnetically coupled through a nonmagnetic spacer. The SyAF reduces the magnetostatic energy, resulting in a reduced switching field for the nanometer scale elements and facilitating the formation of a single domain structure even for lower aspect ratios. In Section 5.2, we will present the domain structures and element width dependence of H_{sw} for SyAFs ($Co_{90}Fe_{10}/Ru/Co_{90}Fe_{10}$) and single films of ($Co_{90}Fe_{10}$).

Another possible method to realize a low switching field is current-induced magnetization switching (CIMS). In 1996, Slonczewski and Berger predicted magnetization reversal induced by spin-polarized current injection in current perpendicular-to-plane (CPP) magnetic multilayers through the spin-transfer effect.[5,6] It is very interesting because CIMS, instead of using a magnetic field to switch, could offer a completely new class of current-controlled memory devices.[7] For example, it could greatly simplify the design of high-density nonvolatile memory. Theoretical[8–10] and experimental[11–16] understandings about the CIMS effect have been primarily established. However, the basic physical processes involved in the switching are not yet fully understood. Until now, almost all of the reported experimental works have concentrated on the CIMS observed in nanometer sized Co/Cu/Co or other pseudo spin-valve (SPV) pillars. It would therefore be a breakthrough if the CIMS phenomenon is observed in exchange-biased SPVs (ESPVs) because ESPVs are always used in magnetic storage devices instead of simple ferromagnet (FM)/Cu/(FM) trilayers in order to precisely control their domain structures and to decrease their coercivities. In Section 5.3, we will report the observation of CIMS in ESPV nanofabricated pillars and demonstrate the critical density reduction by insertion of a thin Ru layer.

Finally, we will present CPP-GMR enhancement in spin valves using a thin Ru layer. As the areal density of the hard disk drive is growing at the rate of 200% a year, the read-back device is required to have higher output.[17–20] The giant magnetoresistance (GMR) elements in current-in-plane spin valve (CIP-SV) mode will have reached its maximum sensitivity for areal densities approaching 100 Gb/in², because the track widths become too narrow. On the other hand, GMR elements in CPP mode have also been studied because it is reported that CPP-GMR is larger than CIP-GMR at room temperature in the case of multilayers,[21,22] and the downsizing of element area makes the resistance change (ΔR) higher without increasing film performance, such as the resistance change and area product (ΔRA). Spin valve (SPV) structures, however, are suitable for a magnetic read head to control the magnetization process. In the CPP-SPV case, there are fewer interfaces than that of multilayers for spin-dependent scattering and there is also a high-resistivity antiferromagnetic layer, used for pinning one ferromagnetic layer, in series with their sensor layers.[18,23] If we can obtain an increase in resistance change or magnetoresistance (MR) ratio (MR = ($\Delta RA/RA$)), the CPP-SPV could be a potential replacement for CIP-SPV in future high area recording density applications. In Section 5.4, we will report CPP-GMR ratio enhancement by inserting a Ru cap layer in the CPP-GMR SPV structure.

5.2 DOMAIN STRUCTURES AND H_{SW} FOR SyAFs ($Co_{90}Fe_{10}/Ru/Co_{90}Fe_{10}$)

Typical magnetic force microscopy (MFM) images are shown in Figure 5.1 for 2 μm × 2 μm bits of (a) $Co_{90}Fe_{10}$ (6 nm)/Ru (0.6 nm)/$Co_{90}Fe_{10}$ (10 nm) SyAF and (b) $Co_{90}Fe_{10}$ (10 nm) single film bits. The separation between the bits is greater than 2 μm for both samples. The direction of the bit length is along that of the applied magnetic field during the deposition. The single film bits exhibit a multidomain structure, indicated by the bright-dark patterns. The SyAF bits, on the other hand, demonstrate single-domain behavior indicated by the dark and bright areas observed at the sides of the SyAF elements. The single-domain structure was also observed for the bit length from 0.5 to 8 μm with aspect ratio 1 for SyAFs. Figure 5.1c and Figure 5.1d show magnetic hysteresis loops for the SyAF bits and single film bits,

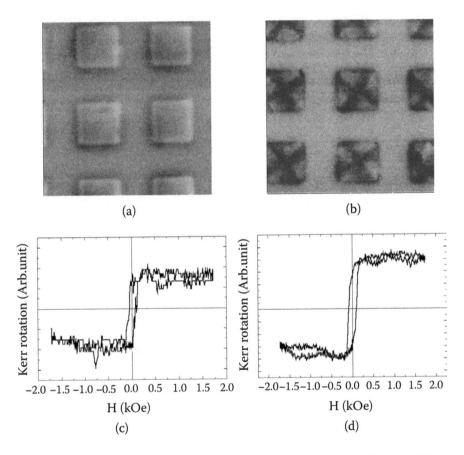

(a)   (b)

(c)   (d)

FIGURE 5.1 MFM images of (a) $Co_{90}Fe_{10}$ (6 nm)/Ru (0.6 nm)/$Co_{90}Fe_{10}$ (10 nm) SyAF and (b) $Co_{90}Fe_{10}$ (10 nm) single film bits with bit length of 2 μm and aspect ratio of 1. Magneto-optic Kerr loops for (c) SyAF and (d) single film bits, which are the same samples as those used for MFM measurement.

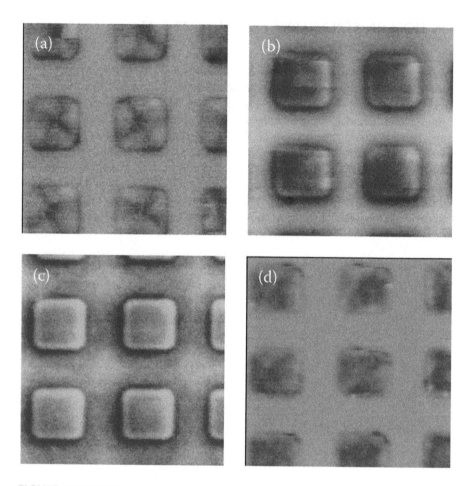

FIGURE 5.2 MFM images of the remnant state for SyAFs of $Co_{90}Fe_{10}$ (t nm)/Ru (0.6 nm) /$Co_{90}Fe_{10}$ (10 nm) for t = (a) 2, (b) 4, (c) 6, and (d) 8.

respectively, which are the same samples as those used for MFM measurements. The remnant magnetization for SyAF is almost the same as the saturation magnetization, while that of the single film is about 30% smaller than the saturation magnetization. The remnant magnetization ratio of 1 for SyAF indicates a single-domain structure, which is consistent with the MFM observations.

Figure 5.2 shows the MFM image for SyAFs with bottom layer thickness (t nm) of (a) 2, (b) 4, (c) 6, and (d) 8 and length 2 µm. In the cases of t = 4 and t = 6, a single-domain structure was observed, and in cases of t = 2 and t = 8, a four-domain closure pattern with 90° walls and a complex multidomain structure were obtained, respectively. Micron-sized SyAF can lead to a single-domain structure due to the reduction of the dipolar stray field. In the case of t = 2, however, the difference of film thickness is too large and as a result the stray field for closing magnetic flux between the top and bottom ferromagnetic films may not be enough. Therefore, a

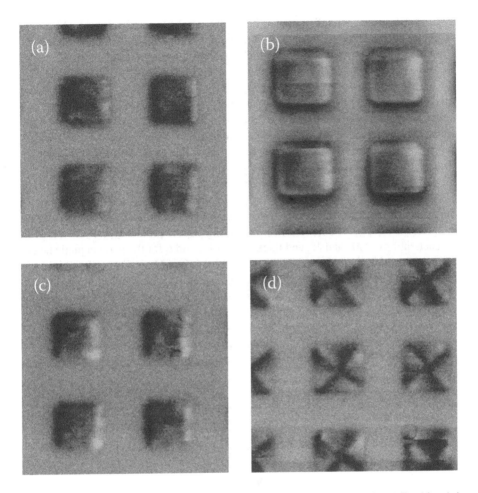

FIGURE 5.3 MFM images of the remnant state for SyAFs of $Co_{90}Fe_{10}$ (6 nm)/Ru (d nm) / $Co_{90}Fe_{10}$ (10 nm) for d = (a) 0.45, (b) 0.6, (c) 0.8, and (d) 1.3.

multidomain structure may be observed. The origin of the multidomain structure for $t = 8$ is not clear yet. It might be explained that as the film thickness increases, topographic magnetostatic coupling occurs due to the morphology or interface roughness.

Figure 5.3 shows MFM images of the remnant state for SyAFs of $Co_{90}Fe_{10}$ (6 nm)/Ru (d nm)/$Co_{90}Fe_{10}$ (10 nm) for d = (a) 0.45, (b) 0.6, (c) 0.8, and (d) 1.3 and length 2 μm. A single-domain structure can be clearly seen in the case of Figure 5.3a, Figure 5.3b, and Figure 5.3c, and a multidomain structure is seen in the case of Figure 5.3d. We measured magnetic hysteresis loops for these SyAF films before patterning. An increase of Ru thickness weakened the antiferromagnetic coupling, from $d = 0.45$ to 1.1 nm and ferromagnetic coupling was observed in the case of $d = 1.3$ nm. Single-domain structures for SyAFs of low-aspect ratio are produced by the antiferromagnetically coupled field between the top and bottom SyAF ferro-magnetic layers. From the results of Figure 5.2 and Figure 5.3, it is found that there is an optimum ferromagnetic film thickness and nonmagnetic layer thickness of SyAF to obtain a single-domain structure for the aspect ratio of 1.

Figure 5.4 shows the coercivity field as a function of bit width of SyAFs and single film bits. For all cases, the magnetic field required to switch the magnetization direction increases systematically as the bit width is decreased as a result of the increased demagnetization field from the magnetic moment. The coercive field is approximately inversely proportional to the bit width. Note also that the coercive field of the SyAF bit is not influenced so much by width compared to that of the single film bit. From the Stoner–Wohlfarth model,[5] the coercivity H_c is given by $H_c \propto C(k)Mt/W$, where $C(k)$ and t are related to the demagnetization factor depending on the aspect ratio k and film thickness t, respectively. $C(k)$ contains complete elliptic integrals and becomes smaller with decreasing aspect ratio. For a SyAF bit with magnetizations of M_1 and M_2 and thicknesses of t_1 and t_2 for the two magnetic layers across the nonmagnetic layer, Mt is replaced by $(M_1t_1 - M_2t_2)$. This reduction of the magnetic moment results in a small size-dependent switching field for SyAFs.

The value of $C(k)$ is zero in the case of $k = 1$, thus, the bits with aspect ratio of 1 and single-domain structure can lead to a size-independent switching field.[24] In the result of Figure 5.4, however, the switching field increased with decreasing bit width for the samples with $k = 1$. From our previous simulation result,[25] it can be expected that coherent rotation while maintaining antiferromagnetic alignment

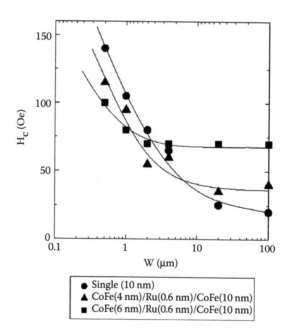

FIGURE 5.4 Coercive field as a function of bit width for (●) $Co_{90}Fe_{10}$ (10 nm), (▲) $Co_{90}Fe_{10}$ (6 nm)/Ru (d nm)/$Co_{90}Fe_{10}$ (10 nm), and (■) $Co_{90}Fe_{10}$ (6 nm)/Ru (0.6 nm)/$Co_{90}Fe_{10}$ (10 nm).

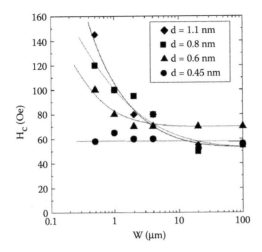

FIGURE 5.5 Coercive field as a function of bit width for SyAF of $Co_{90}Fe_{10}$ (6 nm)/Ru (d nm) /$Co_{90}Fe_{10}$ (10 nm).

during magnetization reversal is necessary to follow the Stoner–Wohlfarth model and obtain the size-independent switching field character for SyAFs with $k = 1$. Thus, it was predicted that a size-independent bit would be obtained only for a SyAF with $k = 1$ and strongly antiferromagnetic coupling fields between the top and bottom ferromagnetic layers. To explore such properties, SyAFs with varied exchange coupling field were fabricated. Figure 5.5 shows the bit width dependence of the coercivity field for SyAFs. The most remarkable result exhibited in Figure 5.5 is the size-independent switching field behavior for SyAF consisting of $Co_{90}Fe_{10}$ (6 nm)/Ru (0.45 nm)/$Co_{90}Fe_{10}$ (10 nm). The coercivity of SyAF with a large exchange coupling field is less affected by bit width. This tendency is equivalent to the result obtained by micromagnetic simulation.

Finally, we investigated the influence of the bit shape and width on coercivity force H_c and fabricated SyAF and monolayer bits, having square and circle shapes with $k = 1$ with bit widths (or diameters) W from 0.2 to 100 μm. Figure 5.6 shows H_c as a function of W for SyAF and monolayer bits. For the monolayer system, H_c is influenced by both W and bit shape. For the SyAF system, however, H_c is not affected by W and bit shape. This demonstrates that SyAFs with $k = 1$ exhibit size- and shape-independent switching fields due to the magnetic closing between two magnetic layers with a strong interlayer antiferromagnetic exchange coupling.

5.3 CIMS IN ESPV NANOFABRICATED PILLARS AND CRITICAL DENSITY REDUCTION BY INSERTION OF A THIN Ru LAYER

The measured CPP-GMR loop of one ESPV pillar ("ESPV 1") with a size A of 280×90 nm², a free layer thickness d_1 of 2.5 nm, and a Ru cap layer thickness d_2 of 0 nm (i.e., no Ru cap layer) is shown in the upper left inset of Figure 5.7. The CPP-GMR of ESPV 1 is around 0.3%. We also measure the resistance switching

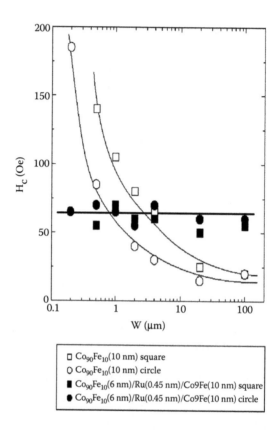

FIGURE 5.6 Coercive force as a function of bit width on diameter for SyAFs and monolayers with squares and circles for $k = 1$.

while sweeping a direct current (DC) current. We start the measurement while keeping the magnetic field at 90 Oe and that makes the antiparallel (AP) state of magnetization in the sample. The sweeping DC current causes the resistance switching in ESPV 1, shown as the closed circles with lines in Figure 5.7. The hysteretic CIMS loop is very similar to previously reported data on Co/Cu/Co nanopillars.[13,15,16] At the beginning, the resistance keeps nearly constant with increasing positive current. When the positive current reaches a critical value $I_P \sim 27$ mA (i.e., the critical current density $J^{AP \to P} = 1.1 \times 10^8$ A/cm^2), the resistance jumps to a lower value, which corresponds to the resistance in the parallel (P) configuration of magnetization. When the current decreases to a critical value $I_{AP} \sim 54$ mA (i.e., the critical current density $J^{P \to AP} = -2.1 \times 10^8$ A/cm^2), the resistance jumps back to the higher value, the AP resistance. We define the average critical current density as $J_C = 1/2(|J^{AP \to P}| + |J^{P \to AP}|)$. The calculated J_C for ESPV 1 is 1.6×10^8 A/cm^2. The abrupt change in resistance by the injected current is the same as the observed CPP-GMR in an external magnetic field. This means that the resistance switching by the injected DC current in ESPV 1 is due to the full reversal of magnetization of the sample through the spin-transfer effect. For ESPV 1, the critical current $|I_P|$ is lower than $|I_{AP}|$, which is in

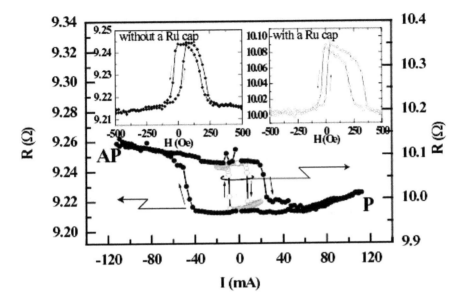

FIGURE 5.7 Resistance (R) vs. sweeping DC current (I) for ESPV 1 (closed circles with lines) and ESPV 2 (open circles with lines). Both curves have been measured under a 90 Oe magnetic field. The insets show the GMR loops of ESPV 1 (closed circles with lines) and ESPV 2 (open circles with lines).

good agreement with the spin-transfer model.[5] We note that the critical current densities here are much higher than the reported ones in the Co/Cu/Co nanopillars,[13] typically of the order of ~10^7 A/cm^2. The thick antiferromagnet (AF) layer of Ru (~10 nm) in ESPV 1 may cause a strong depolarization of the spin current. The reduced spin polarization leads to an even smaller spin torque per unit current applied to the FM free layer, and therefore, a larger current is required to drive the CIMS.

For the sample "ESPV 2," we keep the free layer thickness as d_1 = 2.5 nm and insert a thin Ru layer (d_2 = 0.45 nm) between the free layer and the top Cu layer. The junction size A of ESPV 2 is also 280×90 nm^2. The I-R curve measured under a 90 Oe magnetic field together with its GMR loop are shown as the open circles with line in Figure 5.7 and the inset. A clear resistance switching by an injected DC current can be observed. The CIMS behavior is similar to that of ESPV 1 except that the average critical current density is much lower, 2.2×10^7 A/cm^2. Therefore by inserting a thin Ru cap layer between the free layer and the top electrode, the average critical current density that makes the full reversal of magnetization is effectively decreased.

The insertion of a Ru layer between the ESPV nanopillar and top electrode effectively reduces the critical current for the CIMS. From an application point of view, however, the critical current density of ~10^7 A/cm^2 is still too high for the CIMS to be directly applied in spintronic devices. To further reduce the critical current density, we designed a special ESPV structure (Structure III); Cu (80 nm)/ Co$_{90}$Fe$_{10}$ (5 nm)/Ru (6 nm)/Co$_{90}$Fe$_{10}$ (2.5 nm)/Cu (6 nm)/Co$_{90}$Fe$_{10}$ (5 nm)/IrMn

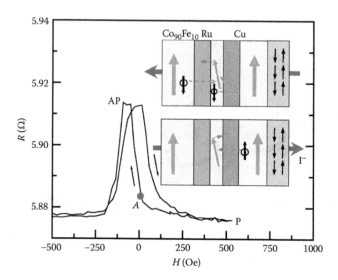

FIGURE 5.8 CPP-GMR loop for one nanopillar with structure III. The area of the pillar is 400×100 nm^2 and the sense current is 0.1 mA. The magnetic configurations of the nanopillar are shown in the inset for both positive and negative current applications when the external magnetic field is in the position A.

(10 nm)/Cu (5 nm)/Ta (2 nm). The CPP-GMR loop for one nanopillar with Structure III and a size of 400×100 nm is shown in Figure 5.8. The nanopillar shows a MR value of around 0.57% at room temperature. As we decrease the external magnetic field from the positive saturation value (1000 Oe) down to zero, that is, at the position A in Figure 5.8, the magnetic configuration of the nanopillar is shown as the inset of Figure 5.8. Because the ion milling etching is controlled to be stopped at the middle of the Ru layer, a half of the Ru spacer layer is left extended, so that there is no dipolar coupling between the two neighboring $Co_{90}Fe_{10}$ layers on the left-hand side in the inset of Figure 5.8 (i.e., the bottom two FM layers). Therefore, the magnetic moment of the middle thin $Co_{90}Fe_{10}$ layer can be easily reversed.

Figure 5.9 shows the CIMS results for the nanopillar with the Structure III at room temperature. The CIMS curves are measured after the magnetic field is decreased from 1000 Oe down to a certain value H. As shown in the insets of Figure 5.9, the injected current clearly drives the magnetization switching in the nanopillar. With increasing magnetic field H, the $I^{P \to AP}$ slightly increases while the $I^{AP \to P}$ largely decreases, which is consistent with the reported results on a Co/Cu/Co pseudo SPV.[26] Therefore, we believe the spin-momentum transfer rather than the current-induced Oested field plays a key role in the CIMS process. The critical current densities are only $J^{AP \to P} = -2 \times 10^6$ A/cm^2 and $J^{P \to AP} = 2.3 \times 10^6$ A/cm^2 without a magnetic field, which are significantly lower than those in the FM/Cu/FM trilayers.[15,16,26] Without a magnetic field, the resistance change due to the injected current is around 0.57% that is the same as the CPP-GMR of the nanopillar, which means a full reversal of the magnetization of the free layer is achieved by the injected current.

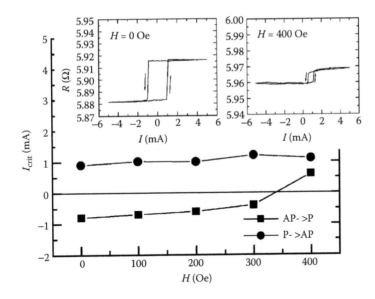

FIGURE 5.9 Magnetic field dependence of the critical current (I_{crit}) for the CIMS. The insets show the corresponding CIMS curves of the nanopillar when the magnetic fields are 0 and 400 Oe.

5.4 CPP-GMR RATIO ENHANCEMENT BY INSERTING Ru CAP LAYER

We fabricated CPP-SPVs with two kinds of free layers. One is a ($Co_{90}Fe_{10}$ t_f nm) single film (conventional structure) and the other is a ($Co_{90}Fe_{10}$ t_f nm/Ru d_{Ru} nm) bilayer. The bottom electrode (Ta 2.5 nm/Cu 20 nm), CPP-SPV film ($Ir_{22}Mn_{78}$ 10 nm/$Co_{90}Fe_{10}$ t_p nm/Cu d_{Cu} nm/free layer), and top contact electrode (Cu 5 nm/Ta 3 nm) are deposited on a thermally oxidized Si wafer in an ultrahigh vacuum sputtering system with a base pressure below 5×10^9 torr. A magnetic field of around 200 Oe was applied during sputtering to induce an easy axis. The bottom electrode and the element of a CPP-SPV were fabricated by using electron beam lithography and ion milling etching techniques, and then they were coated with a SiO_2 film with a contact hole. Lastly, the top electrode of Cu was fabricated using a lift-off process. We grew these structures under the same conditions, and the difference between the conventional and Ru cap structures is only whether or not the thin Ru layer exists. MR curves of the CPP-SPV elements were measured by a four-point probe method in a field from −1000 to 1000 Oe at room temperature. The magnetic field direction was aligned to the easy axis of the films. The measuring current was kept below 1 mA to avoid any other effects induced by the current. To avoid current distribution effects and to obtain negligible contact resistance in our system, the MR ratio was obtained as $A\Delta R/AR$ from size A, resistance R, and resistance change ΔR. Note that our experimental results for the conventional CPP-SPV fit well with the V-F model calculation.[27,28]

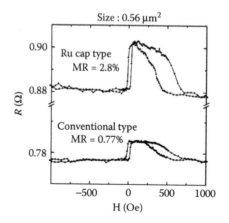

FIGURE 5.10 MR curves for two kinds of CPP-SPV structures when the size of both elements is 0.56 μm².

A resistance vs. field curve is shown in Figure 5.10 for a conventional and Ru cap structures. As is found for the Ru cap structure, inserting a Ru layer between the $Co_{90}Fe_{10}$ and top Cu layers increases the MR ratio by a factor of approximately 3.5. Namely, the structure displays an enhanced MR effect with higher resistance and resistance change. Consequently, the dependence of the MR ratio on the thickness of Ru layer was examined in detail as shown in Figure 5.11. The MR ratio exhibits a maximum at a Ru thickness between 0.2 and 0.4 nm. The slight decrease in MR ratio with increasing Ru layer thickness greater than 0.5 nm can be understood as a dilution effect due to the serial resistance of the Ru layer. More importantly,

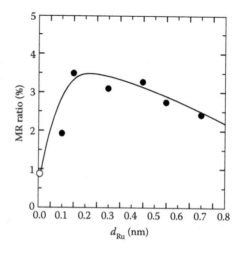

FIGURE 5.11 MR ratio as a function of Ru thickness for Ru cap structure of CPP-GMRs.

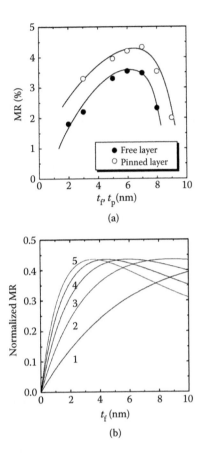

FIGURE 5.12 (a) MR ratio as functions of the free and pinned layer thickness for Ru cap structure of CPP-GMRs. (b) Calculation results of the relative CPP-GMR as a function of ferromagnetic film thickness.

an increase in MR ratio with Ru layer thickness up to 0.2 nm is observed. Thus the importance of interface scattering between CoFe and Ru is suggested.

Figure 5.12A shows the MR ratio as a function of free and pinned ferromagnetic layer thickness for the Ru cap structure. For both cases, the maximum occurs around 7 nm (t_{MAX}). This is a significantly smaller thickness than the spin-diffusion length, whose value is reported to be about 12 nm for $Co_{91}Fe_9$ at 4.2 K.[29] The thickness dependence of MR ratio can be described[30] by

$$MR = (MR)_{max}[1 - \exp(-nt_f/l_{sf})] / [1 + nt_f/t_{rest}]$$

where l_{sf} (= 12 nm) is the spin-diffusion length, t_{rest} is an effective thickness representing the serial resistance of the rest of the structure (= 25 nm), and n is the number of passes through the ferromagnetic layer because of reflection by the Ru cap and IrMn or Cu layer. Because Ru impurities in Co scatter majority spins more strongly than minority ones,[31] we assume that reflection of electron spins occurs at the interface between the $Co_{90}Fe_{10}$ and Ru layer. In Figure 5.12B, the calculation results using this equation are shown. As can be seen from this figure, t_{MAX} is shifted to smaller film thickness with increasing n. Again the t_{MAX} associated with the spin-diffusion length becomes smaller with increasing n, and in this case, t_{MAX} for $n = 3$ corresponds to the experimental result.

To summarize, we have studied CPP-GMR for conventional and Ru cap structures and found the insertion of the Ru layer enhances the MR ratio. The MR ratio enhancement effect decreases with increasing the ferromagnetic layer. Comparison between experimental results and phenomenological calculation suggests the interfacial spin scattering between Ru and $Co_{90}Fe_{10}$ layers.

5.5 SUMMARY

Magnetization switching field (H_{sw}) and magnetic domain structure were investigated for SyAF elements with micron to submicron sizes and different aspect ratios. We found that the strongly antiferromagnetically coupled SyAF with an aspect ratio of 1 showed single-domain structure and size independent H_{sw}. This comes from the zero demagnetization field of SyAF bits with $k = 1$.[32,33]

It was demonstrated that a thin Ru layer has a strong effect on the CIMS behavior. A very high current density, about 10^8 A/cm^2, is needed for CIMS in a single spin valve-type structure; however, the insertion of a Ru layer between the free layer and top electrode decreases the critical current density for magnetization switching by at least one order of magnitude. Furthermore, in a well-designed antisymmetrical structure, this critical current density can be reduced to 10^6 A/cm^2. This effect can also be understood by the majority spin reflection by a Ru layer.[34,35]

We fabricated a single spin valve structure with a SyAF free layer. We observed a CPP-GMR enhancement and doubled the resistance area product for a single spin valve with a SyAF free layer. This enhancement is also produced for a single spin valve structure with a $Co_{90}Fe_{10}$/Ru free layer. We argue that the MR enhancement is probably caused by the spin-dependent scattering due to strong reflection of majority spins at the $Co_{90}Fe_{10}$/Ru interfaces.[36]

ACKNOWLEDGMENT

This research was supported by the information technology program of Research and Revolution 2002 (RR2002), the Priority Area Grant No. 14076202, Scientific Research (A) Grant No. 15206074 from Ministry of Education, Culture, Sports, Science, and Technology (MEXT) Industrial Technology Research Grant Program in '0* from New Energy and Industrial Technology Development Organization (NEDO) of Japan, the Mitsubishi Foundation, and the Housoubunka Foundation.

REFERENCES

1. W.J. Gallagher, S.S.P. Parkin, Y. Lu, X.P. Bian, A. Marley, R.A. Altman, S.A. Rishton, K.P. Roche, C. Jahnes, T.M. Shaw, and G. Xiao, *J. Appl. Phys.*, 81, 3741, 1997.
2. S. Tehrani, J.M. Slaughter, E. Chen, and M. Durlam, *IEEE Trans. Magn.*, 35, 2814, 2000.
3. O. Redon, N. Kasahara, K. Shimazawa, S. Araki, H. Morita, and M. Matsuzaki, *J. Appl. Phys.*, 87, 4688, 2000.
4. E. Girgis, J. Schelten, J. Si, J. Janesky, S. Tehrani, and H. Goronkin, *Appl. Phys. Lett.*, 76, 3780, 2000.
5. J.C. Slonczewski, *J. Magn. Magn. Mater.*, 159, L1, 1996.
6. L. Berger, *Phys. Rev. B*, 54, 9353, 1996.
7. K. Inomata, *IEICE Trans. Electron.*, E84-C(6), 740, 2001.
8. X. Waintal, E.B. Myers, P.W. Brouwer, and D.C. Ralph, *Phys. Rev. B*, 62, 12317, 2000.
9. M.D. Stiles and A. Zangwill, *Phys. Rev. B*, 66, 014407, 2002.
10. S. Zhang, P.M. Levy, and A. Fert, *Phys. Rev. Lett.*, 88, 236601, 2002.
11. E.B. Myers, D.C. Ralph, J.A. Katine, R.N. Louie, and R.A. Buhrman, *Science*, 285, 867, 1999.
12. M. Tsoi, A.G.M. Jansen, J. Bass, W.C. Chiang, V. Tsoi, and P. Wyder, *Nature*, 406, 46, 2000.
13. F.J. Albert, N.C. Emley, E.B. Myers, D.C. Ralph, and R.A. Buhrman, *Phys. Rev. Lett.*, 89, 226802, 2002.
14. K. Bussmann, G.A. Prinz, S.-F. Cheng, and D. Wang, *Appl. Phys. Lett.*, 75, 2476, 1999.
15. F.B. Mancoff, R.W. Dave, N.D. Rizzo, T.C. Eschrich, B.N. Engel, and S. Tehrani, *Appl. Phys. Lett.*, 83, 1596, 2003.
16. S. Urazhdin, N.O. Birge, W.P. Pratt, Jr., and J. Bass, *Appl. Phys. Lett.*, 84, 1516, 2004.
17. R. Rottomayer and J. Zhu, *IEEE Trans. Magn.*, 31, 2597, 1995.
18. A. Tanaka, Y. Shimizu, Y.A. Seyama, K. Nagasaka, R. Kondo, H. Oshima, S. Eguchi, and H. Kanai, *IEEE Trans. Magn.*, 38, 84, 2002.
19. M.A. Seigler, P.A. Van der Heijiden, A.E. Litvinov, and R. Rottmayer, *IEEE Trans. Magn.*, 39, 1855, 2003.
20. H. Sakakima, Y. Sugita, M. Satomi, and Y. Kawawake, *J. Magn. Magn. Mater.*, 198, 9, 1999.
21. M.A.M. Gijs, S.K.J. Lenczowski, and J.B. Giesberg, *Phys. Rev. Lett.*, 70, 3343, 1993.
22. M.A.M. Gijs, M.T. Johnson, A. Reinders, P.E. Husman, R.J.M. van de Veerdonk, S.K.J. Lenczowski, and R.M.J. van Gansewinkel, *Appl. Phys. Lett.*, 66, 1989, 1995.
23. K. Nagasaka, Y. Seyama, L. Varga, Y. Shimizu, and A. Tanaka, *J. Appl. Phys.*, 89, 6943, 2001.
24. N. Tezuka, N. Koike, K. Inomata, and S. Sugimoto, *J. Appl. Phys.*, 93, 7441, 2003.
25. N. Tezuka, E. Kitagawa, K. Inomata, S. Sugimoto, N. Kikuchi, and Y. Shimada, *J. Magn. Magn. Mater.*, 76, 3780, 2000.
26. J.A. Katine, F.J. Albert, R.A. Buhrman, E.B. Myers, and D.C. Ralph, *Phys. Rev. Lett.*, 84, 3149, 2000.
27. Y. Jiang, S. Abe, T. Nozaki, N. Tezuka, and K. Inomata, *Appl. Phys. Lett.*, 83, 2874, 2003.
28. T. Valet and A. Fert, *Phys. Rev. B*, 48, 7099, 1993.
29. A.C. Reilly, W. Park, R. Slater, B. Ouaglal, R. Loloee, W.P. Pratt, Jr., and J. Bass, *J. Magn. Magn. Mater.*, 195, L269, 1999.
30. N. Stelkov, A. Vedyaev, and B. Dieny, *J. Appl. Phys.*, 94, 3278, 2003.

31. I.A. Campbell and A. Fert, *Ferromagnetic Materials*, E.P. Wolforth, Ed., Vol. 3, North-Holland, Amsterdam, 1982, Chap. 9, p. 752.
32. N. Tezuka, N. Koike, K. Inomata, and S. Sugimoto, *Appl. Phys. Lett.,* 82, 604, 2003.
33. K. Inomata, N. Koike, T. Nozaki, S. Abe, and N. Tezuka, *Appl. Phys. Lett.,* 82, 2667, 2003.
34. Y. Jiang, S. Abe, T. Ochiai, T. Nozaki, A. Hirohata, N. Tezuka, and K. Inomata, *Phys. Rev. Lett.,* 92, 167204, 2004.
35. Y. Jiang, T. Nozaki, S. Abe, T. Ochiai, A. Hirohata, N. Tezuka, and K. Inomata, *Nat. Mater.,* 3, 361, 2004.
36. N. Tezuka, S. Abe, Y. Jiang, and K. Inomata, *J. Magn. Magn. Mater.*, 290, 1150, 2005.

6 The Spin-Dependent Interfacial Transparency

Ke Xia

CONTENTS

6.1 INTRODUCTION

The great achievement of the controlled growth of layered structures has opened a new frontier in which interface effects dominate bulk properties. In artificial transition metal materials, the distance between interfaces can be as small as nanometers. A number of important discoveries have been made in these structures, such as oscillatory exchange coupling,[1] giant magnetoresistance (GMR),[2] and current-induced magnetization switching.[3]

It was discovered that in layered structures of ferromagnetic (FM) layers with nonferromagnetic (NM) spacer layers, the alignment of the magnetization of FM layers depends on the thickness of the spacer layer. For certain thicknesses of the NM layers, two successive FM layers are coupled antiferromagnetically. The strength of the exchange coupling in general is very small compared with that in the bulk materials, such as Fe, Co, and Ni. So when a small magnetic field H is applied, the magnetization of the layers can be easily directed toward H. Later experiments showed that there is a large change of resistance as the angle between the magnetization directions of adjacent magnetic films changes, which is known as "giant magnetoresistance." More interestingly, because the spin injected into a magnetic material experiences a torque due to the magnetic moment, it has been argued that passage of a current through adjacent magnetic layers should lead to the transfer of spin-angular momentum from one layer to the other with possible reorientation of the magnetization for sufficiently large currents,[3] now known as current-induced spin torque. The experimental observation of current-induced magnetization reversal effects has been done by a number of groups.[4]

Spin-dependent interfacial transparency plays an important role in all these phenomena. Interlayer exchange coupling can be understood by the spin-dependent interference between the interfaces. For a sandwich formed by two ferromagnetic layers separated by nonmagnetic spacer layers, interlayer exchange coupling is a product of the geometrical properties of the Fermi surface of the spacer layer material and the reflection amplitudes from the interfaces.[5]

The GMR effect can be understood qualitatively in terms of the two current model, originally proposed by Mott, to explain the increase in resistivity of FM metals when heated above the Curie temperature. In this model it is assumed that the electrical conductivity of GMR structures can be split into two independent channels corresponding to the two possible spin orientations of the electrons. It is important to notice that the two channels can only be treated independently if the spins of the electrons are conserved over a long length scale. In other words, the probability of a spin flip must be small or the spin-flip length must be much longer than the layer thickness. The scattering rates of the two different channels can be quite different in these structures. Strongly spin-dependent interface reflection can lead to a large asymmetry for different spin channels. For the case of a current perpendicular to the interfaces, the GMR comes from the spin dependence of the resistance associated with each interface.

In circuit theory,[6] the spin torque and current in a ferromagnetic heterostructure are formulated in terms of (real) spin-up and spin-down conductances, $G^{\uparrow}$ and $G^{\downarrow}$, and spin-mixing conductance, G^{mix}, which is complex. Here $G^{\uparrow}, G^{\downarrow}$ and G^{mix} can all be formulated as the function of the scattering matrix elements at the interface.

The spin-dependent scattering at the ferromagnet/nonmagnetic metal (FM/NM) interface has many important physical consequences in a magnetic layered structure. However, measuring spin-dependent reflectivity is a difficult job, because the spin-up and spin-down electrons flow together in most electronic transport experiments. The most common way to probe the spin polarization is to make use of the two-current model. For example, the spin-dependent interface resistance can be measured by changing the magnetic alignment of the magnetic multilayer and the spin-dependent interface resistance can be obtained in the frame of the two-current model. Andreev reflection spectroscopy at the ferromagnet/superconductor (F/S) interfaces in ballistic point contacts,[7,8] proposed as an alternative way to measure the spin polarization, contains information of the scattering matrix at the interface but is very difficult to resolve.[9] Therefore, there is a demand for a method that can be used as a tool to measure the transmission of different spins through an interface separately.

When the electrons are confined in a thin film, the energy levels will be discrete. These are called quantum well (QW) states. The confinement can be provided by a real barrier (such as a vacuum) or a symmetry gap at the metallic interface. When the confinement is provided by a ferromagnetic metal, the confinement is spin polarized, so the position of the energy levels of the QW states will also be spin-dependent. These QW states can be probed by angle resolved photoemission spectroscopy (ARPES).[10]

Both theory and experiment show that the linewidths of spectroscopy are related to the quasiparticle lifetime and the interfacial reflectivity. According to the Bohr–Sommerfeld quantization rule, the peak width of the spectroscopy depends on reflectivity at the boundaries

$$\delta E = \Gamma \eta \frac{1 - R \exp(-1/\eta)}{\sqrt{R} \exp(-1/2\eta)},$$

where R is the product of the reflection at the interface and the surface, and δE is the spectroscopy linewidth. Gamma is the quasiparticle inverse lifetime, and eta is a parameter related to the thickness of film. By measuring these linewidths, one can get information about the reflectivity of the interface, which provides us with a new way to separately measure the spin-dependent reflectivity of a ferromagnetic and normal metal interface.

This method can be used to demonstrate the idea of measuring the spin-dependent transparency provided the following conditions are satisfied. (i) A high quality interface must be obtained experimentally to observe the QW states. (ii) There must be a "relative gap" at the interface to confine the electron propagation along the direction perpendicular to the interface, so as to form the QW states. (iii) It should contain a NM/FM interface to provide spin-dependent reflectivity.

Our previous study[11] suggested that interface roughness significantly influences transport properties. The transmission can be increased or decreased by interface impurity scattering and the magnitude of the change can be as high as 70% in some systems, such as Fe/Cr bcc(001) (bcc — body-centered cubic). However, experimentally it is rather difficult to find a system that is clean enough to verify the theoretical calculation for the ideal interface. Ag/Fe provides us with a system in which the impurity scattering can be systematically changed from weak to strong.

This chapter is organized as follows. In Section 6.2, the calculation method is introduced. In Section 6.3, the interface resistance for Co/Cu is presented. In Section 6.4, we apply our method to calculate QW states in Fe/Ag, and in Section 6.5, a brief summary is given.

6.2 TRANSPORT CALCULATION BASED ON THE FIRST-PRINCIPLES METHOD

Consider the two-terminal structure sketched in Figure 6.1. The system can be divided into three parts, a left electrode, an intermediate scattering region, and a right electrode. The intermediate region can be any structure, such as a single

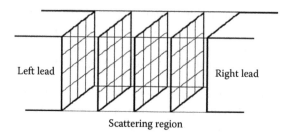

Left lead

Right lead

Scattering region

FIGURE 6.1 Sketch of the configuration used in the Landauer–Büttiker transport formulation to calculate the two-terminal conductance. A scattering region is sandwiched by left- and right-hand leads that have translation symmetry and are partitioned into principal layers perpendicular to the transport direction.

interface, multilayer, or a tunneling junction. The scattering by defects or disconti-nuities in the electronic structure only happens in the intermediate region. The elements of the scattering matrix (i.e., transmission and reflection coefficients) contain the information of how a Bloch wave coming from one electrode is scattered by the intermediate region to other Bloch states in the same or the other electrode. To compute the scattering matrix, we have to specify the Bloch states in the electrodes and solve the scattering problem in the intermediate region. The Bloch states can be found by making use of the translation symmetry of the crystal structure of the ideal leads and the scattering problem can be solved using a Green's function method.[12]

6.2.1 THE TIGHT-BINDING HAMILTONIAN

Let us start with the tight-binding Hamiltonian H based on the first-principles elec-tronic structure calculation[13]

$$H_{RL,R'L'} = \varepsilon_{RL} O_{RLR'L'} + t_{RLR'L'} \tag{6.1}$$

where R is the site index, ε_{RL} is the on site energy of R with the orbit denoted by the angular momentum $L = (lm)$ and $t_{RLR'L'}$ are the hopping matrix elements between RL and $R'L'$.

In the mixed representation of lateral wave vector $k_\parallel$ in the two-dimensional interface Brillouin zone and real space layer index i

$$
\begin{aligned}
&[(\varepsilon O_{ii}(k_\parallel) - \varepsilon_{ii}(k_\parallel)) - t_{ii}(k_\parallel)]u_i + [\varepsilon O_{i,i-1}(k_\parallel) - \varepsilon_{i,i-1}(k_\parallel) - t_{i,i-1}(k_\parallel)]u_{i-1} \\
&+ [\varepsilon O_{i,i+1}(k_\parallel) - \varepsilon_{i,i+1}(k_\parallel) - t_{i,i+1}(k_\parallel)]u_{i+1} = 0
\end{aligned}
\tag{6.2}
$$

where u_i is a $(2l+1)N \times 1$ vector describing the amplitudes of the i-th layer consisting of N sites and $(2l+1)$ bands. $\varepsilon O_{ii}(k_\parallel)$ and $t_{ii}(k_\parallel)$ are $(2l+1)N \times (2l+1)N$ matrices.

Here

$$O_{i,j}(k_\parallel) = \sum_{R \in \{R_{i,j}\}} O(R)\exp(ik_\parallel R) \tag{6.3}$$

$$\varepsilon_{i,j}(k_\parallel) = \sum_{R \in \{R_{i,j}\}} \varepsilon(R)O(R)\exp(ik_\parallel R) \tag{6.4}$$

$$t_{i,j}(k_\parallel) = \sum_{R \in \{R_{i,j}\}} t(R)\exp(ik_\parallel R) \tag{6.5}$$

where $R_{i,j}$ denotes the set of vectors that connect one lattice site in the i-th layer with all lattice sites in the j-th layer. Note that we can apply the formalism described in the following to any electronic structure program with tight-binding form. In this study, we got the Hamiltonian from the tight-binding linear muffin-tin orbital (TB-LMTO) method.[13] For the sake of simplicity of notation, from now on we omit explicit reference to the $k_\parallel$. Rewrite the equation of motion as

$$[\varepsilon O_{ii} - H_{ii}]u_i - H_{i,i-1}u_{i-1} - H_{i,i+1}u_{i+1} = 0 \tag{6.6}$$

here

$$H_{ii} = \varepsilon_{ii}(k_\parallel) + t_{ii}(k_\parallel) \tag{6.7}$$

$$H_{i,i-1} = -[\varepsilon O_{i,i-1}(k_\parallel) - \varepsilon_{i,i-1}(k_\parallel) - t_{i,i-1}(k_\parallel)] \tag{6.8}$$

$$H_{i,i+1} = -[\varepsilon O_{i,i+1}(k_\parallel) - \varepsilon_{i,i+1}(k_\parallel) - t_{i,i+1}(k_\parallel)] \tag{6.9}$$

6.2.2 Eigenstates in the Leads

Now consider the Bloch states in the ideal lead. To obtain linearly independent solutions, we set $C_i = \lambda^i C_o$, because in a periodic potential, the wave function should satisfy the Bloch theorem. The potential function is the same for all sites. The constant structure matrix depends only on the relative positions. The equation of motion then becomes

$$\lambda \begin{pmatrix} u_i \\ u_{i-1} \end{pmatrix} = \begin{pmatrix} H_{01}^{-1}(\varepsilon O_{00} - H_{00}) & -H_{01}^{-1}H_{10} \\ 1 & 0 \end{pmatrix}\begin{pmatrix} u_i \\ u_{i-1} \end{pmatrix} \tag{6.10}$$

where the wave number k is related to the eigenvalue $\lambda = \exp(ika)$, and a is the distance between adjacent layers. This equation has $2(2l+1)N$ eigenvalues and $2(2l+1)N$ eigenvectors, corresponding to $(2l+1)N$ right- and $(2l+1)N$ left-going waves.

Let $u_1(-), \ldots, u_N(-)$ denote the left-going solutions corresponding to eigenvalues $\lambda_1(-), \ldots, \lambda_N(-)$, and $u_1(+), \ldots, u_N(+)$ denote the right-going solutions corresponding to eigenvalues $\lambda_1(+), \ldots, \lambda_N(+)$. Define $U(\pm) = (u_1(\pm), \ldots, u_N(\pm))$ and $\Lambda(\pm)$, the matrix of diagonal elements as $\lambda_i(\pm)$. With $U(\pm)\Lambda(\pm)U^{-1}(\pm) \equiv F(\pm)$, one finds easily that $u_i(\pm) = F^{i-i'}(\pm)u_{i'}(\pm)$. $F^{i-i'}(\pm)$ is the Bloch factor in the leads representing the bases used in the equation of motion.

In a given lead there are $2M$ eigenvalues corresponding to real k vectors, which means there will only be M right-going and M left-going propagating states, the others are evanescent with $M \leq (2l+1)N$.

6.2.3 SCATTERING PROBLEM

Consider a Bloch wave incident from the left lead and scattered by the intermediate region. One part of the incoming wave is reflected back into the left lead and the rest is transmitted through the intermediate region into the right lead. We number the layers in the intermediate region from 1 to N. As there is no translation symmetry in the direction perpendicular to the layered plane, we treat this direction in real space. The Hamiltonian matrix for an infinite system in real space is also infinite, but because the scattering only happens in the intermediate region, we use boundary conditions to represent the ideal electrodes so that we can reduce the number of coupled equations of motion.

First, we separate the amplitude in the zeroth layer into right- and left-going solutions $u_0 = u_0(+) + u_0(-)$. As there is no scattering outside the intermediate region, we can use the Bloch factor to relate the amplitude in the lth layer to that in the zeroth layer as

$$u_{-1} = F_l^{-1}(+)u_0(+) + F_l^{-1}(-)u_0(-) = [F_l^{-1}(+) - F_l^{-1}(-)]u_0(+) + F_l^{-1}(-)u_0 \quad (6.11)$$

The equations of motion for the zeroth layer amplitude therefore involve only that in the first layer

$$\{\varepsilon O_{00} - \tilde{H}_{0,0}\}u_0 - H_{01}u_1 = H_{10}[F_l^{-1}(+) - F_l^{-1}(-)]u_0(+) \quad (6.12)$$

So, we have truncated the equation of motion at layer zero by using a boundary condition for the periodic lead. Here, $\tilde{H}_{0,0} = H_{0,0} + H_{10}F_l^{-1}(-)$, l denotes the left lead, and $-H_{10}F_l^{-1}(-)$ is the "embedding potential" for the left lead.

Only the right-going wave can exist in the $(N+1)$-th layer, so the boundary condition is simply $u_{N+2} = F_r(+)u_{N+1}(+)$. We can truncate the equation that couples the wave functions to the right lead r as

$$(\varepsilon O_{N+1,N+1} - \tilde{H}_{N+1,N+1})u_{N+1} - H_{N+1,N}u_N = 0 \quad (6.13)$$

where $\tilde{H}_{N+1,N+1} = H_{N+1,N+1} + H_{N+1,N+2}F_r(+)$ and $-H_{N+1,N+2}F_r(+)$ is the embedding potential for the right lead.

By using these boundary conditions, the number of equations of motion is finite. The Green's function matrix in real space is the inverse of the (finite) Hamiltonian matrix

$$
G = \begin{pmatrix}
\varepsilon O_{00} - \tilde{H}_{0,0} & H_{01} & 0 & 0 \\
H_{10} & \varepsilon O_{11} - H_{11} & H_{1,2} & 0 \\
0 & H_{2,1} & \cdots & H_{N,N+1} \\
0 & 0 & H_{N+1,N} & \varepsilon O_{N+1,N+1} - \tilde{H}_{N+1,N+1}
\end{pmatrix}^{-1}
\tag{6.14}
$$

In fact, this treatment is very similar to the widely used surface Green's function method. The boundary conditions used above are equivalent to the retarded Green's function boundary condition, of which the imaginary part is really infinitesimal.

The Green's function satisfies an inhomogeneous equation of motion, differing from the homogeneous Schrödinger equation only by a unit matrix. One can relate the wave function in the right electrode to the incoming wave from the left electrode through the Green's function by using the transformation between the eigenstates and the localized basis functions; in this way, we obtain the transmission and reflection matrix elements[12]

$$
t_{\mu\nu} = \left(\frac{v_\mu}{v_\nu}\right)^{1/2} \left\{ U_r^{-1}(+) G_{N+1,0} H_{1,0} \left[F_l^{-1}(+) - F_l^{-1}(-) \right] U_l(+) \right\}_{\mu\nu}
\tag{6.15}
$$

$$
r_{\mu\nu} = \left(\frac{v_\mu}{v_\nu}\right)^{1/2} \left\{ U_r^{-1}(-) < G_{0,0} H_{1,0} \left[F_l^{-1}(+) - F_l^{-1}(-) \right] - 1 > U_l(+) \right\}_{\mu\nu}
\tag{6.16}
$$

where $\mu\nu$ are the Bloch states, $G_{N+1,0}$ is the Green's function element defined above, and v_μ is the group velocity of the Bloch wave.

Let us calculate the group velocity of a Bloch wave

$$
|n> = \sum_{\alpha,i} e^{ik_n x_i} u_\alpha^n |i,\alpha>
\tag{6.17}
$$

where α is the index of the basis and n is an eigenstate index. The current operator can be expressed as:

$$
j_0 = \frac{ie}{\hbar} \sum_{\alpha\beta} H_{i-1,i}^{\alpha\beta} |i-1,\alpha><i,\beta| - H_{i+1,i}^{\beta\alpha} |i+1,\beta><i,\alpha|
\tag{6.18}
$$

with expectation values

$$
<n|j_0|n> = \frac{ie}{\hbar} \sum_{\alpha\beta} H_{i-1,i}^{\alpha\beta} u_\alpha^{n*} u_\beta^n e^{-ik_n a} - H_{i+1,i}^{\beta\alpha} u_\beta^{n*} u_\alpha^n e^{ik_n a}
\tag{6.19}
$$

For a single-band one-dimension chain, $<n \mid j_o \mid n> = 1/\hbar\, 2t \sin ka$, which is the well-known group velocity of a single-band tight-binding model. It is convenient to flux-normalize the states, such that $<n \mid j_o \mid n> = 1$.

6.2.4 INTERFACE RESISTANCE[14]

In current perpendicular to the interface plane (CPP) experiments,[15] the resistance variations as the layer thickness changes can be fitted quite well by a simple resistor model. For a nonmagnetic multilayer composed of materials A and B with resistivities ρ_A and ρ_B and thickness d_A and d_B, respectively, the resistor model for the total multilayer resistance R_T times the cross-sectional area S is

$$SR_T = M[\rho_A d_A + SR_{A/B} + \rho_B d_B + SR_{A/B}] \qquad (6.20)$$

where M is the number of bilayers and $R_{A/B}$ is the resistance of an A/B interface. The total resistance is thus simply a sum of the bulk and interface resistances. If some of the layers are magnetic, the resistor model should be extended to include spin polarization. By measuring the total resistance as a function of the layer thickness, the bulk resistivities and interface resistances can be determined experimentally. The interface resistance turns out to be strongly spin-dependent and it dominates the resistance and magnetoresistance for layer thicknesses that are not too large. Understanding CPP magnetoresistance is then largely a matter of understanding the origin of the interface resistance.

In a diffuse environment in which there is no coherent scattering between adjacent interfaces due to the sufficiently strong bulk scattering, the interface resistance has been expressed explicitly in terms of the transmission matrix T by Schep[14]

$$SR_{A/B} = S\frac{b}{e^2}\left[\frac{1}{\sum T_{\mu\nu}} - \frac{1}{2}\left(\frac{1}{N_A} + \frac{1}{N_B}\right)\right] \qquad (6.21)$$

where S is the interface area. $T_{\mu\nu}$, which describes how the electronic structure mismatch at an interface affects electron transport, is the probability of eigenstate μ in material A to be transmitted through the interface into eigenstate ν in material B, the trace simply sums over all incident and transmitted channels ν and μ at the Fermi level, and $e^2/hN_{A(B)}$ is the Sharvin conductance of material $A(B)$.

6.3 Co/Cu INTERFACE

The Co/Cu interface is perhaps the most extensively studied magnetic multilayer system. The equilibrium lattice constant of Cu is 3.614 Å and Co is 3.549 Å if both are face-centered cubic (fcc) structures. Experimentally, an atomically smooth interface can be obtained by molecular beam epitaxy (MBE) growth and the electronic transport properties have also been measured accurately by the Michigan State University group.[15]

TABLE 6.1
Cu/Co Interface, a = 3.549 Å, Basis Set *spd*

Position of the Layer	Cu (4)	Cu (3)	Cu (2)	Cu (Interface)	Co (Interface)	Co (2)	Co (3)	Co (4)
m(μ_B)	0.001	−0.002	−0.005	0.005	1.57	1.65	1.64	1.64
			T_{maj} = 0.589				T_{min} = 0.504	

TABLE 6.2
Cu/Co Interface, a = 3.549 Å, Basis Set *spdf*

Position of the Layer	Cu (4)	Cu (3)	Cu (2)	Cu (Interface)	Co (Interface)	Co (2)	Co (3)	Co (4)
m(μ_B)	0.001	−0.002	−0.005	0.001	1.53	1.61	1.59	1.59
			T_{maj} = 0.553				T_{min} = 0.504	

There is a 1.6% lattice constant mismatch for the Co/Cu interface; to accurately determine the interface structure of the Co/Cu interface is beyond our ability. We calculate the magnetic moments of Cu/Co interface with three different lattice constants, these magnetic moments along with the transmission amplitudes are shown in Table 6.1 through Table 6.4. The calculations are carried out with the structure: Cu(bulk)/four layers of Cu/four layers of Co/Co(bulk). The potential of bulk Cu and

TABLE 6.3
Cu/Co Interface, a = 3.581 Å, Basis Set *spd*

Position of the Layer	Cu (4)	Cu (3)	Cu (2)	Cu (Interface)	Co (Interface)	Co (2)	Co (3)	Co (4)
m(μ_B)	0.001	0.001	−0.005	0.002	1.60	1.67	1.66	1.66
			T_{maj} = 0.601				T_{min} = 0.501	

TABLE 6.4
Cu/Co Interface, a = 3.614 Å, Basis Set *spd*

Position of the Layer	Cu (4)	Cu (3)	Cu (2)	Cu (Interface)	Co (Interface)	Co (2)	Co (3)	Co (4)
m(μ_B)	0.001	0.002	−0.005	0.006	1.63	1.69	1.68	1.68
			T_{maj} = 0.615				T_{min} = 0.501	

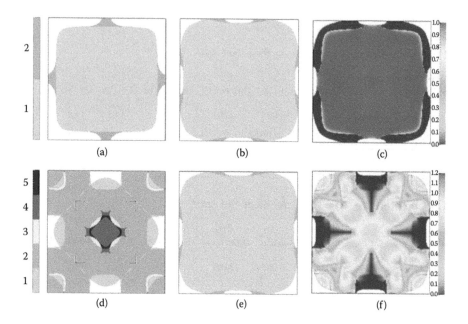

FIGURE 6.2 Transmission amplitude in two-dimensional BZ of both majority (Panel c) and minority (Panel f) spin channels for an fcc Cu/Co(100) interface. In the left and middle panels are shown the projections of the Cu (Panel b and Panel e) and Co (Panel a and Panel d) Fermi surfaces onto a plane perpendicular to the (100) direction.

Co are calculated from separate calculations. The potential of four layers of Cu/four layers of Co as the intermediate region are calculated self-consistently by a TB-LMTO program. From the calculation results, we can see that four layers away from the interface the magnetic moments are quite close to the bulk materials. In general, the slight changes in lattice constants do not have a significant effect on our calculation results. We see that the magnetic moments of Co and the induced magnetic moments in Cu can be changed by the lattice constants used in the calculation, but not by more than 5%. The different basis sets used in the calculation can also introduce effects of a similar size.

In the following, 3.549 Å is used as the lattice constant of fcc Co/Cu. We calculate the interface resistance of Co/Cu interfaces for both (100) and (111) orientations. We calculate the transmission matrix of a Cu/Co fcc(111) or ideal interface using a 1×1 cell. The relative conductance difference is as small as 0.2% when the $k_\parallel$ mesh points in the interfacial two-dimensional Brillouin zone (IBZ) change from 3,600 to 14,400. Consequently, the following calculation will be carried out with a $k_\parallel$ mesh density of 3,600 $k_\parallel$ mesh points in the IBZ for a 1×1 cell.

For the fcc(001) orientation, the Fermi surface projection of Cu and Co onto the IBZ are shown in Figure 6.2. The left and middle panels show the projections of the Cu (Panel B and Panel E) and Co (Panel A and Panel D) Fermi surfaces onto

TABLE 6.5
Results of Calculations for Different Systems

System	Roughness	$SR^S_{maj}(f\Omega m^2)$	$SR^S_{min}(f\Omega m^2)$	$SR^{LB}_{maj}(f\Omega m^2)$	$SR^{LB}_{maj}(f\Omega m^2)$
Co/Cu(100)	Clean	0.33	1.79	2.27	3.11
Co$_{hcp}$/Cu(111)	Clean	0.60	2.24	2.67	3.65
Co/Cu(111)	Clean	0.39	1.46	2.39	2.80
Co/Cu(111)	two-layer 50-50 alloy	0.41	1.82 ± 0.03	2.40	3.14
Co/Cu(111)	Exp. [MSU][15]	0.26 ± 0.06	1.84 ± 0.14		
Fe/Cr(100)	Clean	2.82	0.50	3.51	1.45
Fe/Cr(100)	two-layer 50-50 alloy	0.99	0.50	1.68	1.45
Fe/Cr(110)	Clean	2.74	1.05	4.22	3.17
Fe/Cr(110)	Clean [Stiles][16]	2.11	0.81		
Fe/Cr(110)	two-layer 50-50 alloy	2.05	1.10	3.53	3.22
Fe/Cr(110)	Exp. [MSU][15]	2.71	0.48		

the IBZ perpendicular to the (100) direction. Panel C and Panel F are the calculated number of channels of the Co/Cu interface. If we define the reflection from the Cu side as the difference between the number of channels of the Co/Cu interface and the Fermi surface projection of Cu onto the IBZ, we will find the reflection for majority spin and minority spin are not very different. In fact, the spin-dependent interface resistance observed by MSU group[15] cannot be explained by the normal Landauer–Büttiker conductance, as we can see in Table 6.5. According to Schep,[14] the interface resistance should be expressed as Equation 6.21. We define $\frac{h}{e^2 N_A}$ as R^{LB}, the inverse of the Sharvin conductance of material $A(B)$. For Co/Cu(001), we can see the main spin dependence comes from the R^{LB} of Co.

Similar plots are also shown in Figure 6.3 for the fcc(111) orientation. For comparison, we plot Cu(111)/Co(hcp) (hcp — hexagonal-close packed) interfaces in Figure 6.4 and Figure 6.5, and we can see that the interface structure has an observable effect on the Co/Cu interface. The electronic structures of hcp and fcc Co are different, and for a clean interface between hcp and fcc cobalt, we predict an interface resistance of 0.14 fΩm^2 and 0.39 fΩm^2 for majority and minority spins, respectively. For a clean Co(0001)hcp/Cu(111)fcc interface, both majority and minority spin resistances are substantially larger than for the fcc case and even larger than the experimental values (Table 6.5). In view of the substantial difference predicted between these two interfaces, it would be interesting to test this experimentally.

Next we use a lateral supercell to model the effect of interfacial disorder. We check the two-layer interface alloy using an 8 × 8 supercell for Co$_{0.5}$Cu$_{0.5}$ in a Co/Cu interface and calculate some configurations. The largest fluctuation is from the

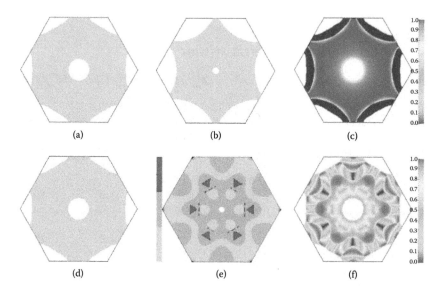

FIGURE 6.3 Transmission amplitude in a two-dimensional BZ of both the majority (Panel c) and minority (Panel f) spin channel for a fcc Cu/Co(111) interface. In the left and middle panels are shown the projections of the Cu (Panel a and Panel d) and Co (Panel b and Panel e) Fermi surfaces onto a plane perpendicular to the (111) direction.

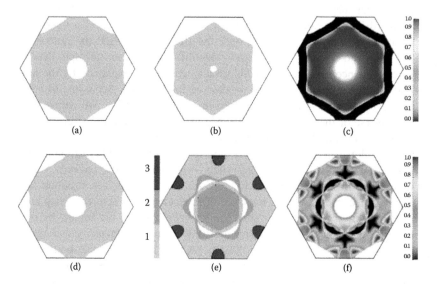

FIGURE 6.4 Transmission amplitude in a two-dimensional BZ of both majority (Panel c) and minority (Panel f) spin channel for a fcc Cu(111)/Co$_{hcp}$(0001) interface. In the left and middle panels are shown the projections of the Cu (Panel a and Panel d) and Co (Panel b and Panel e) Fermi surfaces onto a plane perpendicular to the (111) direction. The interface structure is Cu(ABC)/Co(ABAB), here a, b, and c are three different possible sites in the fcc(111) orientation.

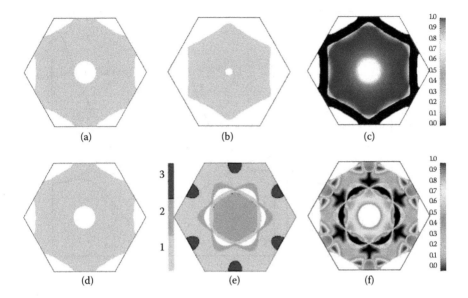

FIGURE 6.5 Transmission amplitude in a two-dimensional BZ of both majority (Panel c) and minority (Panel f) spin channel for a fcc Cu/Co$_{hcp}$(111) interface. In the left and middle panels are shown the projections of the Cu (Panel a and Panel d) and Co (Panel b and Panel e) Fermi surfaces onto a plane perpendicular to the (111) direction. The interface structure is Cu(ABC)/Co(ACAC), a, b, and c are three different possible sites in the fcc(111) orientation.

minority spin that produces an interface resistance uncertainty of about ±5%. This uncertainty is smaller than the experimental error bar, so, we chose an 8 × 8 supercell to model the disorder in the interface. The results and corresponding experimental data are summarized in Table 6.5.

In summary, the experimentally measured interface resistance can be understood by Schep's formula. For the Co/Cu interface, the main spin dependence comes from the spin-dependent Sharvin conductance of Co. Whereas for the Fe/Cr interface, the main spin dependence comes from the mismatch of the Bloch state wave functions at the interface.

6.4 Fe/Ag ARPES

The lattice constant of bulk Ag is a_{Ag} = 4.09 Å, while the lattice constant of bulk Fe is a_{Fe} = 2.87 Å. So if the lattice of bcc Fe is rotated by 45°, it can fit onto the fcc Ag lattice with a lattice mismatch of 0.8%. And more importantly, experimentally very high-quality silver films can be grown on the top of iron crystal whiskers.[10] The calculated energy bands of fcc Ag and bcc ferromagnetic Fe bulk along the Γ-X and Γ-H directions of the respective IBZ are shown in Figure 6.6.

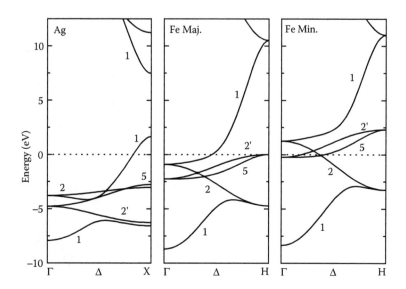

FIGURE 6.6 The band structure of fcc Ag along Γ-X and bcc ferromagnetic Fe bulk along the Γ-H direction.

At the Ag/Fe interface, a symmetry gap between Δ_1 states of Ag and the states of the minority spin electrons of Fe is formed near Fermi energy. The QW state will be formed in the Ag film. By measuring the linewidth of this QW state, we can get information that relates to the electronic transport properties across the Ag/Fe interface.

In our calculation, we keep the lattice constant of Fe the same as the bulk material. The interfacial area of Ag has to be the same as that of Fe and the lattice constant of Ag in the direction perpendicular to the interface is elongated to $a_{Ag} =$ 4.15 Å. The Ag lattice is elongated by 1.5% in the perpendicular direction. In the calculation, we let the charge density of the six principal layers around the Ag/Fe(100) interface be calculated self-consistently. When the number of k points in the IBZ increases from 576 to 1600, the maximum difference for magnetic moment of Fe is about 0.1%. The change in the spin polarization of the Ag atoms is larger, but the absolute value of the spin polarization of the Ag is rather small, so its effect on the transmission and reflection is negligible. In the following calculation, we take 576 k points in the IBZ for the electronic structure calculation. The conductance changes less than 1% when the number of k points increases from 14,400 to 57,600 in the IBZ in the calculation; so in the following calculations, we used 14,400 k points in the IBZ or similar k points density in the supercell calculation.

The calculated magnetic moments of Fe are 2.26 μ_B per atom away from the interface and 2.91 μ_B per atom at the interface. Reflection in the IBZ of both majority and minority (right-hand panel) spin channels are shown in the Figure 6.7. In the left and middle panels are shown the projections of the Ag and Fe Fermi surfaces onto a plane perpendicular to the (100) direction. For the majority spin, electrons

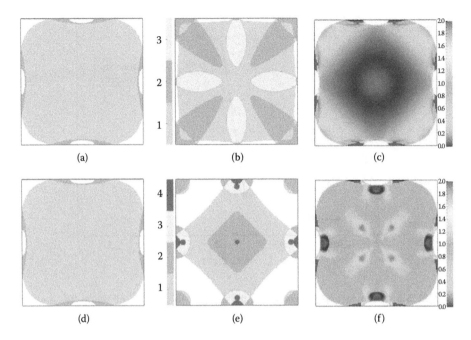

FIGURE 6.7 Reflection in the IBZ of both majority and minority (right-hand panel) spin channels for a fcc Ag/Fe(100) interface. In the left and middle panels are shown the projections of the Ag and Fe Fermi surfaces onto a plane perpendicular to the (100) direction.

can transmit through the interface across most of the area of the IBZ, only at the edge of the IBZ are electrons reflected. There will be no confinement of the majority electrons at the Ag/Fe interface. Although for the minority spins, the electrons are completely reflected in most of the first Brillouin zone (BZ), and electrons can only transmit through the interface at a few points. The confinement of minority electron at the Ag/Fe interface provides the opportunity for the formation of QW states.

Only the electrons near the Fermi level can contribute to the electronic transport in a low bias experiment. So, in principle, we only need the information about the *sp* band of Ag. Most of the published data on the ARPES of the QW states of the Ag film in the Ag/Fe system are for normal emission, because the measurement in this direction is relatively accurate. These results should be compared with the $\bar{\Gamma}$ point of the IBZ of Ag; so we will focus on the $\bar{\Gamma}$. The coupling between the Δ_1 state of bcc Fe and the Δ_1 state of fcc Ag is large, and the $\Delta_{2'}$ state of bcc Fe and Δ_1 state of fcc Ag is small. As a result, the electrons at the Δ_1 state of fcc Ag are easily transported to the Δ_1 state of bcc Fe.

In our calculation, we calculate the reflectivity of Δ_1 states at the $\bar{\Gamma}$ point. To study the reflectivity of a particular state of Ag at the Ag/Fe interface, we have to identify states in a supercell calculation with the states in the IBZ of Ag. In our calculation, we calculate the velocity of the states in the IBZ of Ag and determine the corresponding velocity in the supercell calculation to identify the states we want.

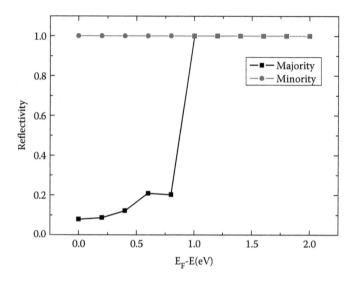

FIGURE 6.8 The reflection probability of the Δ_1 band as function of energy for a clean interface.

There is only one state at the Fermi level for Ag (let us denote it as $\mu = 1$). Then we calculate the transmission matrices through the Ag/Fe(100) interface, denoted as T_{1v}. The reflectivity can be expressed as R_{11}. For the supercell calculation, we use the group velocity of the state to distinguish the states in IBZ.

The energy-dependent reflectivity at the Ag/Fe interface of the Ag sp-like band is shown in Figure 6.8. The reflectivity for the minority spin is near unity in our calculation as expected. For the majority spin, there is a leap at the energy of 1.0 eV. This can be explained from the energy band of Fe. At this energy, the Δ_1 state of the majority spin of Fe has disappeared. The state at this energy is a d-like state. Because of the mismatch of the band structure, the transmission is decreased on a large scale (nearly to zero). At this time, experimenters do not have a complete analysis for the majority spin states, the problem is that under most experimental conditions, the majority states have very low photoemission cross sections, making them difficult to measure.

Experimentally, the reflectivity of the sp-like band is found to be about 0.86 near the Fermi level at 100 K. Electron-phonon (e-ph) coupling is the main reason for this discrepancy. The effect of e-ph interactions on the linewidth of the ARPES has been intensively investigated experimentally.[17] In fact, ARPES is an important tool in the study of the e-ph interaction strength. E-ph interactions complicate our interpretation, even at zero temperature, e-ph interactions still exist and broaden the linewidth of the spectrum. What we are interested in is the reflection at the interface without e-ph interactions. This difficulty can be removed by looking at the temperature-dependent linewidth. Paggel et al.[17] had measured the impurity scattering self-energy by assuming the impurity scattering is independent of the energy. A similar treatment

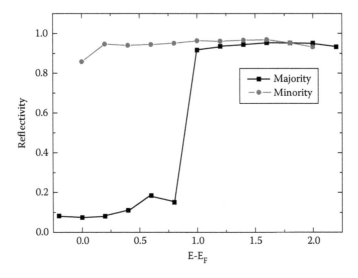

FIGURE 6.9 The reflection probability of the Δ_1 band as function of energy for one Ag impurity at the interface.

can be applied to the interfacial reflectivity and the broadening due to the finite transmission caused by the impurity scattering can in this way be measured.

The reflection of minority spin electrons at the Ag/Fe interface can be reduced in the presence of diffuse scattering centers. Let us now study the effect of disorder on reflectivity of Ag/vacuum/Ag sandwich to compare with experimental results. We treat the interface disorder by putting Ag ad atoms on the top of the Ag layer with a 8×8 lateral supercell. We randomly distribute the appropriate concentration of Ag atoms within this lateral supercell. When we compare the result of adding 1.6% impurity with the clean case, we can find that for the minority spin, the order of magnitude of transmission changes from 1.0×10^{-5} up to 1.0×10^{-1}. For the minority spin, when 6.25% impurity is added, the reflectivity is 36% at Fermi level. While when 18.75% impurities are added, the reflectivity is 16%. The energy-dependent reflectivities of the *sp* state as a function of energy of Ag/Fe interface are shown in Figure 6.9.

At the Fermi energy, we plot the reflectivity of the minority spin as a function of the number of Ag impurities in Figure 6.10. The error bar caused by different configurations is also shown. The shape of the curve coincides with the experiment results[10] even though direct comparison is unjustified. The diffuse scattering center in the experiment is on the surface of silver film, and in our calculation, the diffuse scattering center is at the interface.

The interface resistance as a function of the number of impurities is shown in Figure 6.11. We see a huge drop of the interface resistance for the minority spin channel as the interfacial impurity concentration increases. So, although there is no experiment to measure electronic transport properties across the Fe/Ag interface as the interface roughness changes, ARPES should be able to probe such effects.

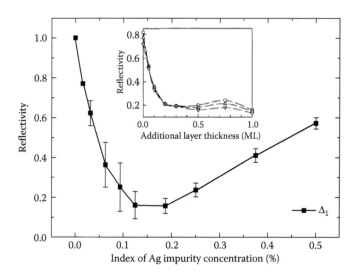

FIGURE 6.10 The reflection probability of the Δ_1 band for the minority spin as function of impurity concentration at Fermi level. The inset is the experimental results. (See Reference 10 for details.)

6.5 SUMMARY AND OUTLOOK

Using first-principles electronic structure calculations, we have developed a method to study the electronic transport through a FM/NM interface. The interface resistance of Co/Cu and Fe/Ag has been calculated for different interface structures, and

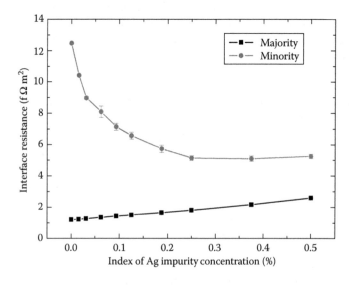

FIGURE 6.11 The interface resistance as a function of impurity concentration.

comparison is made with experimental data. Finally, we proposed a way to measure the spin-dependent transparency using the linewidth of the QW states in ARPES.

ACKNOWLEDGMENT

The author would like to thank Professor P.J. Kelly, Professor G.E.W. Bauer, Dr. M. Zwierzycki, and P.X. Xu for collaboration. This work was supported in part by National Science Foundation of China (Grant No. 90303014).

REFERENCES

1. P. Grünberg, R. Schreiber, Y. Pang, M.B. Brodsky, and H. Sowers, *Phys. Rev. Lett.,* 57, 2442, 1986.
2. M.N. Baibich, J.M. Broto, A. Fert, F. Nguyen Van Dau, F. Petroff, P. Etienne, G. Creuzet, A. Friederich, and J. Chazelas, *Phys. Rev. Lett.,* 61, 2472, 1988.
3. J.C. Slonczewski, *J. Magn. Magn. Mater.,* 159, L1, 1996.
4. M. Tsoi, A.G.M. Jansen, J. Bass, W.-C. Chiang, M. Seck, V. Tsoi, and P. Wyder, *Phys. Rev. Lett.,* 80, 4281, 1998; J.-E. Wegrowe, D. Kelly, Y. Jaccard, Ph. Guittienne, and J.-Ph. Ansermet, *Europhys. Lett.,* 45, 626, 1999; J.Z. Sun, *J. Magn. Magn. Mater.,* 202, 157, 1999; E.B. Myers, D.C. Ralph, J.A. Katine, R.N. Louie, and R.A. Buhrman, *Science,* 285, 867, 1999.
5. M.D. Stiles, *J. Appl. Phys.,* 79, 5805, 1996.
6. A. Brataas, Yu.V. Nazarov, and G.E.W. Bauer, *Phys. Rev. Lett.,* 84, 2481, 2000.
7. M.J.M. de Jong and C.W.J. Beenakker, *Phys. Rev. Lett.,* 74, 1657, 1995.
8. S.K. Upadhyay et al., *Phys. Rev. Lett.,* 81, 3247, 1998; *Appl. Phys. Lett.,* 74, 3881, 1999; R.J. Soulen et al., *Science,* 282, 85, 1998; B. Nadgorny et al., *Phys. Rev. B,* 61, R3788, 2000.
9. K. Xia et al., *Phys. Rev. Lett.,* 89, 166603, 2002.
10. T.-C. Chiang, *Surf. Sci. Rep.,* 39, 181, 2000; J.J. Paggel, T. Miller, and T.-C. Chiang, *Science,* 283, 1709, 1999.
11. K. Xia, P.J. Kelly, G.E.W. Bauer, I. Turek, J. Kudrnovsky, and V. Drchal, *Phys. Rev. B,* 63, 064407, 2001.
12. T. Ando, *Phys. Rev. B,* 44, 8017, 1991.
13. O.K. Andersen, O. Jepsen, and D. Glötzel, in *Highlights in Condensed Matter Theory,* F. Bassani, F. Fumi, and M.P. Tosi, Eds., North-Holland, Amsterdam, 1985, p. 59.
14. K. Schep, P. Kelly, and G. Bauer, *Phys. Rev. B,* 56, 10805, 1997.
15. W.P. Pratt, Jr., S.-F. Lee, J.M. Slaughter, R. Loloee, P.A. Schroeder, and J. Bass, *Phys. Rev. Lett.,* 66, 3060, 1991.
16. M.D. Stiles and D.R. Penn, *Phys. Rev. B,* 61, 3200, 2000.
17. J.J. Paggel, T. Miller, and T.-C. Chiang, *Phys. Rev. Lett.,* 83, 1415, 1999.

Part II

Spin Torque and Domain Wall Magneto Resistance

7 Current-Driven Switching of Magnetization: Theory and Experiment

D.M. Edwards and F. Federici

CONTENTS

7.1 INTRODUCTION

Recently there has been a lot of interest in magnetic nanopillars of 10 to 100 nm in diameter. The pillar is a metallic layered structure with two ferromagnetic layers, usually of cobalt, separated by a nonmagnetic spacer layer, normally of copper. Nonmagnetic leads are attached to the magnetic layers so that an electric current may be passed through the structure. In the simplest case, the pillar may exist in two states, with the magnetization of the two magnetic layers being parallel or antiparallel. The state of a pillar can be read by measuring its resistance, this being smaller in the parallel state than in the antiparallel one. This dependence of the resistance on magnetic configuration is the giant magnetoresistance (GMR) effect.[1] A dense array of these nanopillars could form a magnetic memory for a computer. Normally one of the magnetic layers in a pillar is relatively thick and its magnetization direction is fixed. To write into the memory, the magnetization direction of the second, thinner layer must be switched. This might be achieved by a local magnetic field of suitable strength and methods have been proposed[2] for providing such a local field by currents in a crisscross array of conducting strips. However an alternative, and potentially more efficient, method proposed by Slonczewski[3] makes use of a current passing up the pillar itself. Slonczewski's effect relies on "spin transfer" and not on the magnetic field produced by the current, which in the nanopillar geometry is ineffective. The idea of spin transfer is as follows. In a ferromagnet, there are more electrons of one spin orientation than of the other so that current passing through the thick magnetic layer (the polarizing magnet) becomes spin polarized. In general, its state of spin polarization changes as it passes through the second (switching) magnet so that spin-angular momentum is transferred to the switching magnet. This transfer of spin-angular momentum is called spin-transfer torque, and if the current exceeds a critical value, it may be sufficient to switch the direction of magnetization of the switching magnet. This is called current-induced switching.

In Section 7.2, we show how to calculate the spin-transfer torque for a simple model.

7.2 SPIN-TRANSFER TORQUE IN A SIMPLE MODEL

For simplicity, we consider a structure of the type shown in Figure 7.1, where **p** and **m** are unit vectors in the direction of the magnetizations. This models the layered structure of the pillars used in experiments but the atomic planes shown are considered to be unbounded instead of having the finite cross section of the pillar. This means that there is translational symmetry in the x and z directions. The structure consists of a thick (semi-infinite) left magnetic layer (polarizing magnet), a nonmagnetic metallic spacer layer, a thin second magnet (switching magnet) and a semi-infinite nonmagnetic lead. In the simplest model, we assume the atoms form a simple cubic lattice with lattice constant a, and we adopt a one-band tight-binding model with hopping Hamiltonian

$$H_0 = t \sum_{\mathbf{k}_\parallel \sigma} \sum_n c^\dagger_{\mathbf{k}_\parallel n \sigma} c_{\mathbf{k}_\parallel n - 1 \sigma} + \text{h.c.} \tag{7.1}$$

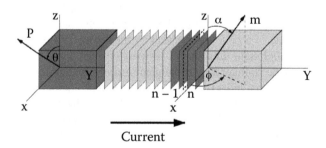

FIGURE 7.1 Schematic picture of a magnetic layer structure for current-induced switching (magnetic layers are darker; nonmagnetic layers are lighter).

Here $c^\dagger_{k_\parallel n\sigma}$ creates an electron on plane n with two-dimensional wave vector $\mathbf{k}_\parallel$ and spin σ, and t is the nearest-neighbor hopping integral.

In the tight-binding description, the operator for spin-angular momentum current between planes $n-1$ and n, which we require to calculate spin-transfer torque, is given by

$$\mathbf{j}_{n-1} = -\frac{it}{2}\sum_{\mathbf{k}_\parallel}\left(c^\dagger_{k_\parallel,n,\uparrow},c^\dagger_{k_\parallel,n,\downarrow}\right)\boldsymbol{\sigma}\left(c_{k_\parallel,n-1,\uparrow},c_{k_\parallel,n-1,\downarrow}\right)^{\mathrm{T}} + \mathrm{h.c.} \qquad (7.2)$$

Here $\boldsymbol{\sigma} = (\sigma_x,\sigma_y,\sigma_z)$ where the components are Pauli matrices. Equation 7.2 yields the charge current j^c_{n-1} if $\frac{1}{2}\boldsymbol{\sigma}$ is replaced by a unit matrix multiplied by the number $e/\hbar$, where e is the electronic charge (negative). All currents flow in the y direction, perpendicular to the layers, and the components of the vector $\mathbf{j}$ correspond to transport of x, y, and z components of spin. The justification of Equation 7.2 for $\mathbf{j}_{n-1}$ relies on an equation of continuity, as will be pointed out in Section 7.4.

To define the present model completely, we must supplement the hopping Hamiltonian H_0 by specifying the on-site potentials in the various layers. For simplicity, we assume the on-site potential for both spins in nonmagnetic layers, and for majority spin in ferromagnetic layers, is zero. We assume an infinite exchange splitting in the ferromagnets so that the minority spin potential in these layers is infinite. Thus minority spin electrons are completely excluded from the ferromagnets. Clearly the definition of majority and minority spin relate to spin quantization in the direction of the local magnetization. We take $\alpha = 0$, so that the magnetization of the switching magnet is in the z direction and take $\theta = \psi$, where ψ is the angle between the magnetizations.

To describe spin transport in the structure, we adopt the generalized Landauer approach of Waintal et al.[4] Thus the structure is placed between two reservoirs, one on the left and one on the right, with electron distributions characterized by Fermi functions $f(\omega-\mu_L)$, $f(\omega-\mu_R)$, respectively. The system is then subject to a bias voltage V_b given by $eV_b = \mu_L - \mu_R$, which is the difference between the chemical potentials. We discuss the ballistic limit where scattering occurs only at interfaces, the effect of impurities being negligible. We label the atomic planes so that $n=0$ corresponds to the last atomic plane of the polarizing magnet. The planes of the spacer layer correspond to $n = 1, 2, \cdots, N$ and $n = N+1$ is the first plane of the switching magnet.

Consider first an electron incident from the left with wave function $|k,maj\rangle$, where $k > 0$, which corresponds to a Bloch wave $|k\rangle = \sum_n e^{ikna}|\mathbf{k}_\parallel n\rangle$ with majority spin in the polarizing magnet. In this notation, the label $\mathbf{k}_\parallel$ is suppressed. The particle is partially reflected by the structure and finally emerges as a partially transmitted wave in the lead, with spin $\uparrow$ corresponding to majority spin in the switching magnet. Thus the wave function is of the form

$$|P_k\rangle = |k,maj\rangle + B|-k,maj\rangle \qquad (7.3)$$

in the polarizer and

$$\left|L_{k}\right\rangle = F\left|k,\uparrow\right\rangle \tag{7.4}$$

in the lead. A majority spin in either ferromagnet enters or leaves the spacer without scattering, since in our simple model there is no potential step. Also the minority spin wave function entering a ferromagnet is zero. Therefore, the spacer wave function may be written in two ways

$$\left|S_{k}\right\rangle = F\left|k,\uparrow\right\rangle + E\left(e^{-ik(N+1)a}\left|k,\downarrow\right\rangle - e^{ik(N+1)a}\left|-k,\downarrow\right\rangle\right) \tag{7.5}$$

or

$$\left|S_{k}\right\rangle = \left|k, maj\right\rangle + B\left|-k, maj\right\rangle + D\left(\left|k, min\right\rangle - \left|-k, min\right\rangle\right)$$

$$= \cos(\psi/2)\left|k,\uparrow\right\rangle + \sin(\psi/2)\left|k,\downarrow\right\rangle + B\left[\cos(\psi/2)\left|-k,\uparrow\right\rangle + \sin(\psi/2)\left|-k,\downarrow\right\rangle\right]$$

$$+ D\left[-\sin(\psi/2)\left|k,\uparrow\right\rangle + \cos(\psi/2)\left|k,\downarrow\right\rangle + \sin(\psi/2)\left|-k,\uparrow\right\rangle - \cos(\psi/2)\left|-k,\downarrow\right\rangle\right].$$

$$\tag{7.6}$$

On equating coefficients of $\left|k,\uparrow\right\rangle$, $\left|k,\downarrow\right\rangle$, $\left|-k,\uparrow\right\rangle$, and $\left|-k,\downarrow\right\rangle$ in Equation 7.5 and Equation 7.6, we have four equations that may be solved for B, D, E, and F. In particular, the transmission coefficient T is given by

$$T = \left|F\right|^{2} = \frac{4\cos^{2}(\psi/2)\sin^{2}k(N+1)a}{\sin^{4}(\psi/2) + 4\cos^{2}(\psi/2)\sin^{2}k(N+1)a}. \tag{7.7}$$

Similarly an electron incident from the right with wave function $\left|-k,\uparrow\right\rangle$ in the lead is partially reflected and finally emerges as a partially transmitted wave $F'\left|-k, maj\right\rangle$ in the polarizing magnet. It is found that $F' = F$ so that the transmission coefficient is the same for particles from left or right.

The spin-angular momentum current in a particular layer, which we shall denote by S although it need not be the spacer layer, is the sum of currents carried by left- and right-moving electrons. Thus we have a Landauer-type formula[5]

$$\mathbf{j}_{s} = \frac{a}{2\pi}\sum_{\mathbf{k}_{\parallel}}\left\{\int_{k>0}dk\left[\left\langle S_{k}\left|\mathbf{j}_{n-1}\right|S_{k}\right\rangle f(\omega - \mu_{L}) + \left\langle S_{-k}\left|\mathbf{j}_{n-1}\right|S_{-k}\right\rangle f(\omega - \mu_{R})\right]\right\}$$

$$\tag{7.8}$$

where $\left|S_k\right\rangle, \left|S_{-k}\right\rangle$ are wave functions in the layer considered corresponding to electrons incident from left and right, respectively. Here ω, the energy of the Bloch wave k, is given by the tight-binding formula

$$\omega = u_{\mathbf{k}_\parallel} + 2t\cos ka \qquad (7.9)$$

where $u_{\mathbf{k}_\parallel} = 2t(\cos k_x a + \cos k_z a)$. We take $t < 0$ so that positive k corresponds to positive velocity $\hbar^{-1}\partial\omega/\partial k$ as we have assumed. The current $\mathbf{j}_s$ in layer S calculated by Equation 7.8 does not depend on the particular planes $n-1$, n between which it is calculated. On changing the integration variable in Equation 7.8 we find

$$\mathbf{j}_s = \frac{1}{2\pi}\sum_{\mathbf{k}_\parallel}\int d\omega[\mathbf{J}_+ f(\omega - \mu_L) + \mathbf{J}_- f(\omega - \mu_R)] \qquad (7.10)$$

where

$$\mathbf{J}_\pm = \frac{\left\langle S_{\pm k}\left|\mathbf{j}_{n-1}\right|S_{\pm k}\right\rangle}{-2t\sin ka}. \qquad (7.11)$$

Here $k = k(\omega, \mathbf{k}_\parallel)$ is the positive root of Equation 7.9. Equation 7.10 may be written as

$$\mathbf{j}_s = \frac{1}{4\pi}\sum_{\mathbf{k}_\parallel}\int d\omega\{(\mathbf{J}_+ + \mathbf{J}_-)[f(\omega - \mu_L) + f(\omega - \mu_R)]$$

$$+ (\mathbf{J}_+ - \mathbf{J}_-)[f(\omega - \mu_L) - f(\omega - \mu_R)]\}. \qquad (7.12)$$

Before discussing this spin current, we briefly consider the charge current j^c, and we denote the analogs of $\mathbf{J}_\pm$ by $J^c_\pm$. Because the charge current is conserved throughout the structure J^c_+ and j^c can be calculated in different ways, for example, in the lead for J^c_+ and in the polarizer for J^c_-. Because $T = |F|^2 = |F'|^2$, we find $J^c_+ + J^c_- = 0$, and for small bias $eV_b = \mu_L - \mu_R$, the charge current is given by

$$j^c = \frac{2e^2 V_b}{h}\sum_{\mathbf{k}_\parallel} T \qquad (7.13)$$

where the transmission coefficient T is given by Equation 7.7 with $k = k(\mu, \mathbf{k}_\parallel)$, μ being the common chemical potential as $V_B \to 0$. This is the well-known Landauer formula.[5]

The spin-transfer torque on the switching magnet is given by

$$\mathbf{T}^{s-t} = \left\langle \mathbf{j}_{spacer} \right\rangle - \left\langle \mathbf{j}_{lead} \right\rangle \qquad (7.14)$$

where $\left\langle \mathbf{j}_{spacer} \right\rangle$ and $\left\langle \mathbf{j}_{lead} \right\rangle$ are spin currents in the spacer and lead, respectively. For zero bias ($\mu_L = \mu_R$), there is clearly no charge current in the structure and straightforward calculation shows that all components of spin current in the spacer and the lead vanish, except for a nonzero y-spin current in the spacer. Therefore, there is a nonzero y component of spin-transfer torque acting on the switching magnet for zero bias, and its dependence on the angle ψ between the magnetizations is found to be approximately $\sin \psi$. This torque is due to exchange coupling, analogous to a Ruderman–Kittel–Kasuya–Yosida (RKKY) coupling, between the two magnetic layers. This coupling oscillates as a function of spacer thickness and tends to zero as the thickness tends to infinity. For finite bias V_B, the second term in the integrand of Equation 7.12 comes into play. In general, this leads to finite x and y components of $\mathbf{T}^{s-t}$ proportional to V_b (for small V_b), whereas $T_z^{s-t} = 0$. However for the special model considered here, with infinite exchange splitting in both ferromagnets, it turns out that $T_y^{s-t} = 0$. For this model, the only nonzero component of $\mathbf{T}^{s-t}$ proportional to V_b is found to be

$$T_x^{s-t} = \frac{\hbar j^c}{2|e|} \tan \frac{\psi}{2} \qquad (7.15)$$

where j^c is the charge current given by Equation 7.13.

Slonczewski[3] originally obtained this result for the analogous parabolic band model. From Equation 7.15, Equation 7.13, and Equation 7.7, it follows that T_x^{s-t} contains an important factor $\sin \psi$, although this does not represent the whole angle dependence. Clearly, from Equation 7.15, the torque proportional to bias remains finite for arbitrarily large spacer thickness in the ballistic limit. For this model, with infinite exchange splitting, the torque is independent of the thickness of the switching magnet.

From the results of this simple model, we can infer a general form of the spin-transfer torque $\mathbf{T}^{s-t}$ which is independent of the choice of coordinate axes. Thus, we write

$$\mathbf{T}^{s-t} = \mathbf{T}_\perp + \mathbf{T}_\| \qquad (7.16)$$

where

$$\mathbf{T}_\perp = (g^{ex} + g_\perp e V_b)(\mathbf{m} \times \mathbf{p})$$
$$\mathbf{T}_\| = g_\| e V_b \mathbf{m} \times (\mathbf{p} \times \mathbf{m}) \qquad (7.17)$$

With the choice of axes in Figure 7.1, $T_\parallel$ corresponds to the x component of torque, which is the component parallel to the plane containing the magnetization directions **m** and **p**. Similarly, $T_\perp$ corresponds to the y component of torque, this being perpendicular to the plane of **m** and **p**. The modulus of both the vectors $\mathbf{m} \times \mathbf{p}$ and $\mathbf{m} \times (\mathbf{p} \times \mathbf{m})$ is $\sin\psi$, so that the factors g^{ex}, $g_\perp$, and $g_\parallel$ are functions of ψ which contain deviations from the simple $\sin\psi$ behavior. The bias-independent term g^{ex} corresponds to the interlayer exchange coupling, as discussed above, and henceforth we assume that the spacer is thick enough for this term to be negligible. Sometimes the $\sin\psi$ factor accounts for most of the angular dependence of $T_\perp$ and $T_\parallel$ so that $g_\perp$ and $g_\parallel$ may be regarded as constant parameters for the given structure. In Section 7.3, we use Equation (7.17) for the spin-transfer torque in a phenomenological theory of current-induced switching of magnetization. This phenomenological treatment enables us to understand most of the available experimental data. It is more usual in experimental works to relate spin-transfer torque to current rather than bias. However in theoretical work, based on the Landauer or Keldysh approach, bias is more natural. In practice, the resistance of the system considered is rather constant (the GMR ratio is only a few percent) so that bias and current are in a constant ratio.

7.3 PHENOMENOLOGICAL TREATMENT OF CURRENT-INDUCED SWITCHING OF MAGNETIZATION

In this section, we explore the consequences of the spin-transfer torque acting on a switching magnet using a phenomenological Landau–Lifshitz equation with Gilbert damping (LLG equation). This is essentially a generalization of the approach used originally by Slonczewski[3] and Sun.[6] We assume that there is a polarizing magnet whose magnetization is pinned in the xz plane in the direction of a unit vector **p**, which is at general fixed angle θ to the z axis as shown in Figure 7.1. The pinning of the magnetization of the polarizing magnet can be due to its large coercivity (thick magnet) or a strong uniaxial anisotropy. The role of the polarizing magnet is to produce a stream of spin-polarized electrons, that is, a spin current, that is going to exert a torque on the magnetization of the switching magnet whose magnetization lies in the general direction of a unit vector **m**. The orientation of the vector **m** is defined by the polar angles α, ϕ shown in Figure 7.1. There is a nonmagnetic metallic layer inserted between the two magnets whose role is merely to separate magnetically the two magnetic layers and allow a strong charge current to pass. The total thickness of the whole trilayer sandwiched between two nonmagnetic leads must be smaller than the spin-diffusion length l_{sf} so that there are no spin flips due to impurities or spin-orbit coupling. A typical junction in which current-induced switching is studied experimentally[7] is shown schematically in Figure 7.2. The thickness of the polarizing magnet is 40 nm, that of the switching magnet is 2.5 nm, and the nonmagnetic spacer is 6 nm thick. The materials for the two magnets and the spacer are cobalt and copper, respectively, which are those most commonly used. The junction cross section is oval shaped with dimensions 60 × 130 nm. A small diameter

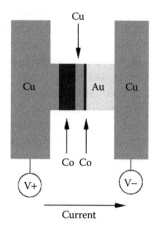

FIGURE 7.2 Schematic picture of a junction in which current-induced switching is studied experimentally.

is necessary so that the torque due to the Oersted field generated by a charge current of 10^7 to 10^8 A/cm^2, required for current-induced switching, is much smaller than the spin-transfer torque we are interested in.

The aim of most experiments is to determine the orientation of the switching magnet moment as a function of the current (applied bias) in the junction. Sudden jumps of the magnetization direction, that is, current-induced switching, are of particular interest. The orientation of the switching magnet moment **m** relative to that of the polarizing magnet **p**, which is fixed, is determined by measuring the resistance of the junction. Because of the GMR effect, the resistance of the junction is higher when the magnetizations of the two magnets are antiparallel than when they are parallel. In other words, what is observed are hysteresis loops of resistance vs. current. A typical experimental hysteresis loop of this type[8] is reproduced in Figure 7.3. It can be seen from Figure 7.3 that, for any given current, the switching magnet moment is stationary (the junction resistance has a well-defined value), that is, the system is in a steady state. This holds everywhere on the hysteresis loop except for the two discontinuities where current-induced switching occurs. As indicated by the arrows, jumps from the parallel (P) to antiparallel (AP) configurations of the magnetization and from AP to P configurations occur at different currents. It follows that in order to interpret experiments that exhibit such hysteresis behavior, the first task of the theory is to determine from the LLG equation all the possible states and then investigate their dynamic stability. At the point of instability, the system seeks out a new steady state, that is, a discontinuous transition to a new steady state with the switched magnetization occurs. We have tacitly assumed that there is always a steady state available for the system to jump to. There is now experimental evidence that this is not always the case. In the absence of any stable steady state, the switching magnetic moment remains permanently in the time-dependent state. This interesting case is implicit in the phenomenological LLG treatment, and we shall discuss it in detail later.

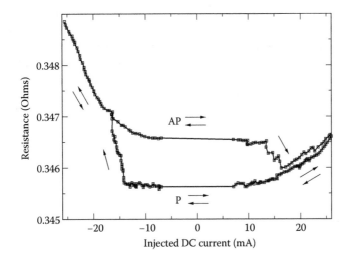

FIGURE 7.3 Resistance vs. current hysteresis loop. (After J. Grollier et al., *Appl. Phys. Lett.*, 78, 3663, 2001. With permission.)

In describing the switching magnet by a unique unit vector **m**, we assume that it remains uniformly magnetized during the switching process. This is only strictly true when the exchange stiffness of the switching magnet is infinitely large. It is generally a good approximation as long as the switching magnet is small enough to remain a single domain, so that the switching occurs purely by rotation of the magnetization as in the Stoner–Wohlfarth theory[9] of field switching. This seems to be the case in many experiments.[7,8,10,11]

Before we can apply the LLG equation to study the time evolution of the unit vector **m** in the direction of the magnetization of the switching magnet, we need to determine all the contributions to the torque acting on the switching magnet. First, there is the spin-transfer torque $\mathbf{T}^{s-t}$ discussed in Section 7.2. Second, there is a torque due to the uniaxial in-plane and easy-plane (shape) anisotropies. The easy-plane shape anisotropy torque arises because the switching magnet is a thin layer typically only a few nanometers thick. The in-plane uniaxial anisotropy is usually also a shape anisotropy arising from an elongated cross section of the switching magnet.[7] We take the uniaxial anisotropy axis of the switching magnet to be parallel to the z axis of the coordinate system shown in Figure 7.1. Because the switching magnet lies in the xz plane, we can write the total anisotropy field as

$$\mathbf{H}_A = \mathbf{H}_u + \mathbf{H}_p \tag{7.18}$$

where $\mathbf{H}_u$ and $\mathbf{H}_p$ are given by

$$\mathbf{H}_u = H_{u0}(\mathbf{m} \cdot \mathbf{e}_z)\mathbf{e}_z \tag{7.19}$$

$$\mathbf{H}_p = -H_{p0}(\mathbf{m} \cdot \mathbf{e}_y)\mathbf{e}_y. \tag{7.20}$$

Here $\mathbf{e}_x$, $\mathbf{e}_y$, and $\mathbf{e}_z$ are unit vectors in the directions of the axes shown in Figure 7.1. If we write the energy of the switching magnet in the anisotropy field as $-\mathbf{H}_A \cdot \langle \mathbf{S}_{tot} \rangle$, where $\langle \mathbf{S}_{tot} \rangle$ is the total spin-angular momentum of the switching magnet, then H_{u0}, H_{p0} which measure the strengths of the uniaxial and easy-plane anisotropies have dimensions of frequency. These quantities may be converted to a field in tesla by multiplying by $\hbar / 2\mu_B = 5.69 \times 10^{-12}$.

We are now ready to study the time evolution of the unit vector $\mathbf{m}$ in the direction of the switching magnet moment. The LLG equation takes the usual form

$$\frac{d\mathbf{m}}{dt} + \gamma \mathbf{m} \times \frac{d\mathbf{m}}{dt} = \mathbf{\Gamma} \tag{7.21}$$

where the reduced total torque $\mathbf{\Gamma}$ acting on the switching magnet is given by

$$\mathbf{\Gamma} = [-(\mathbf{H}_A + \mathbf{H}_{ext}) \times \langle \mathbf{S}_{tot} \rangle + \mathbf{T}_\perp + \mathbf{T}_\parallel]/ |\langle \mathbf{S}_{tot} \rangle|. \tag{7.22}$$

Here $\mathbf{H}_{ext}$ is an external field, in the same frequency units as $\mathbf{H}_A$, and γ is the Gilbert damping parameter. Following Sun,[6] Equation 7.21 may be written more conveniently as

$$(1 + \gamma^2) \frac{d\mathbf{m}}{dt} = \mathbf{\Gamma} - \gamma \mathbf{m} \times \mathbf{\Gamma}. \tag{7.23}$$

It is also useful to measure the strengths of all the torques in units of the strength of the uniaxial anisotropy.[6] We shall, therefore, write the total reduced torque $\mathbf{\Gamma}$ in the form

$$\mathbf{\Gamma} = H_{u0}\{(\mathbf{m} \cdot \mathbf{e}_z)\mathbf{m} \times \mathbf{e}_z - h_p(\mathbf{m} \cdot \mathbf{e}_y)\mathbf{m} \times \mathbf{e}_y + v_\parallel(\psi)\mathbf{m} \times (\mathbf{p} \times \mathbf{m}) + [v_\perp(\psi) + h_{ext}]\mathbf{m} \times \mathbf{p}\}$$

$$\tag{7.24}$$

where the relative strength of the easy plane anisotropy $h_p = H_{p0}/H_{u0}$ and $v_\parallel(\psi) = v g_\parallel(\psi)$, $v_\perp(\psi) = v g_\perp(\psi)$ measure the strengths of the torques $\mathbf{T}_\parallel$ and $\mathbf{T}_\perp$. The reduced bias is defined by $v = eV_b / (|\langle \mathbf{S}_{tot} \rangle| H_{u0})$ and has the opposite sign from the bias voltage because e is negative. Thus positive v implies a flow of electrons from the polarizing to the switching magnet. The last contribution to the torque in Equation 7.24 is due to the external field H_{ext} with $h_{ext} = H_{ext}/H_{u0}$. The external field is taken in the direction of the magnetization of the polarizing magnet, as is the case in most experimental situations.

It follows from Equation 7.21 that in a steady state, $\mathbf{\Gamma} = 0$. We shall first consider some cases of experimental importance where the steady state solutions are trivial and the important physics are concerned entirely with their stability. To discuss stability, we linearize Equation 7.23, using Equation 7.24, about a steady state solution $\mathbf{m} = \mathbf{m}_0$. Thus

$$\mathbf{m} = \mathbf{m}_0 + \xi \mathbf{e}_\alpha + \eta \mathbf{e}_\phi \qquad (7.25)$$

where $\mathbf{e}_\alpha$, $\mathbf{e}_\phi$ are unit vectors in the direction $\mathbf{m}$ moves when α and ϕ are increased independently. The linearized equation may be written in the form

$$\frac{d\xi}{d\tau} = A\xi + B\eta, \quad \frac{d\eta}{d\tau} = C\xi + D\eta. \qquad (7.26)$$

Following Sun,[6] we have introduced the natural dimensionless time variable $\tau = tH_{u0}/(1 + \gamma^2)$. The conditions for the steady state to be stable are

$$F = A + D \leq 0, \quad G = AD - BC \geq 0 \qquad (7.27)$$

excluding $F = G = 0$.[12] For simplicity, we give these conditions explicitly only for the case where either $v_\parallel'(\psi_0) = v_\perp'(\psi_0) = 0$, with $\psi_0 = \cos^{-1}(\mathbf{p} \cdot \mathbf{m}_0)$, or $\mathbf{m}_0 = \pm \mathbf{p}$. The case $\mathbf{m}_0 = \pm \mathbf{p}$ is very common experimentally as is discussed below. The stability condition $G \geq 0$ may be written

$$Q^2 v_\parallel^2 + (Qh + \cos 2\alpha_0)(Qh + \cos^2 \alpha_0) + h_p\{Qh(1 - 3\sin^2 \phi_0 \sin^2 \alpha_0)$$

$$+ \cos 2\alpha_0(1 - 2\sin^2 \alpha_0 \sin^2 \phi_0)\} - h_p^2 \sin^2 \alpha_0 \sin^2 \phi_0(1 - 2\sin^2 \phi_0 \sin^2 \alpha_0) \geq 0$$

$$(7.28)$$

where $v_\parallel = v_\parallel(\psi_0)$, $h = v_\perp(\psi_0) + h_{ext}$, and $Q = \cos \psi_0$. The condition $F \leq 0$ takes the form

$$-2(v_\parallel + \gamma h)Q - \gamma(\cos 2\alpha_0 + \cos^2 \alpha_0) - \gamma h_p(1 - 3\sin^2 \phi_0 \sin^2 \alpha_0) \leq 0. \quad (7.29)$$

We now discuss several interesting examples, the first of these relating to experiments of Grollier et al.[10] and others. In these experiments, the magnetization of the polarizing magnet, the uniaxial anisotropy axis, and the external field are all collinear (along the in-plane z axis in our convention). In this case, the equation $\Gamma = 0$, with Γ given by Equation 7.24, shows immediately that possible steady states are given by $\mathbf{m}_0 = \pm \mathbf{p}(\alpha_0 = 0, \pi)$, which corresponds to the switching magnet moment along the z axis. These are the only solutions when $h_p = 0$. For $h_p \neq 0$, other steady state solutions may exist, but in the parameter regime that has been investigated, they are always unstable.[13] We shall assume this is always the case and concentrate on the solutions $\mathbf{m}_0 = \pm \mathbf{p}$. In the state of parallel magnetization (P) $\mathbf{m}_0 = \mathbf{p}$, we have

$v_{\parallel} = vg_{\parallel}(0)$, $h = vg_{\perp}(0) + h_{ext}$, $\alpha_0 = 0$, and $Q = 1$. The stability conditions Equation 7.28 and Equation 7.29 become

$$[g_{\parallel}(0)]^2 v^2 + [vg_{\perp}(0) + h_{ext} + 1]^2 + h_p[vg_{\perp}(0) + h_{ext} + 1] \geq 0 \qquad (7.30)$$

$$g_{\parallel}(0)v + \gamma\left[vg_{\perp}(0) + h_{ext} + 1 + \frac{1}{2}h_p\right] \geq 0. \qquad (7.31)$$

In the state of antiparallel magnetization (AP), $\mathbf{m}_0 = -\mathbf{p}$, we have $v_{\parallel} = vg_{\parallel}(\pi)$, $h = vg_{\perp}(\pi) + h_{ext}$, $\alpha_0 = \pi$, and $Q = -1$. The stability conditions for the AP state are thus

$$[g_{\parallel}(\pi)]^2 v^2 + [-vg_{\perp}(\pi) - h_{ext} + 1]^2 + h_p[-vg_{\perp}(\pi) - h_{ext} + 1] \geq 0 \qquad (7.32)$$

$$g_{\parallel}(\pi)v + \gamma\left[vg_{\perp}(\pi) + h_{ext} - 1 - \frac{1}{2}h_p\right] \leq 0. \qquad (7.33)$$

In the regime of low external field ($h_{ext} \approx 1$, that is, $H_{ext} \approx H_{u0}$), we have $H_p \gg H_{ext}$ ($h_p \approx 100$). Equation 7.30 and Equation 7.32 may be then approximated by

$$vg_{\perp}(0) + h_{ext} + 1 > 0 \qquad (7.34)$$

$$vg_{\perp}(\pi) + h_{ext} - 1 < 0. \qquad (7.35)$$

Equation 7.34 corresponds to P stability and Equation 7.35 to AP stability. It is convenient to define scalar quantities $T_{\perp}$, $T_{\parallel}$ by $T_{\perp} = g_{\perp}(\psi)\sin\psi$, $T_{\parallel} = g_{\parallel}(\psi)\sin\psi$, these being scalar components of spin-transfer torque in units of eV_b (Equation 7.17). Then $g_i(0) = [dT_i/d\psi]_{\psi=0}$ and $g_i(\pi) = -[dT_i/d\psi]_{\psi=\pi}$ with $i = \perp, \parallel$. Model calculations[13] show that both $g_{\perp}$ and $g_{\parallel}$ can be of either sign, although positive values are more common. Also there is no general rule about the relative magnitude of $g_i(0)$ and $g_i(\pi)$.

We now illustrate the consequences of the above stability conditions by considering two limiting cases. We first consider the case $g_{\perp}(\psi) = 0$, $g_{\parallel} > 0$, as assumed by Grollier et al.[10] in the analysis of their data. In Figure 7.4, we plot the regions of P and AP stability deduced from Equation 7.31, Equation 7.33 to Equation 7.35 in the (v, h_{ext}) plane. Grollier et al. plot current instead of bias but this should not change the form of the figure. Theirs is rather more complicated, owing to a less transparent stability analysis with unnecessary approximation. The only approximations made above, to obtain Equation 7.34 and Equation 7.35, can easily be removed,

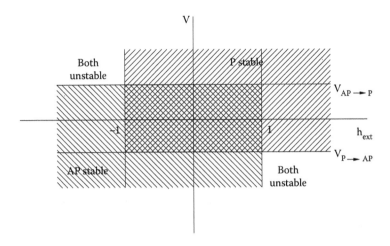

FIGURE 7.4 Bias-field stability diagram for $g_\perp(\psi) = 0, g_\parallel(\psi) > 0$. A small downward slope of the lines $V_{AP \to P}, V_{P \to AP}$ (see Equation 7.36) is not shown.

which results in the critical field lines $h_{ext} = \pm 1$ acquiring a very slight curvature given by $h_{ext} \approx 1 + [vg_\parallel(\pi)]^2/h_p$ and $h_{ext} \approx -1 - [vg_\parallel(0)]^2/h_p$. The critical biases in the figure are given by

$$v_{AP \to P} = \gamma \left[1 + \frac{1}{2}h_p - h_{ext} \right] \Big/ g_\parallel(\pi)$$

$$v_{P \to AP} = -\gamma \left[1 + \frac{1}{2}h_p + h_{ext} \right] \Big/ g_\parallel(0).$$

$$(7.36)$$

A downward slope from left to right of the corresponding lines in Figure 7.4 is not shown there. Since the damping parameter γ is small ($\gamma \approx 0.01$), this downward slope of the critical bias lines is also small. From Figure 7.4 we can deduce the behavior of resistance vs. bias in the external field regimes $|h_{ext}| < 1$ and $|h_{ext}| > 1$.

Consider first the case $|h_{ext}| < 1$. Suppose we start in the AP state with a bias $v = 0$, which is gradually increased to $v_{AP \to P}$. At this point, the AP state becomes unstable and the system switches to the P state as v increases further. On reducing v, the hysteresis loop is completed via a switch back to the AP state at the negative bias $v_{P \to AP}$. The hysteresis loop is shown in Figure 7.5a. The increase in resistance R between the P and AP states is the same as would be produced by varying the applied field in a GMR experiment. Now consider the case $h_{ext} < -1$. Starting again in the AP state at $v = 0$, we see from Figure 7.4 that, on increasing v to $v_{AP \to P}$, the AP state becomes unstable but there is no stable P state to switch to. This point is marked by an asterisk in Figure 7.5b. For $v > v_{AP \to P}$, the moment of the switching magnet is in a persistently time-dependent state. However, if v is now decreased below $v_{AP \to P}$ the system homes in on the stable AP state and the overall behavior

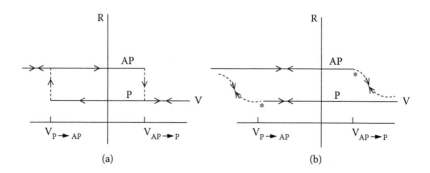

FIGURE 7.5 (a) Hysteresis loop of resistance vs. bias for $|h_{ext}| < 1$; (b) Reversible behavior (no hysteresis) for $h_{ext} < -1$ (upper curve) $h_{ext} > 1$ (lower curve). The dashed lines represent hypothetical behavior of average resistance in regions of Figure 7.4 marked "both unstable" where no steady states exist.

is reversible, i.e., no switching and no hysteresis occur. When $h_{ext} > 1$, similar behavior, now involving the P state, occurs at negative bias, as shown in Figure 7.5b. The dashed curves in Figure 7.5b show a hypothetical time-averaged resistance in the regions of time-dependent magnetization. As discussed later, time-resolved measurements of resistance suggest that several different types of dynamics can occur in these regions.

It is clear from Figure 7.5a that the jump AP → P always occurs for positive bias v, which corresponds to flow of electrons from the polarizing to the switching magnet. This result depends on the assumption that $g_{\parallel} > 0$; if $g_{\parallel} < 0$, it is easy to see that the sense of the hysteresis loop is reversed and the jump P → AP occurs for positive v. To our knowledge, this reverse jump has never been observed, although $g_{\parallel} < 0$ can occur in principle and is predicted theoretically[13] for the Co/Cu/Co(111) system with a switching magnet consisting of a single atomic plane of Co. It follows from Equation 7.36 that $|v_{P \to AP}/v_{AP \to P}| = |g_{\parallel}(\pi)/g_{\parallel}(0)|$ in zero external field. Experimentally this ratio, essentially the same as the ratio of critical currents, may be considerably less than 1 (e.g., < 0.5),[7] greater than 1 (e.g., ≈ 2)[11] or close to 1.[8] Usually the field dependence of the critical current is found to be stronger than that predicted by Equation 7.36.[7,8]

We now discuss the reversible behavior shown in Figure 7.5(b), which occurs for $|h_{ext}| > 1$. The transition from hysteretic to reversible behavior at a critical external field seems to have been first seen in pillar structures by Katine et al.[14] Curves similar to the lower one in Figure 7.5b are reported with $|v_{P \to AP}|$ increasing with increasing h_{ext}, as expected from Equation 7.36. Plots of the differential resistance dV/dI show a peak near the point of maximum gradient of the dashed curve. Similar behavior has been reported by several groups.[15–17] It is particularly clear in the work of Kiselev et al.[15] that the transition from hysteretic behavior (as in Figure 7.5a) to reversible behavior with peaks in dV/dI occurs at the coercive field 600 Oe of the switching layer ($h_{ext} = 1$). The important point about the peaks in dV/dI is that for a given sign of h_{ext}, they only occur for one sign of the bias. This clearly shows that this effect is due to spin transfer and not to Oersted fields. Myers et al.[18] show a current-field stability diagram similar to the bias-field one of Figure 7.4 with a critical field of 1500 Oe. They examine the time dependence of the resistance at room temperature with the field and current adjusted so that the system is in the "both unstable" region in the fourth quadrant of Figure 7.4 but very close to its top left-hand corner. They observe telegraph-noise-type switching between approximately P and AP states with slow switching times in the range of 0.1 to 10 s. Similar telegraph noise with faster switching times was observed by Urazhdin et al[16] at a current and field close to a peak in dV/dI. In the region of P and AP instability, Kiselev et al.[15] and Pufall et al.[17] report various types of dynamics of precessional type and random telegraph switching type in the microwave GHz regime. Kiselev et al.[15] propose that systems of the sort considered here might serve as nanoscale microwave sources or oscillators, tunable by current and field over a wide frequency range.

We now return to the stability conditions Equation 7.31, Equation 7.33 to Equation 7.35 and consider the case of $g_\perp(\psi) \neq 0$ but $h_{ext} = 0$. These conditions of stability of the P state may be written approximately, remembering that $\gamma \ll 1$, $h_p \gg 1$, as

$$vg_\perp(0) > -1, \quad vg_\parallel(0) > -\frac{1}{2}\gamma h_p. \tag{7.37}$$

The conditions for stability of the AP state are

$$vg_\perp(\pi) < 1, \quad vg_\parallel(\pi) < \frac{1}{2}\gamma h_p. \tag{7.38}$$

In Figure 7.6, we plot the regions of P and AP stability, assuming $g_\perp(0) = g_\perp(\pi) = g_\perp$ and $g_\parallel(0) = g_\parallel(\pi) = g_\parallel$ for simplicity. We also put $r = g_\perp / g_\parallel$. For $r > 0$, we find the normal hysteresis loop, as in Figure 7.5a, if we plot R against $vg_\parallel$ (valid for either sign of $g_\parallel$). In Figure 7.7, we plot the hysteresis loops for the cases $r_c < r < 0$ and $r < r_c$, where $r_c = -2/(\gamma h_p)$ is the value of r at the point X in Figure 7.6. The points labeled by asterisks have the same significance as in Figure 7.5b. If in Figure 7.7a, we increase $vg_\parallel$ beyond its value indicated by the right-hand asterisk, we move into the "both unstable" region where the magnetization direction of the switching magnet is perpetually in a time-dependent state. Thus negative r introduces behavior in the zero applied field which is similar to that found when the applied field exceeds the coercive field of the switching magnet for $r = 0$. This behavior was predicted by Edwards et al.,[13] in particular for a Co/Cu/Co(111) system with the switching magnet consisting of a Co monolayer. Zimmler et al.[19] use methods similar to the ones described here to analyze their data on a Co/Cu/Co nanopillar and deduce that $g_\parallel > 0$, $r = g_\perp/g_\parallel \approx -0.2$. It would be interesting to carry out time-resolved resistance measurements on this system at a large current density (corresponding to $vg_\perp < -1$) and zero external field.

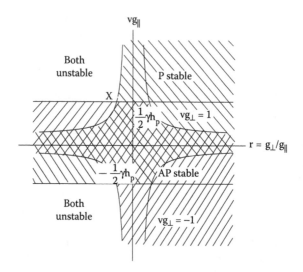

FIGURE 7.6 Stability diagram for $h_{ext} = 0$.

So far we have considered the low-field regime ($H_{ext} \approx$ coercive field of switching magnet) with both magnetizations and the external field in plane. There is another class of experiments in which a high field, greater than the demagnetizing field ($> 2T$), is applied perpendicular to the plane of the layers. The magnetization of the polarizing magnet is then also perpendicular to the plane. This is the situation in the early experiments where a point contact was employed to inject high current densities into magnetic multilayers.[20–22] In this high-field regime, a peak in the differential resistance dV/dI at a critical current was interpreted as the onset of current-induced excitation of spin waves in which the spin-transfer torque leads to uniform precession of the magnetization.[20,21,23] No hysteretic magnetization reversal was observed and it seemed that the effect of spin-polarized current on the magnetization is quite different in the low- and high-field regimes. Recently, however, Özyilmaz et al.[24] have studied Co/Cu/Co nanopillars (≈ 100 nm in diameter) at

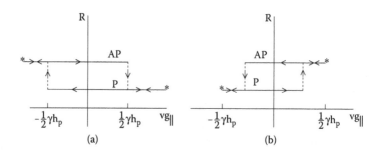

FIGURE 7.7 Hysteresis loop for (a) $r_c < r < 0$; (b) $r < r_c$.

$T = 4.2$ K for large applied fields perpendicular to the layers. They observe hysteretic magnetization reversal and interpret their results using the Landau–Lifshitz equation. We now give a similar discussion within the framework of this section.

Following Özyilmaz et al., we neglect the uniaxial anisotropy term in Equation 7.24 for the reduced torque Γ while retaining H_{u0} as a scalar factor. Hence

$$\Gamma = H_{u0}\{[h_{ext} + v_{\perp}(\psi) - h_p \cos\psi]\mathbf{m} \times \mathbf{p} + v_{\parallel}(\psi)\mathbf{m} \times (\mathbf{p} \times \mathbf{m})\} \qquad (7.39)$$

where $\mathbf{p}$ is the unit vector perpendicular to the plane. When $v_{\parallel}(\psi) \neq 0$, the only possible steady state solutions of $\Gamma = 0$ are $\mathbf{m}_0 = \pm\mathbf{p}$. On linearizing Equation 7.23 about $\mathbf{m}_0$ as before, we find that the condition $G \geq 0$ is always satisfied. The second stability condition $F < 0$ becomes

$$\{v_{\parallel}(\psi_0) + \gamma[v_{\perp}(\psi_0) + h_{ext} - h_p]\}\cos\psi_0 > 0 \qquad (7.40)$$

where $\psi_0 = \cos^{-1}(\mathbf{m}_0 \cdot \mathbf{p})$. Applying this to the P state ($\psi_0 = 0$), and the AP state ($\psi_0 = \pi$), we obtain the conditions

$$v > \gamma(h_p - h_{ext})/g(0) \qquad (7.41)$$

$$v < -\gamma(h_p + h_{ext})/g(\pi) \qquad (7.42)$$

where the first condition applies to the P stability and the second to the AP stability. Here $g(\psi) = g_{\parallel}(\psi) + \gamma g_{\perp}(\psi)$. The corresponding stability diagram is shown in Figure 7.8, where we have assumed $g(\pi) > g(0) > 0$ for definiteness. The boundary

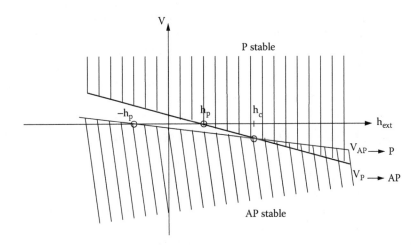

FIGURE 7.8 Bias-field stability diagram for large external field ($h_{ext} > h_p$) perpendicular to the layers.

lines cross at $h_{ext} = h_c$, where $h_c = h_p[g(\pi) + g(0)]/[g(\pi) - g(0)]$. This analysis is only valid for fields larger than the demagnetizing field ($h_{ext} > h_p$), and we see from Figure 7.8 that for $h_{ext} > h_c$ hysteretic switching occurs. This takes place for only one sign of the bias (current) and the critical biases (currents) increase linearly with h_{ext} as does the width of the hysteresis loop $|v_{P \to AP} - v_{AP \to P}|$. This accords with the observations of Özyilmaz et al. The critical currents are not larger than those in the low-field or zero-field regimes (c.f. Equation 7.41, Equation 7.42 with Equation 7.36) and yet the magnetization of the switching magnet can be switched against a very large external field. However, in this case, the AP state is only stabilized by maintaining the current.

The experiments on spin transfer discussed above have mainly been carried out at constant temperature, typically 4.2 K or room temperature. The effect on current-driven switching of varying the temperature has recently been studied by several groups.[16,18,25] The standard Néel–Brown theory of thermal switching[26] does not apply because the Slonczewski in-plane torque is not derivable from an energy function. Li and Zhang[27] have generalized the standard stochastic Landau–Lifschitz equation, which includes white noise in the effective applied field, to include spin-transfer torque. In this way, they have successfully interpreted some of the experimental data. A full discussion of this work is outside the scope of the present review. However, it should be pointed out that in addition to the classical effect of white noise, there is an intrinsic temperature dependence of quantum origin. This arises from the Fermi distribution functions that appear in expressions for the spin-transfer torque (see Equation 7.14 and Equation 7.12).

So far we have discussed steady-state solutions of the LLG equation (Equation 7.23). It is important to study the magnetization dynamics of the switching layer in the situation during the jumps AP $\to$ P and P $\to$ AP of the hysteresis curve in zero external field, and secondly under conditions where only time-dependent solutions are possible, for example in the regions of sufficiently strong current and external field marked "both unstable" in Figure 7.4. The first situation has been studied by Sun,[6] assuming single-domain behavior of the switching magnet, and by Miltat et al.[28] with more general micromagnetic configurations. Both situations have been considered by Li and Zhang.[29] In the second case, they find precessional states, and the possibility of "telegraph noise" at room temperature, as seen experimentally in References 15 and 17. Switching times (AP $\to$ P and P $\to$ AP) are estimated to be of the order 1 ns. Micromagnetic simulations[28] indicate that the Oersted field cannot be completely ignored for typical pillars with diameter of the order of 100 nm.

Finally, in this section, we briefly discuss some practical considerations that may ultimately decide whether current-induced switching is useful in spintronics. Sharp switching, with nearly rectangular hysteresis loops, is obviously desirable and this demands single-domain behavior. In experiments on nanopillars of a circular cross section,[14] multidomain behavior was observed with the switching transition spread over a range of currents. Subsequently the same group[7] found sharp switching in pillars whose cross section was an elongated hexagon, which introduces strong uniaxial in-plane shape anisotropy. It was known from earlier magnetization studies of nanomagnet arrays[30] that such a shape anisotropy can result in single-domain behavior. A complex switching transition need not necessarily indicate multidomain behavior. It could also arise from a marked departure of $T_\perp(\psi)$ and or $T_{\parallel}(\psi)$ from sinusoidal behavior, such as occurs near $\psi = \pi$ in calculations for Co/Cu/Co(111) with two atomic planes of Co in the switching magnet (see Figure 7.9b). In the calculations of the corresponding hysteresis loops (Figure 7.11), the torques were approximated by sine curves but an accurate treatment would certainly complicate the AP $\rightarrow$ P transition that occurs at negative bias in Figure 7.11b. Studies of this effect are planned.

The critical current density for switching is clearly an important parameter. From Equation 7.36, the critical reduced bias for the P $\rightarrow$ AP transition is to a good approximation given by $-\gamma h_p/[2g_{\parallel}(0)]$. Using the definitions of reduced quantities given after Equation 7.24, we may write the actual critical bias in volts as

$$V_{P \rightarrow AP} = M\gamma M_s H_d/[2g_{\parallel}(0)\,|\,e\,|], \qquad (7.43)$$

where M is the number of atomic planes in the switching magnet, M_s is the average moment (J/T) of the switching magnet per atomic plane per unit area, and $H_d = \hbar H_{p0}/(2\mu_B)$ is the easy-plane anisotropy field in tesla. As expressed earlier $g_{\parallel}(0) = (dT_{\parallel}/d\psi)_{\psi=0}$, where the torque $T_{\parallel}$ is per unit area in units of eV_B. (The calculated torques in Figure 7.9 and Figure 7.10 of Section 7.5 are per surface atom so that if these are used to determine $g_{\parallel}(0)$ in Equation 7.43, M_s must be taken per surface atom.)

An obvious way to reduce the critical bias, and hence the critical current, is to reduce M, the thickness of the switching magnet. Calculations show[13] (see also Figure 7.10) that $g_{\parallel}$ does not decrease with M and may, in fact, increase for small values such as $M = 2$. Careful design of the device might also increase $g_{\parallel}(0)$ beyond the values (< 0.01 per surface atom), which seem to be obtainable in simple trilayers.[13] Jiang et al.[31,32] have studied various structures in which the polarizing magnet is pinned by an adjacent antiferromagnet (exchange biasing) and in which a thin Ru layer is incorporated between the switching layer and the lead. Critical current densities of 2×10^6 A cm^{-2} have been obtained that are substantially lower than those in Co/Cu/Co trilayers. Such structures can quite easily be investigated theoretically by the methods of Section 7.5.

Decreasing the magnetization M_s, and hence the demagnetizing field ($\propto H_d$), would be favorable but $g_{\parallel}$ then tends to decrease also.[13] A possible way of decreasing H_d without decreasing local magnetic moments in the system is to use a synthetic ferrimagnet as the switching magnet.[33] The Gilbert damping factor γ is another crucial parameter but it is uncertain whether this can be decreased significantly. However, the work of Capelle and Gyorffy[34] is an interesting theoretical development. The search for structures with critical current densities low enough for use in spintronic devices (10^5 A cm^{-2} perhaps)[35] is an enterprise where experiment and quantitative calculations[13] should complement each other fruitfully.

7.4 QUANTITATIVE THEORY OF SPIN-TRANSFER TORQUE

7.4.1 GENERAL PRINCIPLES

To put the phenomenological treatment of Section 7.3 on a first-principle quantitative basis, we must calculate the spin-transfer torques (Equaion 7.17) in a steady state for real systems. For this purpose it is convenient to describe the magnetic and nonmagnetic layers of Figure 7.1 by tight-binding models, in general multiorbital with s, p, and d orbitals, whose one-electron parameters are fitted to first-principle bulk band structure.[36] The Hamiltonian is therefore of the form

$$H = H_0 + H_{int} + H_{anis} \tag{7.44}$$

where the one-electron hopping term H_0 is given by

$$H_0 = \sum_{k_\parallel \sigma} \sum_{m\mu,n\upsilon} t_{m\mu,n\upsilon}(\mathbf{k}_\parallel) c^\dagger_{k_\parallel m\mu\sigma} c_{k_\parallel n\upsilon\sigma} \tag{7.45}$$

where $c^\dagger_{k_\parallel m\mu\sigma}$ creates an electron in a Bloch state, with in-plane wave vector $\mathbf{k}_\parallel$ and spin σ, formed from a given atomic orbital μ in plane m. Equation 7.45 generalizes the single orbital Equation 7.1. H_{int} is an on-site interaction between electrons in d orbitals that leads to an exchange splitting of the bands in the ferromagnets and is neglected in the spacer and lead. Finally, H_{anis} contains anisotropy fields in the switching magnet and is given by

$$H_{anis} = -\sum_n \mathbf{S}_n \cdot \mathbf{H}_A \tag{7.46}$$

where $\mathbf{S}_n$ is the operator of the total spin-angular momentum of plane n and $\mathbf{H}_A$ is given by Equation 7.18 to Equation 7.20 with the unit vector $\mathbf{m}$ in the direction of $\sum_n \langle \mathbf{S}_n \rangle$, where $\mathbf{S}_n$ is the thermal average of $\mathbf{S}_n$. We assume here that the anisotropy fields H_{u0}, H_p are uniform throughout the switching magnet but we could generalize to include, for example, a surface anisotropy.

In the tight-binding description, the spin-angular momentum operator $\mathbf{S}_n$ is given by

$$\mathbf{S}_n = \frac{1}{2}\hbar \sum_{k_\parallel \mu} (c^\dagger_{k_\parallel n\mu\uparrow}, c^\dagger_{k_\parallel n\mu\downarrow}) \sigma (c_{k_\parallel n\mu\uparrow}, c_{k_\parallel n\mu\downarrow})^T \tag{7.47}$$

and the corresponding operator for the spin-angular momentum current between planes $n-1$ and n is

$$\mathbf{j}_{n-1} = -\frac{1}{2} i \sum_{k_\parallel \mu \upsilon} t(\mathbf{k}_\parallel)_{n\upsilon,n-1\mu} (c^\dagger_{k_\parallel n\upsilon\uparrow}, c^\dagger_{k_\parallel n\upsilon\downarrow}) \sigma (c_{k_\parallel n-1\mu\uparrow}, c_{k_\parallel n-1\mu\downarrow})^T + \text{H.c.} \tag{7.48}$$

which generalizes the single-orbital expression Equation 7.2. The rate of change of $\mathbf{S}_n$ in the switching magnet is given by

$$i\hbar \frac{d\mathbf{S}_n}{dt} = [\mathbf{S}_n, H_0] + [\mathbf{S}_n, H_{anis}]. \tag{7.49}$$

This results holds because the spin operator commutes with the interaction Hamiltonian H_{int}.

It is straightforward to show that

$$[\mathbf{S}_n, H_0] = i\hbar(\mathbf{j}_{n-1} - \mathbf{j}_n) \tag{7.50}$$

and

$$[\mathbf{S}_n, H_{anis}] = -i\hbar(\mathbf{H}_A \times \mathbf{S}_n). \tag{7.51}$$

On taking the thermal average, Equation 7.49 becomes

$$\left\langle \frac{d\mathbf{S}_n}{dt} \right\rangle = \langle \mathbf{j}_{n-1} \rangle - \langle \mathbf{j}_n \rangle - \mathbf{H}_A \times \langle \mathbf{S}_{tot} \rangle. \tag{7.52}$$

This corresponds to an equation of continuity, stating that the rate of change of spin-angular momentum on plane n is equal to the difference between the rate of flow of this quantity onto and off the plane, plus the rate of change due to precession around the field $\mathbf{H}_A$. When Equation 7.52 is summed over all planes in the switching magnet, we have

$$\frac{d}{dt} \langle \mathbf{S}_{tot} \rangle = \mathbf{T}^{s-t} - \mathbf{H}_A \times \langle \mathbf{S}_{tot} \rangle \tag{7.53}$$

where the total spin-transfer torque $\mathbf{T}^{s-t}$ is given by Equation 7.14 and $\langle \mathbf{S}_{tot} \rangle$ is the total spin-angular momentum of the switching magnet. Equation 7.53 is equivalent to Equation 7.21, for zero external field, in the absence of damping. Equation 7.14 shows how $\mathbf{T}^{s-t}$ required for the phenomenological treatment of Section 7.3 is to be determined from the calculated spin currents in the spacer and lead. As discussed in Section 7.3, the magnetization of a single-domain sample is essentially uniform and the spin-transfer torque $\mathbf{T}^{s-t}$ depends on the angle ψ between the magnetizations of the polarizing and switching magnets.

To consider time-dependent solutions of Equation 7.21, it is necessary to calculate $\mathbf{T}^{s-t}$ for arbitrary angle ψ, and for this purpose, $\mathbf{H}_A$ can be neglected. To reduce the calculation of the spin-transfer torque to effectively a one-electron problem, we replace H_{int} by a self-consistent exchange field term $-\Sigma_n \mathbf{S}_n \times \mathbf{\Delta}_n$, where the exchange field $\mathbf{\Delta}_n$ should be determined self-consistently in the spirit of an unrestricted Hartree–Fock (HF) or local spin density (LSD) approximation. The essential self-consistency condition in any HF or LSD calculation is that the local moment $\langle \mathbf{S}_n \rangle$ in a steady state is in the same direction as $\mathbf{\Delta}_n$. Thus we require

$$\mathbf{\Delta}_n \times \langle \mathbf{S}_n \rangle = 0 \tag{7.54}$$

for each atomic plane of the switching magnet. It is useful to consider first the situation when there is no applied bias and the polarizing and switching magnets are separated by a spacer that is so thick that the zero-bias oscillatory exchange coupling[37] is negligible. In that case, we have two independent magnets and the self-consistent exchange field in every atomic plane of the switching magnet is parallel to its total magnetization which is uniform and assumed to be along the z axis. Referring to Figure 7.1, the self-consistent solution therefore corresponds to uniform exchange fields in the polarizing and switching magnets that are at an assumed angle $\psi = \theta$ with respect to one another.

When a bias V_b is applied, and a uniform exchange field $\mathbf{\Delta} = \Delta \mathbf{e}_z$ in the switching magnet imposed, the calculated local moments $\langle \mathbf{S}_n \rangle$ will deviate from the z-direction so that the solution is not self-consistent. To prepare a self-consistent state with $\mathbf{\Delta}$ and all $\langle \mathbf{S}_n \rangle = \langle \mathbf{S} \rangle$ in the z-direction, it is necessary to apply fictitious constraining fields $\mathbf{H}_n$ of magnitude proportional to V_b. The local field for plane n is thus $\mathbf{\Delta} + \mathbf{H}_n$, but to calculate the spin currents in the spacer and lead, and hence $\mathbf{T}^{s-t}$ from Equation 7.14, the fields $\mathbf{H}_n$, of the order of V_b, may be neglected compared with $\mathbf{\Delta}$. Although the fictitious constraining fields $\mathbf{H}_n$ need therefore never be calculated, it is interesting to see that they are in fact related to $\mathbf{T}^{s-t}$. For the constrained self-consistent steady state ($\langle \mathbf{S}_n \rangle = \langle \mathbf{S} \rangle$, $\langle \dot{\mathbf{S}}_n \rangle = 0$) in the presence of the constraining fields, with $\mathbf{H}_A$ neglected as discussed above, it follows from Equation 7.52 that

$$\langle \mathbf{j}_{n-1} \rangle - \langle \mathbf{j}_n \rangle = (\mathbf{\Delta} + \mathbf{H}_n) \times \langle \mathbf{S} \rangle = \mathbf{H}_n \times \langle \mathbf{S} \rangle, \tag{7.55}$$

where the local field $\mathbf{\Delta} + \mathbf{H}_n$ replaces $\mathbf{H}_A$. On summing over all atomic planes n in the switching magnet, we have

$$\mathbf{T}^{s-t} = \langle \mathbf{j}_{spacer} \rangle - \langle \mathbf{j}_{lead} \rangle = \sum_n \mathbf{H}_n \times \langle \mathbf{S} \rangle. \tag{7.56}$$

Thus, as expected, in the prepared state with a given angle ψ between the magnetizations of the magnetic layers, the spin-transfer torque is balanced by the total torque due to the constraining fields.

In the simple model of Section 7.2, with infinite exchange splitting in the magnets, the local moment is constrained to be in the direction of the exchange field so the question of self-consistency is not raised.

The main conclusion of this section is that the spin-transfer torque for a given angle ψ between magnetizations may be calculated using uniform exchange fields making the same angle with one another. Such calculations are described in Section 7.2 and Section 7.5. The use of this spin-transfer torque in the LLG equation of Section 7.3 completes what we shall call the "standard model" (SM). It underlies the original work of Slonczewski[3] and most subsequent work. The spin-transfer torque calculated in this way should be appropriate even for time-dependent solutions of the LLG equation. This is based on the reasonable assumption that the time for the electronic system to attain a "constrained steady state" with given ψ is short compared with the timescale (≈ 1 ns) of the macroscopic motion of the switching magnet moment.

Although the SM is a satisfactory way of calculating the spin-transfer torque, its lack of self-consistency leads to some nonphysical concepts. The first of these is the "transverse spin accumulation" in the switching magnet.[38,39] This refers to the deviations of local moments $\langle \mathbf{S}_n \rangle$ from the direction of the exchange field, assumed to be uniform in the SM. In a self-consistent treatment such deviations do not occur because the exchange field is always in the direction of the local moment. A related nonphysical concept is the "spin decoherence length" over which the spin accumulation is supposed to decay.[38,39] More detailed critiques of these concepts are given elsewhere.[13,40]

7.4.2 KELDYSH FORMALISM FOR FULLY REALISTIC CALCULATIONS OF THE SPIN-TRANSFER TORQUE

The wave-function approach to spin-transfer torque described in Section 7.2 is difficult to apply to realistic multiorbital systems. For this purpose, Green's functions are much more convenient and Keldysh[41] developed a Green's function approach to the nonequilibrium problem of electron transport. In this section, we apply this method to calculate spin currents in a magnetic layer structure, following Edwards et al.[13]

The structure we consider is shown schematically in Figure 7.1. It consists of a thick (semi-infinite) left magnetic layer (polarizing magnet), a nonmagnetic metallic spacer layer of N atomic planes, a thin switching magnet of M atomic planes, and a semi-infinite lead. The broken line between the atomic planes $n-1$ and n represents a cleavage plane separating the system into two independent parts so that charge carriers cannot move between the two surface planes $n-1$ and n. It will be seen that our ability to cleave the whole system in this way is essential for the implementation of the Keldysh formalism. This can be easily done with a tight-binding parametrization of the band structure by simply switching off the matrix of hopping integrals $t_{n\upsilon,n-1\mu}$ between atomic orbitals υ, μ localized in planes $n-1$ and n. We therefore adopt the tight-binding description with the Hamiltonian defined by Equation 7.44 to Equation 7.47.

To use the Keldysh formalism[41,42,43] to calculate the charge or spin currents flowing between the planes $n-1$ and n, we consider an initial state at time $\tau = -\infty$ in which the hopping integral $t_{n\upsilon,n-1\mu}$ between planes $n-1$ and n is switched off. Then both sides of the system are in equilibrium but with different chemical potentials

μ_L on the left and μ_R on the right, where $\mu_L - \mu_R = eV_b$. The interplane hopping is then turned on adiabatically and the system evolves to a steady state. The cleavage plane, across which the hopping is initially switched off, may be taken in either the spacer or in one of the magnets or in the lead. In principle, the Keldysh method is valid for arbitrary bias V_b, but here we restrict ourselves to small bias corresponding to linear response. This is always reasonable for a metallic system. For larger bias, which might occur with a semiconductor or insulator as spacer, electrons would be injected into the right part of the system far above the Fermi level and many-body processes neglected here would be important. Following Keldysh,[41,42] we define a two-time matrix

$$G^+_{RL}(\tau,\tau') = i\langle c^\dagger_L(\tau')c_R(\tau)\rangle \qquad (7.57)$$

where $R \equiv (n, \nu, \sigma')$ and $L \equiv (n-1, \mu, \sigma)$, and we suppress the $k_\parallel$ label. The thermal average in Equation 7.57 is calculated for the steady state of the coupled system. The matrix G^+_{RL} has dimensions $2m \times 2m$, where m is the number of orbitals on each atomic site, and is written so that the $m \times m$ upper diagonal block contains matrix elements between up ($\uparrow$) spin orbitals and the $m \times m$ lower diagonal block relates to down ($\downarrow$) spin. $2m \times 2m$ hopping matrices t_{LR} and t_{RL} are written similarly and in this case only the diagonal blocks are nonzero. If we denote t_{LR} by t, then $t_{RL} = t^\dagger$. We also generalize the definition of σ so that its components are now direct products of the 2×2 Pauli matrices σ_x, σ_y, σ_z, and the $m \times m$ unit matrix. The thermal average of the spin-current operator, given by Equation 7.49, may now be expressed as

$$\langle \mathbf{j}_{n-1}\rangle = \frac{1}{2}\sum_{\mathbf{k}_\parallel} \mathrm{Tr}\{[G^+_{RL}(\tau,\tau)t - G^+_{LR}(\tau,\tau)t^\dagger]\sigma\}. \qquad (7.58)$$

Introducing the Fourier transform $G^+(\omega)$ of $G^+(\tau,\tau')$, which is a function of $\tau - \tau'$, we have

$$\langle \mathbf{j}_{n-1}\rangle = \frac{1}{2}\sum_{\mathbf{k}_\parallel}\int \frac{d\omega}{2\pi} \mathrm{Tr}\{[G^+_{RL}(\omega)t - G^+_{LR}(\omega)t^\dagger]\sigma\} \qquad (7.59)$$

The charge current is given by Equation 7.59 with $\frac{1}{2}\sigma$ replaced by the unit matrix multiplied by $e/\hbar$.

Following Keldysh,[41,42] we now write

$$G^+_{AB}(\omega) = \frac{1}{2}(F_{AB} + G^a_{AB} - G^r_{AB}) \qquad (7.60)$$

where the suffices A and B are either R or L. $F_{AB}(\omega)$ is the Fourier transform of

$$F_{AB}(\tau,\tau') = -i\langle [c_A(\tau), c^\dagger_B(\tau')]_-\rangle \qquad (7.61)$$

and G^a, G^r are the usual advanced and retarded Green's functions.[44] Note that in Reference 41 and Reference 42, the definitions of G^a and G^r are interchanged, and that in the Green's function matrix defined by these authors, G^+ and G^- should be interchanged.

Charge and spin current are related by Equation 7.59 and Equation 7.60 to the quantities G^a, G^r and F_{AB}. The latter are calculated for the coupled system by starting with decoupled left and right systems, each in equilibrium, and turning on the hopping between planes L and R as a perturbation. Hence, we express G^a, G^r, and F_{AB} in terms of retarded surface Green's functions $g_L \equiv g_{LL}$, $g_R \equiv g_{RR}$ for the decoupled equilibrium system. It is then found[13] that the spin current between the planes $n-1$ and n can be written as the sum $\langle \mathbf{j}_{n-1} \rangle = \langle \mathbf{j}_{n-1} \rangle_1 + \langle \mathbf{j}_{n-1} \rangle_2$, where the two contributions to the spin current $\langle \mathbf{j}_{n-1} \rangle_1$, $\langle \mathbf{j}_{n-1} \rangle_2$ are given by

$$\langle \mathbf{j}_{n-1} \rangle_1 = \frac{1}{4\pi} \sum_{\mathbf{k}_\parallel} \int d\omega \Re \operatorname{Tr}[(B-A)\sigma][f(\omega - \mu_L) + f(\omega - \mu_R)] \qquad (7.62)$$

$$\langle \mathbf{j}_{n-1} \rangle_2 = \frac{1}{2\pi} \sum_{\mathbf{k}_\parallel} \int d\omega \Re \operatorname{Tr} \left\{ [g_L t A B g_R^\dagger t^\dagger - AB + \frac{1}{2}(A+B)]\sigma \right\} \qquad (7.63)$$

$$[f(\omega - \mu_L) - f(\omega - \mu_R)].$$

Here, $A = [1 - g_R t^\dagger g_L t]^{-1}$, $B = [1 - g_R^\dagger t^\dagger g_L^\dagger t]^{-1}$, and as in Section 7.2, $f(\omega - \mu)$ is the Fermi function with chemical potential μ and $\mu_L - \mu_R = eV_b$. In the linear-response case of small bias that we are considering, the Fermi functions in Equation 7.63 are expanded to first order in V_b. Hence the energy integral is avoided, being equivalent to multiplying the integrand by eV_b and evaluating it at the common zero-bias chemical potential μ_0.

It can be seen that Equation 7.62 and Equation 7.63, which determine the spin and the charge currents, depend on just two quantities: the surface retarded one-electron Green's functions for a system cleaved between two neighboring atomic planes. The surface Green's functions can be determined without any approximations by the standard adlayer method (see, e.g. , Reference 36, Reference 37) for a fully realistic band structure.

We first note that there is a close correspondence between Equation 7.62, Equation 7.63, and the generalized Landauer formula (Equation 7.12). The first term in Equation 7.12 corresponds to the zero-bias spin current $\langle \mathbf{j}_{n-1} \rangle_1$ given by Equation 7.62. When the cleavage plane is taken in the spacer, the spin current $\langle \mathbf{j}_{n-1} \rangle_1$ determines the oscillatory exchange coupling between the two magnets and it is easy to verify that the formula for the exchange coupling obtained from Equation 7.62 is equivalent to the formula used in previous total energy calculations of this effect.[36,37] The contribution to the transport spin current given by Equation 7.63 clearly corresponds to the second term in the Landauer formula (Equation 7.12), which is proportional to the bias in the linear response limit. Placing the cleavage plane first between any two neighboring atomic planes in the spacer and then between any two neighboring planes in the lead, we obtain from Equation 7.63 the total spin-transfer torque $\mathbf{T}^{s-t}$ of Equation 7.14 in Section 7.2.

The equivalence of the Keldysh and Landauer methods has been demonstrated by calculating the currents in Equation 7.62 and Equation 7.63 analytically for the simple single-orbital model of Section 7.2. The results of that section, such as Equation 7.15 are reproduced.[13]

7.5   QUANTITATIVE RESULTS FOR Co/Cu/Co(111)

We now discuss the application of the Keldysh formalism to a real system. In particular, we consider a realistic multiorbital model of fcc Co/Cu/Co(111) with tight-binding parameters fitted to the results of the first-principles band structure calculations, as described previously.[36,37]

Referring to Figure 7.1, the system considered by Edwards et al.[13] consists of a semi-infinite slab of Co (polarizing magnet), the spacer of 20 atomic planes of Cu, the switching magnet containing M atomic planes of Co, and the lead that is semi-infinite Cu. The spacer thickness of 20 atomic planes of Cu was chosen so that the contribution of the oscillatory exchange coupling term is so small that it can be neglected. The spin currents in the right lead and in the spacer were determined from Equation 7.63. Figure 7.9 shows the angular dependences of $T_{\parallel}$, $T_{\perp}$ for the cases $M = 1$ (a) and $M = 2$ (b), respectively. For the monolayer switching magnet, the torques $T_{\parallel}$ and $T_{\perp}$ are equal in magnitude and they have opposite sign. However, for $M = 2$, the torques have the same sign and $T_{\perp}$ is somewhat smaller than $T_{\parallel}$. A negative sign of the ratio of the two torque components has important and unexpected consequences for hysteresis loops as already discussed in Section 7.3. It can be seen that the angular dependence of both torque components is dominated by a $\sin \psi$ factor but distortions from this dependence are clearly visible. In particular, the slopes at $\psi = 0$ and $\psi = \pi$ are quite different. As pointed out in Section 7.3, this

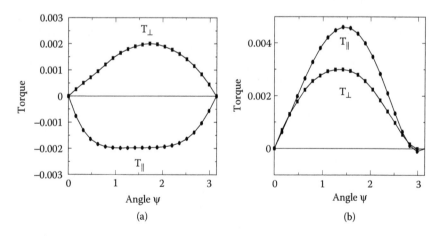

FIGURE 7.9 Dependence of the spin-transfer torque $T_{\parallel}$ and $T_{\perp}$ for Co/Cu/Co(111) on the angle ψ. The torques per surface atom are in units of eV_b. Panel a is for $M = 1$, and Panel b for $M = 2$ monolayers of Co in the switching magnet.

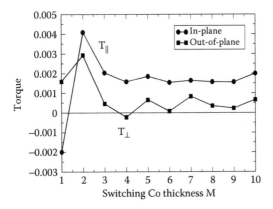

FIGURE 7.10 Dependence of the spin-transfer torque $T_\parallel$ and $T_\perp$ for Co/Cu/Co(111) on the thickness of the switching magnet M for $\psi = \pi/3$. The torques are in units of eV_b.

is important in the discussion of the stability of steady states and leads to quite different magnitudes of the critical biases $V_P \rightarrow V_{AP}$ and $V_{AP} \rightarrow V_P$.

In Figure 7.10, we reproduce the dependence of $T_\perp$ and $T_\parallel$ on the thickness of the Co switching magnet. It can be seen that the out-of-plane torque $T_\perp$ becomes smaller than $T_\parallel$ for thicker switching magnets. However, $T_\perp$ is by no means negligible (27% of $T_\parallel$) even for a typical experimental thickness of the switching Co layer of 10 atomic planes. It is also interesting that beyond the monolayer thickness, the ratio of the two torques is positive with the exception of $M = 4$.

The microscopically calculated spin-transfer torques for Co/Cu/Co(111) were used by Edwards et al.[13] as an input into the phenomenological LLG equation. For simplicity, the torques as functions of ψ were approximated by sine curves but this is not essential. The LLG equation was first solved numerically to determine all the steady states and then the stability discussion outlined in the phenomenological section was applied to determine the critical bias for which instabilities occur. Finally, the ballistic resistance of the structure was evaluated from the real-space Kubo formula at every point of the steady state path. Such a calculation for the realistic Co/Cu system then gives hysteresis loops of the resistance vs. bias which can be compared with the observed hysteresis loops. The LLG equation was solved including a strong easy-plane anisotropy with $h_p = 100$. If we take $H_{u0} = 1.86 \times 10^9$ sec^{-1}, corresponding to a uniaxial anisotropy field of about 0.01 T, this value of h_p corresponds to the shape anisotropy for a magnetization of 1.6×10^6 A/m, similar to that of Co.[6] Also a realistic value[6] of the Gilbert damping parameter $\gamma = 0.01$ was used. Finally, referring to the geometry of Figure 7.1, two different values of the angle θ were employed in these calculations: $\theta = 2$ rad and $\theta = 3$ rad, the latter value being close to the value of π, which is realized in most experiments.

We first reproduce in Figure 7.11, the hysteresis loops for the case of Co switching magnet consisting of two atomic planes. We note that the ratio $r = T_\perp / T_\parallel \approx 0.65$ deduced from Figure 7.9 is positive in this case. Figure 7.11a shows

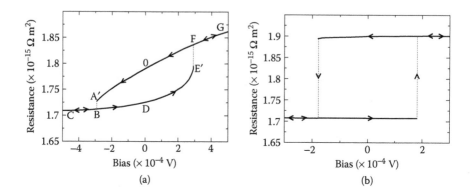

FIGURE 7.11 Resistance of the Co/Cu/Co(111) junction as a function of the applied bias, with $M = 2$ monolayers of Co in the switching magnet. Panel a is for $\theta = 2$ rad, and Panel b is for $\theta = 3$ rad.

the hysteresis loop for $\theta = 2$ and Figure 7.11b that for $\theta = 3$. The hysteresis loop for $\theta = 3$ shown in Figure 7.11b is an illustration of the stability scenario in zero applied field with $r > 0$ discussed in Section 7.3. As pointed out there, the hysteresis curve is that of Figure 7.5a, which agrees with Figure 7.11b when we remember that the reduced bias used in Figure 7.5 has the opposite sign from the bias in volts used in Figure 7.11. It is rather interesting that the critical bias for switching is ≈ 0.2 mV both for $\theta = 2$ and $\theta = 3$. When this bias is converted to the current density using the calculated ballistic resistance of the junction, it is found[13] that the critical current for switching is $\approx 10^7$ A/cm², which is in very good agreement with experiments.[7]

The hysteresis loops for the case of the Co switching magnet consisting of a single atomic plane are reproduced in Figure 7.12. The values of h_p, γ, H_{u0}, and

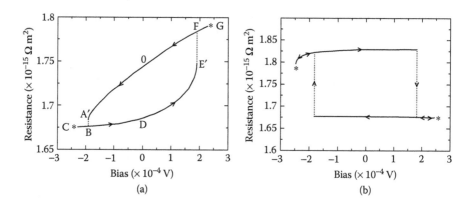

FIGURE 7.12 Resistance of the Co/Cu/Co(111) junction as a function of the applied current, with $M = 1$ monolayer of Co in the switching magnet. Panel a is for $\theta = 2$ rad, and Panel b is for $\theta = 3$ rad.

θ are the same as in the previous example. However, the ratio $r \approx -1$ is now negative and the hysteresis loops in Figure 7.12 illustrate the interesting behavior discussed in Section 7.3 when the system subjected to a bias higher than a critical bias moves to the "both unstable" region shown in Figure 7.6. As in Figure 7.7, the points on the hysteresis loop in Figure 7.12 corresponding to the critical bias are labeled by asterisks. Figure 7.12b and Figure 7.7a are in close correspondence because Figure 7.7a is for $r_c < r < 0$ and in the present case $r = -1$, $r_c = -2/(\gamma h_p) = -2$. Also, from Figure 9.7a, $g_\parallel < 0$ so that $v g_\parallel$ in Figure 7.7a has the same sign as the voltage V in Figure 7.12b.

7.6 SUMMARY

Spin-transfer torque is responsible for current-driven switching of magnetization in magnetic layered structures. The simplest theoretical scheme for calculating spin-transfer torque is a generalized Landauer method, and this is used in Section 7.2 to obtain analytical results for a simple model. The general phenomenological form of spin-transfer torque is deduced in Section 7.3 and this is introduced into the Landau–Lifshitz-Gilbert equation, together with torques due to anisotropy fields. This describes the motion of the magnetization of the switching magnet and the stability of the steady states (constant current and stationary magnetization direction) is studied under different experimental conditions, with and without external field. This leads to hysteretic and reversible behavior in resistance vs. bias (or current) plots in agreement with a wide range of experimental observations. In Section 7.4, the general principles of a self-consistent treatment of spin-transfer torque are discussed and the Keldysh formalism for quantitative calculations is introduced. This approach to the nonequilibrium problem of electron transport uses Green's functions, which are very convenient to calculate for a realistic multiorbital tight-binding model of the layered structure. In Section 7.5, quantitative calculations for Co/Cu/Co(111) systems are presented that yield switching currents of the observed magnitude.

ACKNOWLEDGMENTS

This study of current-driven switching of magnetization was carried out in collaboration with J. Mathon and A. Umerski, and financial support was provided by the U.K. Engineering and Physical Sciences Research Council.

REFERENCES

1. P. Grünberg, R. Schreiber, Y. Pang, M.B. Brodsky, and H. Sower, *Phys. Rev. Lett.*, 57, 2442, 1986; M.N. Baibich, J.M. Broto, A. Fert, Van Dau Nguyen, F. Petroff, P. Etienne, G. Creuset, A. Friederich, and J. Chazelas, *Phys. Rev. Lett.*, 61, 2472, 1998.
2. S.S.P. Parkin et al., *J. Appl. Phys.*, 85, 5828, 1999.
3. J.C. Slonczewski, *J. Magn. Magn. Mater.*, 159, L1 1996.
4. X. Waintal, E.B. Myers, P.W. Brouwer, and D.C. Ralph, *Phys. Rev. B*, 62, 12317, 2000.

5. D.K. Ferry and S.M. Goodnick, *Transport in Nanostructures,* Cambridge University Press, Cambridge, 1997.

6. J.Z. Sun, *Phys. Rev. B*, 62, 570, 2000.

7. F.J. Albert, J.A. Katine, R.A. Buhrman, and D.C. Ralph, *Appl. Phys. Lett.*, 77, 3809 2000.

8. J. Grollier, V. Cross, A. Hamzic, J.M. George, H. Jaffres, A. Fert, G. Faini, J. Ben Youssef, and H. Le Gall, *Appl. Phys. Lett.*, 78, 3663 2001.

9. E.C. Stoner and E.P. Wohlfarth, *Phil., Trans. Roy. Soc. A*, 240, 599, 1948.

10. J. Grollier, V. Cross, H. Jaffres, A. Hamzic, J.M. George, G. Faini, J. Ben Youssef, H. Le Gall, and A. Fert, *Phys. Rev. B*, 67, 174402, 2003.

11. F.J. Albert, N.C. Emley, E.B. Myers, D.C. Ralph, and R.A. Buhrman, *Phys. Rev. Lett.*, 89, 226802, 2002.

12. D.W. Jordan and P. Smith, *Nonlinear Ordinary Differential Equations*, Clarendon Press, Oxford, 1977.

13. D.M. Edwards, F. Federici, J. Mathon, and A. Umerski, *Phys. Rev. B*, 71, 054407, 2005.

14. J.A. Katine, F.J. Albert, R.A. Buhrman, E B. Myers, and D.C. Ralph, *Phys. Rev. Lett.*, 84, 3149, 2000.

15. S.I. Kiselev, J.C. Sankey, I.N. Krivorotov, N.C. Emley, R.J. Schoelkopf, R.A. Buhrman, and D.C. Ralph, *Nature*, 425, 380, 2003.

16. S. Urazhdin, N.O. Birge, W.P. Pratt, Jr., and J. Bass, *Phys. Rev. Lett.*, 91, 146803, 2003.

17. M.R. Pufall, W.H. Rippard, S. Kaka, S.E. Russek, T.J. Silva, J. Katine, and M. Carey, *Phy. Rev. Lett.*, 69, 214409, 2004.

18. E.B. Myers, F.J. Albert, J.C. Sankey, E. Bonet, R.A. Buhrman, and D.C. Ralph, *Phys. Rev. Lett.,* 89, 196801, 2002.

19. M.A. Zimmler, B. Özyilmaz, W. Chen, A.D. Kent, J.Z. Sun, M.J. Rooks, and R. H. Koch, *Phys. Rev. B*, 70, 184438, 2004.

20. M. Tsoi, A.G.M. Jansen, J. Bass, W.C. Chiang, M. Seck, V. Tsoi, and P. Wyder, *Phys. Rev. Lett.*, 80, 4281, 1998.

21. M. Tsoi, A.G.M. Jansen, J. Bass, W.C. Chiang. V. Tsoi, and P. Wyder, *Nature*, 406, 46, 2000.

22. E.B. Myers, D.C. Ralph, J.A. Katine, R.N. Louie, and R.A. Buhrman, *Science,* 285, 867, 1999.

23. J.C. Slonczewski, *J. Magn. Magn. Mater.*, 195, L261 1999; 247, 324, 2002.

24. B. Özyilmaz, A.D. Kent, D. Monsma, J.Z. Sun, M.J. Rooks, and R.H. Koch, *Phys. Rev. Lett.*, 91, 067203, 2003.

25. M. Tsoi, J.Z. Sun, M.J. Rooks, R.H. Koch, and S.S.P. Parkin, *Phys. Rev. B*, 69, 100406(R), 2004.

26. W.F. Brown, *Phys. Rev. B*, 130, 1677, 1963.

27. Z. Li and S. Zhang, *Phys. Rev. B*, 69, 134416, 2004.

28. J. Miltat, G. Albuquerque, A. Thiaville, and C. Vouille, *J. Appl. Phys.*, 89, 6982, 2001.

29. Z. Li and S. Zhang, *Phys. Rev. B*, 68, 024404, 2003.

30. C.R.P. Cowburn, C.K. Koltsov, A.O. Adeyeye, and M.E. Welland, *Phys. Rev. Lett.*, 83, 1042, 1999.

31. Y. Jiang, S. Abe, T. Ochiai, T. Nozaki, A. Hirohata, N. Tezuka, and K. Inomata, *Phys. Rev. Lett.*, 92, 167204, 2004.

32. Y. Jiang, T. Nozaki, S. Abe, T. Ochiai, A. Hirohata, N. Tezuka, and K. Inomata, *Nat. Materi.*, 3, 361, 2004.

33. N. Tezuka, private communication.

34. K. Capelle and B.L. Gyorffy, *Europhys. Lett.*, 61, 354, 2003.

35. J. Sun, *Nature*, 424, 359, 2003.
36. J. Mathon, Murielle Villeret, A. Umerski, R B. Muniz, J. d'Albuquerque e Castro, and D.M. Edwards, *Phys. Rev B,* 56, 11797, 1997.
37. J. Mathon, Murielle Villeret, R.B. Muniz, J. d'Albuquerque e Castro, and D.M. Edwards, *Phys. Rev. Lett.*, 74, 3696, 1995.
38. S. Zhang, P.M. Levy, and A. Fert, *Phys. Rev. Lett.*, 88, 236601, 2002.
39. A.A. Kovalev, A. Brataas, and G.E. W. Bauer, *Phys. Rev. B*, 66, 224424, 2002.
40. D.M. Edwards and J. Mathon, in *Nanomagnetism: Multilayers, Ultrathin Films and Textured Media,* J. A.C. Bland and D.L. Mills, Eds., Elsevier, New York, 2006.
41. L.V. Keldysh, *Sov. Phys. JETP*, 20, 1018, 1965.
42. C. Caroli, R. Combescot, P. Nozieres, and D. Saint-James, *J. Phys. C*, 4, 916, 1971.
43. D.M. Edwards in *Exotic States in Quantum Nanostructures,* S. Sarkar, Ed., Kluwer Academic Press, Norwell, MA, 2002.
44. G.D. Mahan, Many Particle Physics, 2nd ed., Plenum Press, New York, 1990.

8 Domain Wall Scattering and Current-Induced Switching in Ferromagnetic Wires—Experiment

Serban Lepadatu and Yongbing Xu

CONTENTS

8.1 INTRODUCTION

Research into spintronics and magnetoelectronic devices has seen growing interest in obtaining devices such as magnetic random access memory (MRAM)[1] and magnetic logic gates,[2] which can be controlled directly by use of applied voltages and currents rather than magnetic fields. This is partly due to the drawbacks associated with the use of magnetic fields, which include cross talk, high power consumption, and decreased integration densities.[3] The current design of MRAM chips involves the use of magnetic field-generating current lines to switch the magnetization state of a spin valve, a multilayered structure consisting of a layering of ferromagnetic and nonmagnetic films whose resistance states can be changed between antiferromagnetic (AFM) coupling — high resistance — and ferromagnetic (FM) coupling — low resistance — by use of applied magnetic fields.[4] The difference in resistance between the AFM and FM coupling states is understood in terms of the giant magnetoresistance (GMR) effect.[5] In MRAM chips, spin valves are used in an array as bit storage elements. Because a spin valve can be switched between the two

179

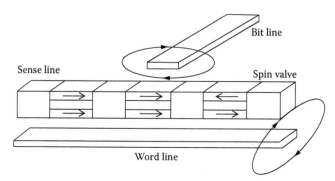

FIGURE 8.1 MRAM with field-generating current lines.

states — AFM coupled or FM coupled — this makes it a natural device to differentiate between logic 0 and logic 1. The writing process involves switching the magnetization of a single spin valve from FM coupled to AFM coupled (and vice versa) using two current lines to generate magnetic fields, called the bit and word lines, as shown in Figure 8.1. The spin valve is tailored so that one of the ferromagnetic layers is strongly pinned, requiring high magnetic fields to switch its magnetization; whereas the second layer is weakly pinned, called a soft layer, requiring lower values of magnetic fields to switch its magnetization. An individual spin valve in an array is accessed by first applying a current pulse on the word line to rotate the magnetic moments in the soft layer of all the spin valves on a row by 90°, followed by a current pulse on the bit line, which performs a similar function for all the spin valves on the respective column. Consequently, at the intersection of the word and bit lines, the magnetization of the soft layer is rotated by 180°, which gives rise to a resistance change as the coupling state of the spin valve is changed. The reading process uses a "destroy and restore" scheme to sense the resistance of an individual spin valve element. To detect the resistance, and hence the logical state, of an individual element, current is run through the "sense line," whose resistance is the sum of the individual elements. Amplifiers at the ends of the sense lines can detect changes in resistance as the spin valve is switched from either FM to AFM or AFM to FM coupled. The use of magnetic field-generating current lines does not lend itself to the high integration densities that would be expected from the next generation of random access memory chips, mainly due to the problems associated with cross talk. Furthermore, if spintronics devices are to be interfaced with existing semiconductor-based devices, the use of voltages and currents alone is essential. One method of achieving this goal is by use of current-induced switching in single- and multilayer magnetic devices, which is the focus of extensive research.[6–14] Following theoretical predictions by Slonczewski[6] and Berger,[7] current-induced magnetization reversal has been observed in multilayered devices.[8] A spin-polarized current is passed perpendicular to two magnetic thin films separated by a metallic spacer, resulting in a rotation of the magnetization of the free layer, which was explained by a spin-transfer torque mechanism.[8] This has important implications for MRAM based on the GMR effect. It is expected that by use of current-induced switching to control the coupling state of the spin valves, the need for field-generating current lines will be removed. Another method of switching the magnetic configuration was demonstrated recently,[9] where current-induced domain wall motion was used to unpin a domain wall from a constriction in a spin-valve structure. This effect was predicted theoretically by Berger[10] and confirmed experimentally.[11] In ferromagnetic metals, the interaction between itinerant electrons and a domain wall can give rise to domain wall motion due to the *s-d* exchange torque exerted by the current-carrying electrons on the domain wall magnetic configuration. Gan et al.[12] have demonstrated this effect in NiFe thin films, by using magnetic force microscopy (MFM) imaging to show the displacement of Bloch walls when direct current (DC) current pulses are applied. It was found that current densities of the order 10^7 A/cm^2 are required to displace a domain wall and its motion is always in the direction of the current carriers.

The definition of artificial pinning sites, such as lateral constrictions, allows for a single domain wall to be trapped and characterized. In a recent experiment,[14] discontinuous resistance changes in current-voltage (*I-V*) measurements of ferromagnetic wires of $Ni_{80}Fe_{20}$ and Ni patterned with a nanoconstriction have been observed. A trapped domain wall at the constriction is removed by current-induced domain wall movement. The difference in resistance between the states with and without a domain wall trapped at the constriction allows for a direct measurement of its resistance contribution. The study of domain wall contributions to magnetoresistance (DWMR) has seen a growing interest, fueled by progress in nanofabrication techniques. In recent studies, negative resistance contributions were observed in microfabricated ferromagnetic wires.[15–17] On the other hand, a positive contribution was observed for an in-plane domain wall in a Gd layer sandwiched between two NiFe layers.[18] Other positive contributions of domain walls to magnetoresistance (MR) have also been shown recently,[19–23] which were discussed in terms of a spin-dependent scattering mechanism. A summary of the main DWMR experimental studies and results is shown in Table 8.1. There are currently two main theoretical treatments for DWMR: for positive contributions, a model based on spin-dependent impurity scattering was proposed by Levy and Zhang,[24] and negative contributions are accounted for by the loss of weak localization of the electrons as shown by Tatara and Fukuyama.[25] On the other hand, large MR changes and nonlinear *I-V* characteristics have been shown recently in nanocontacts of ferrimagnetic crystals[26] and ballistic nanocontacts,[27] where point contacts of only a few atoms give rise to the so-called ballistic magnetoresistance (BMR). This was explained by domain wall scattering due to the inability of the electron spin to accommodate itself adiabatically from one side of the domain wall to the other when the Fermi wavelength, λ_F, is comparable to the domain wall width.[28]

In this chapter, we will discuss the current-induced domain wall motion mechanism and domain wall scattering in single-layer patterned ferromagnetic wires. In Section 8.2, we will discuss some of the experimental techniques used in the fabrication of devices on the micrometer and nanometer scales. In Section 8.3, a method is described by which direct measurements of the domain wall contribution to resistance at zero fields are made. In Section 8.4, the domain wall structure at constrictions and junctions of patterned devices is studied using micromagnetic simulations and MFM imaging. The current-induced domain wall movement is

TABLE 8.1
Summary of Main DWMR Studies

$\Delta\rho$	Temperature	DWMR	Samples	Reference
–	RT	1.4×10^{-3}	NiFe wires	15
–	5 K	5×10^{-4}	Co zigzag wires	16
–	65 K	1×10^{-3}	Fe wires	17
+	RT	1.7×10^{-2}	NiFe constricted wires	14
+	RT	1.82×10^{-2}	Ni constricted wires	14
+	77 K	0.23	NiFe/Gd/NiFe structure	18
+	RT	2×10^{-2}	NiFe crosses	19
+	RT	3.4×10^{-3}	NiFe zigzag wires	20
+	RT	3×10^{-5}	Ni films	21
+	77 K	$1–3 \times 10^{-4}$	Co wires	22
+	RT	5×10^{-3}	Co films	23

RT — room temperature.

studied in Section 8.5 and domain wall scattering in Section 8.6. The interaction between applied measurement currents and magnetic fields in single-layer ferromagnetic wires patterned with a nanoconstriction is studied in Section 8.7. Finally, Section 8.8 contains a brief summary.

8.2 NANOFABRICATION

In this section, we will discuss the experimental techniques used to obtain the devices whose properties we study in subsequent sections. Lithography combined with liftoff is one of the most widely used techniques for obtaining small structures due to the very wide range of devices that can be obtained. There are two types of lithographical processes namely optical and electron beam (e-beam) lithography. With optical lithography, device sizes on the micrometer scale are obtained, limited mainly by the wavelength of light used, whereas with e-beam lithography, devices on the nanometer scale are obtainable. The principle is the same for both processes. A substrate, usually a semiconductor substrate such as Si or GaAs, is coated with a substance known as a resist. The resist is then exposed in a controlled manner using either light or an electron beam source. This is followed by development of the resist, which results in the exposed parts of the resist being washed away (positive resist) or, inversely, not washed away (negative resist). Deposition of a material, usually by thermal evaporation or molecular beam epitaxy (MBE) growth, results in a layered structure with some of the material in direct contact with the substrate and the rest on top of the resist. Using a substance, usually acetone, completely washes away the resist together with the material on top of it leaving behind a patterned structure. This process is shown in Figure 8.2 for clarity. The above procedure may be repeated several times to obtained multilayered devices. In a recent experiment, a set of necked $Ni_{80}Fe_{20}$ and Ni wires have been fabricated using a

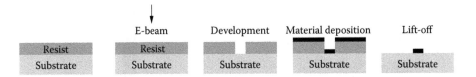

FIGURE 8.2 E-beam lithography and liftoff technique.

standard e-beam lithography and liftoff process.[14] The wires were defined as being 400 μm long, 1 μm wide, 30 nm thick, and with a bow-tie constriction in the center with the width varying from 50 to 350 nm in 50 nm steps. A typical wire with a 50 nm constriction width is shown in Figure 8.3 together with the electrical measurement pads. Thus, for the structures shown in Figure 8.3, the necked wires were obtained by thermal evaporation of $Ni_{80}Fe_{20}$ and Ni, 30 nm thick, using a first level of e-beam lithography combined with ultrasonic assisted liftoff in acetone. To allow for electrical measurements, in a second level of e-beam lithography, the measurement pads were defined with the geometry shown in Figure 8.3a, consisting of thermally evaporated Al, 150 nm thick. A good alignment of the measurement pads on top of the necked wires obtained in the first level was required as shown in Figure 8.3b. This is achieved by defining marks in the first level of lithography, usually known as alignment crosses, which can be scanned and used to precisely align the exposure position in the second level of lithography. Finally electrical contacts are made to the measurement pads to allow for electrical measurements to be carried out.

In another recent experiment, the properties of micrometer and submicrometer cross structures were investigated.[19] The typical geometry of the devices used is shown in Figure 8.4, displaying both the cross structures and the measurement pads used.

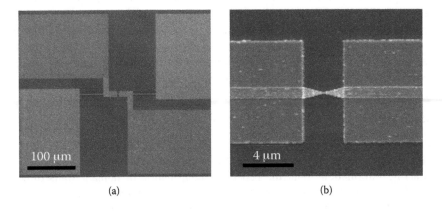

(a)   (b)

FIGURE 8.3 Scanning electron microscopy images of necked ferromagnetic wires showing (a) measurement pads geometry and (b) 50 nm constriction width. (From S. Lepadatu and Y.B. Xu, *Phys. Rev. Lett.,* 92, 127201, 2004. With permission.)

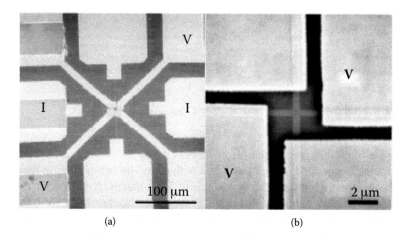

FIGURE 8.4 SEM image of cross structures showing (a) measurement pads geometry and (b) junction area. (From Y.B. Xu et al., *Phys. Rev. B*, 61, 14901, 2000. With permission.)

The crossed wires are composed of $Ni_{80}Fe_{20}$, 30 nm thick, the length of the arm of the crossed wires is fixed at 100 μm and the width varies from 0.2 to 10 μm. The measurement pads are placed very close to the junction area (<2 μm away), as shown in Figure 8.4B. These structures were obtained using a pattern transfer technique similar to the liftoff technique of Figure 8.2. Thus, instead of first depositing a resist on top of the substrate, the material is deposited directly on the substrate using MBE growth. As a side note, the main reason for using pattern transfer rather than liftoff is the possibility of growing high-quality thin films in an ultrahigh vacuum (UHV) chamber using MBE growth, an option not available with the simple liftoff technique as any resist substance will impair the quality of the growth process in the UHV chamber. After the thin film is grown on the substrate, lithography combined with liftoff is used to define a structure known as a mask on top of the thin film. Using a selective etching process, the part of the thin film not covered by the mask is etched away, leaving a layered structure consisting of the thin film covered by the mask. Using another etching process, usually ion beam etching, the mask is also removed, leaving the desired patterned structure on the substrate.

8.3 NONLINEAR *I-V* AT ZERO MAGNETIC FIELD

In this section, the *I-V* characteristics of the necked wires in zero applied magnetic fields are investigated. The main motivation of this study was the possibility of extracting information about the resistance contribution of domain walls directly from *I-V* measurements. It is assumed that due to the necked geometry, a domain wall will be formed at the constriction at the zero magnetic field after reversal from saturation. This assumption will be discussed in detail in Section 8.4. The trapped domain wall is then unpinned from the constriction using a current-induced domain wall movement mechanism at zero field, discussed in Section 8.6. The difference in

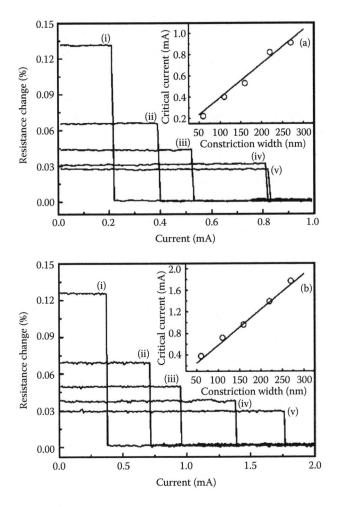

FIGURE 8.5 *I-V* measurements for (a) necked $Ni_{80}Fe_{20}$ wires and (b) necked Ni wires with constriction widths of (i) 50 nm, (ii) 100 nm, (iii) 150 nm, (iv) 200 nm, (iv) 250 nm. In the insets to a and b, the average critical current density is plotted as a function of constriction width for $Ni_{80}Fe_{20}$ and Ni, respectively. (From S. Lepadatu and Y.B. Xu, *Phys. Rev. Lett.*, 92, 127201, 2004. With permission.)

resistance between the state where a domain wall is pinned at the constriction and the state where the domain wall is removed can be used to obtain a direct measurement of its resistance contribution. This has the advantage of canceling any additional contributions, such as anisotropic magnetoresistance (AMR) and Hall effect contributions, which can lead to misinterpreted domain wall scattering in DWMR studies. Figure 8.5a and Figure 8.5b show the resistance vs. applied current for the necked wires with point contact widths 50 to 250 nm for $Ni_{80}Fe_{20}$ and Ni, respectively. The *I-V* measurements were performed at zero magnetic field after reversal from saturation in the longitudinal geometry (magnetic field parallel to current direction, hence parallel to the wire) in all cases. A sharp drop in resistance is observed for the necked wires with point contact width in the range of 50 to 250 nm at a critical current termed the switching current; whereas for the wires with point contact width of 300 and 350 nm, no drop in resistance occurs, showing simply an ohmic I-V characteristic as for a simple wire (the latter not shown). By comparing the switching currents with the cross-sectional areas at the constriction, it is found that the current density required to change the resistance is constant and of the order $10^7 \, A/cm^2$ as shown on the insets to Figure 8.5a and Figure 8.5b, namely $1.1 \times 10^7 \, A/cm^2$ for $Ni_{80}Fe_{20}$ and $2.2 \times 10^7 \, A/cm^2$ for Ni, respectively. Also, the percentage changes in resistance are seen to decrease monotonically with increasing point contact width, namely from 0.13 to 0.024% for $Ni_{80}Fe_{20}$ and from 0.124 to 0.028% for Ni, for constriction widths from 50 to 250 nm, respectively. Upon reversal of current direction, similar behavior is observed showing that the current direction as used in the I-V measurements is not essential, as might be expected from the symmetry of the necked wires. The curves in Figure 8.5 show a flat characteristic, above and below the switching current, with resistance values within the noise margins of the equipment, so that we may conclude that self-heating does not occur to any measurable extent, as a monotonic increase in resistance would be observed with larger applied current if self-heating did occur. The nonlinear I-V characteristics observed for the wires with constriction width in the range of 50 to 250 nm are in marked contrast with the I-V measurements observed for constriction widths above 250 nm. This behavior is best explained as a domain wall contribution to resistance at zero magnetic fields, associated with current-induced domain wall movement in the constriction area. However before this mechanism is detailed, we must first test and discuss the assumption of domain wall pinning at the zero magnetic field.

8.4 DOMAIN WALLS IN PATTERNED STRUCTURES

The formation of a domain wall at constrictions of necked wires after reversal from saturation is tested and demonstrated using micromagnetic simulations and MFM imaging, respectively. Extensive micromagnetic simulations have been performed on the geometry of the samples for both $Ni_{80}Fe_{20}$ and Ni using the Object Oriented Micro-Magnetics Framework (OOMMF) software.[29] In the simulations, a cell size of 30 nm was used and the wire length was kept fixed at 13 μm, rather than 400 μm as for the actual wires, to allow for reasonable simulation times. All the other dimensions have been kept the same, namely 1 μm wide, 30 nm thick, and a constriction in the center with width varying from 50 to 350 nm in 50 nm steps. The simulations have shown the formation of a 180° domain wall at the narrowest part of the constriction at the zero applied magnetic field as shown in Figure 8.6 for the $Ni_{80}Fe_{20}$ wires with constriction widths in the range of 50 to 250 nm. Similar results have also been obtained for the simulated Ni wires. For the necked wires with constriction width 300 and 350 nm, no domain wall is formed, showing simply a parallel alignment of magnetization in the arms either side of the constriction. The rotation of magnetization occurs in the plane, as for a Néel wall, because the thickness of the ferromagnetic layer prevents any Bloch walls from forming. This magnetic configuration was found to be stable for constriction widths up to 250 nm,

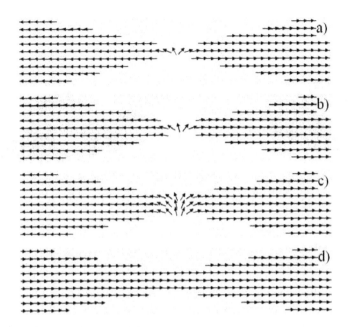

FIGURE 8.6 Micromagnetic simulations for necked $Ni_{80}Fe_{20}$ wires with constriction widths of (a) 50 nm, (b) 100 nm, (c) 250 nm, and (d) 300 nm. (From S. Lepadatu and Y.B. Xu, *Phys. Rev. Lett.*, 92, 127201, 2004. With permission.)

for both $Ni_{80}Fe_{20}$ and Ni, and any further increase in constriction width simply resulted in a parallel alignment of the magnetization in both arms, with no domain walls pinned at the constriction. This is reflected in the results obtained from *I-V* measurements (Figure 8.5), where a change in resistance was observed for samples with nominal constriction width of 50 to 250 nm, whereas the samples with constriction widths of 300 and 350 nm were characterized by a linear *I-V*. The formation of a domain wall in the constriction area at zero field is also observed directly using MFM images. In Figure 8.7, a MFM image for a 50 nm constriction of $Ni_{80}Fe_{20}$ is shown, taken at zero magnetic fields after reversal from saturation in the longitudinal geometry. Thus first a large enough magnetic field is applied along the wire to ensure alignment of the magnetic moments in the direction of the applied field, following which the magnitude of the magnetic field is reduced to zero. The sharp contrast observed at the smallest part of the constriction reveals the presence of a domain wall at zero fields. The formation of a domain wall at low magnetic fields upon reversal from saturation is perhaps surprising, as the wires investigated are symmetrical on either side of the constriction. However, as shown by Bruno,[30] the constriction acts as a pinning potential. To minimize the total energy, a domain wall will naturally be attracted toward the center of the constriction. As the magnitude of the magnetic field is reduced from saturation toward zero, the magnetic configuration is relaxed, giving rise to vortex formations at the constriction because of the necked geometry.[31] Due to the pinning potential created, any such vortex formation will slide toward the center of the constriction, giving rise to a domain wall.

FIGURE 8.7 MFM image of necked $Ni_{80}Fe_{20}$ wire with a 50 nm constriction.

The controlled pinning of a domain wall has also been realized in the $Ni_{80}Fe_{20}$ cross structures.[19] The domain structures of the crosses have been imaged at the zero magnetic field using MFM. The images are reproducible and represent the domain structures of the stable configurations at minimum magnetic energies. In Figure 8.8a and Figure 8.8b, the MFM images of the 0.5 and 1 μm crosses are shown, respectively. In the wire region, a single domain state can be identified from the absence of any contrast. This is because the spins in the wire region are aligned along the wire direction due to the strong shape anisotropy. However, the images show significant contrast across the junction area with diagonal patterns visible. This result indicates the confinement of domain walls at the junction of the crosses. Micromagnetic calculations using the OOMMF software have been performed to simulate the images observed and to determine the detailed domain configuration around the junction. The simulated region is 10×10 μm, i.e., the full length of the arms from one side to the other is 10 μm. A cell size of 30 nm was used, with the starting condition such that the initial magnetization in the junction region of about 3×3 μm² is completely random and the magnetization in the middle of the arms is preferentially aligned along the wire direction. This is justifiable as in the real samples with arms of 100 μm in length, the strong shape anisotropy will keep the magnetization aligned along the wire direction. In Figure 8.8c and Figure 8.8d, the simulations for the 0.5 and 1 μm crosses around the junction are shown. The detailed magnetization configuration is shown in Figure 8.9. A parallel state, in which the spins of the arms on opposite sides of the junction are oriented parallel, was found to be the most stable state. Both the simulated images in Figure 8.8c and Figure 8.8d show a rather similar pattern of change in contrast across a diagonal line, which agrees well with the key feature of the MFM images observed. As a further check, in Figure 8.8e, a large-scale MFM image of a 1 μm wide and 10 μm long cross structure is shown. The charges observed at the wire ends confirm the simulations. Figure 8.9 shows that the magnetization rotates by 45° from the arms to the junction, and all the spins within the junction are aligned approximately along the diagonal

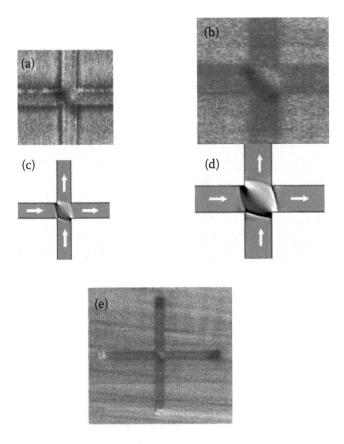

FIGURE 8.8 MFM images of cross structures with arm widths of (a) 0.5 μm and (b) 1 μm and simulated MFM patterns for (c) 0.5 μm and (d) 1 μm crosses. A large-scale MFM image for a 1 μm cross with 10 μm arm lengths is shown in (e). (From Y.B. Xu et al., *Phys. Rev. B*, 61, 14901, 2000. With permission.)

direction. The junction can thus be described as a giant coherent spin block (CSB), essentially forming a 45° domain wall between the arm and the junction.

8.5 CURRENT-INDUCED DOMAIN WALL MOVEMENT

As predicted by Berger[10] and confirmed experimentally,[11] domain walls can be displaced under the influence of a spin-polarized current of sufficient current density. This effect is due to the exchange torque exerted by electrons in the conduction band on the constituent electrons of the domain wall. The net effect is a displacement of the domain wall in the same direction as that of the current carriers, known as current-induced *s-d* exchange interaction. For the necked ferromagnetic wires of Figure 8.3, the difference between the two resistance states observed in the *I-V* measurements is therefore attributed to the removal of the domain wall from the constriction at a critical current density of the order 10^7 A/cm^2 (see Figure 8.5).

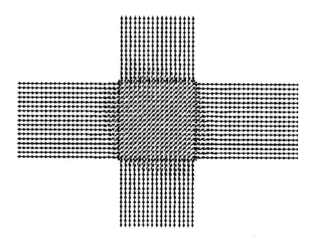

FIGURE 8.9 Detailed domain configuration for a 0.5 μm cross obtained from micromagnetic simulations. (From Y.B. Xu et al., *Phys. Rev. B,* 61, 14901, 2000. With permission.)

In the low resistance state, no domain wall is pinned at the constriction; whereas in the high resistance state, the presence of a domain wall must be responsible for the additional resistance observed. Thus the difference between the two states can be used to obtain a direct measurement of the domain wall resistance, without the use of an applied magnetic field. Three mechanisms have been proposed for current-induced domain wall motion.[12] The first is known as the hydromagnetic domain wall drag, which is based on the Hall effect.[32] The direction of domain wall displacement is dependent on the anomalous Hall coefficient and the direction of the applied current. For $Ni_{80}Fe_{20}$ and Ni at room temperature, the anomalous Hall coefficient assumes a positive value resulting in domain wall displacement in the same direction as that of the current carriers. This effect is strongly dependent on the thickness of the material and Gan et al.[12] have shown that it becomes the dominant effect responsible for domain wall drag in NiFe samples with a thickness greater than 1 μm. This reduces to zero for very thin samples and its effect is negligible in the wires considered here, which have a thickness of 30 nm. The second mechanism is due to the current-induced magnetic field or Oersted field, which runs in closed loops perpendicular to the direction of current flow, studied by Hung and Berger for NiFe thin films.[33] They have shown that this mechanism is not present in films with a thickness smaller than 35 nm. The third mechanism that can give rise to domain wall movement, discussed above, is due to the exchange interaction between $3d$ electrons in the material and $4s$ electrons in the conduction band. A spin-polarized current will exert a torque on the electrons in the domain wall, effectively resulting in a displacement of the domain wall, which is in the same direction as the current flow and its effect is independent of sample thickness. Experimental investigations of this effect,[12] carried by applying current pulses to thin film samples and observing the motion of domain walls by Kerr microscopy, have shown that the current density required to move a domain wall is of the order 10^7 A/cm^2, which is in excellent agreement with the results discussed here. In this case it was not necessary to apply current pulses, as the required current density can be reached by applying DC currents without heating the wires, due to the small cross-sectional area at the constriction.

The current-induced domain wall movement due to s-d exchange interaction was also used to explain the results obtained for CoO/Co/Cu/NiFe spin-valve stripes patterned with a notch.[9] The stripes have been defined with a fixed length of 20 μm and a notch positioned two-thirds along the stripe, dividing the stripe into two arms of different lengths on either side of the notch. By taking a magnetic field loop, with the field direction along the stripe, a head-to-head or tail-to-tail 180° domain wall is pinned at the notch depending on the magnetic field history. Thus starting from an antiparallel state, the magnetic field is increased until the magnetization direction in the longer arm switches first causing a tail-to-tail domain wall to be trapped at the constriction. Further increasing the magnetic field switches the spin-valve completely into the parallel state, as the magnetization in the shorter arm also rotates. Similarly a tail-to-tail domain wall is formed as the spin valve is switched from the parallel to the antiparallel state when the magnetic field is decreased from positive to negative values. A difference in resistance between the parallel and antiparallel states is observed due to the GMR effect. As the domain wall is trapped at the notch, current-induced domain wall movement is used to displace the domain wall. This is reflected in the discontinuous resistance changes observed in the I-V measurements, as the spin valve is switched from the antiparallel to the parallel state (resistance drop) and from the parallel to the antiparallel state (resistance increase). It was again found that the current density required to displace the domain walls is about 10^7 A/cm^2, but one interesting observation that is yet to be explained is the conclusion that the direction of domain wall displacement is independent of current direction, which is contrary to theoretical predictions.[10]

8.6 DOMAIN WALL SCATTERING

The model proposed by Levy and Zhang,[24] suggests that positive contributions of domain walls to resistance may be accounted for by a spin-dependent impurity scattering mechanism. The scattering of electrons in domain walls produces a mixing of the currents in the spin-up and spin-down conduction channels and the additional resistance introduced by domain walls is then due to the different resistivities experienced by spin-up and spin-down electrons. For a current perpendicular to the domain wall, Equation 8.1 may be used to calculate the resistivity ratio[24]

$$R_{CPW} = \frac{\xi^2}{5} \frac{(\rho_0^\uparrow - \rho_0^\downarrow)^2}{\rho_0^\uparrow \rho_0^\downarrow} \left(3 + \frac{10\sqrt{\rho_0^\uparrow \rho_0^\downarrow}}{\rho_0^\uparrow + \rho_0^\downarrow} \right) \tag{8.1}$$

In the above equation, $\xi = h^2 k_F / 16\pi m d J$, where d is the thickness of the domain wall. Taking commonly accepted values of $k_F = 1$ Å^1, $J = 0.5$ eV, $d = 30$ nm, and $\rho_0^\uparrow / \rho_0^\downarrow = 5 - 20$ for typical ferromagnetic materials of NiFe and Ni at room temperature,[24] we

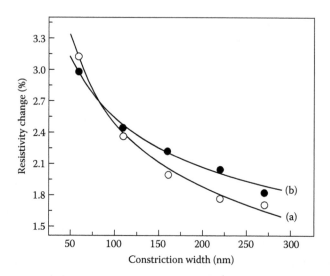

FIGURE 8.10 Percentage change in resistivity vs. constriction width for (A) $Ni_{80}Fe_{20}$ and (B) Ni. The dots are the experimental results and the lines are the simulations. (From S. Lepadatu and Y.B. Xu, *Phys. Rev. Lett.*, 92, 127201, 2004. With permission.)

find this ratio ranges from 0.7 to 3%. The resistance ratios $\Delta R/R_0$, that is, resistance change (high-state resistance minus low-state resistance value) divided by the low-state resistance value, were obtained directly from the *I-V* data in Figure 8.5. To compare the results obtained from *I-V* measurements with the theoretical predictions from spin scattering theory, it is necessary to convert the resistance ratios into resistivity ratios $\Delta \rho/\rho_0$ by solving for the particular current distribution in the geometry of the wires for both NiFe and Ni, assuming a domain wall thickness of 30 nm.[34] The results obtained are shown in Figure 8.10A and Figure 8.10B. Empirical numerical relations may be obtained by curve-fitting these points using Equation 8.2, where S is the cross-sectional area at the constriction. For NiFe and Ni, the constant K takes the values of 1.98×10^{-8} and 1.12×10^{-6} respectively, and the exponent α has values of 0.42 and 0.3, respectively

$$\frac{\Delta \rho}{\rho_0} = \frac{K}{S^\alpha} \qquad (8.2)$$

For the largest constriction widths, the percentage changes in resistivity are within the range predicted by spin scattering theory. As the constriction width is decreased, however, the changes in resistivity with the presence of a domain wall are seen to increase rapidly. For constriction widths of only a few atoms across, the percentage changes would then be expected to be very large, as found by Versluijs et al.[26] and García et al.[27] in nanocontacts using BMR. For a contact area of 1 nm^2, the percentage changes in resistivity due to domain wall scattering, as predicted by the empirical relations of Equation 8.2, would then be ~70% for NiFe and ~30% for Ni. These predictions are in good agreement with the large values found for ballistic nanocontacts[27] at room temperature for a nominal contact area of 1 nm². Possible links between domain wall scattering and BMR have been discussed theoretically in a previous work.[28] In particular, Tatara et al.[28] have shown that a domain wall trapped in a constriction region will strongly scatter electrons if the Fermi wavelength, λ_F, is comparable to the domain wall width. This is due to the inability of the electron spin to accommodate itself adiabatically from one side of the domain wall to the other. These experiments show that there is a correlation between BMR and spin-dependent impurity scattering. Figure 8.10 shows that as the constriction width is increased, the percentage changes in resistivity are in good agreement with the theoretical predictions of spin-dependent impurity scattering; whereas for very small constriction widths, they approach the values obtained from BMR experiments. The region in between these two extremes must then be explained as a combination of domain wall scattering and BMR effects. By considering the samples with a constriction width of 250 nm where the contributions due to BMR are negligible, we obtain values of DWMR of 1.7% for $Ni_{80}Fe_{20}$ and 1.82% for Ni with a 180° domain wall.

The domain wall trapped at the junction of the cross structures studied in Section 8.4 has also been found to give rise to a positive contribution to MR due to domain wall scattering. In Figure 8.11, longitudinal MR measurements of the 0.5 μm cross sample are shown. The magnetic field is swept between −600 and 600 Oe in Figure 8.11a. Two striking features can be seen from this figure: (i) the resistance at points A, B, B′, and D is approximately the same, and (ii) the curve shows two jumps, namely B-C, and C′-D in the reversal magnetization process. When the field is reduced to zero from saturation, the spins of the vertical arms rotate from the horizontal direction at point A to the vertical direction at point B, that is, along the wire direction, as shown by a schematic diagram in Figure 8.11b. The similar resistance of points A and B shows that the switching of the vertical arms contributes little to the resistance change and the CSB in the junction maintains alignment along the wire. When the reversal field is increased to ~150 Oe, a first drop in the resistance occurs in going from B to C, indicating the switching of the CSB close to the diagonal direction. In the field range C to C′, the spin block switches between the two diagonal directions. When the field is further increased to ~400 Oe, the spin block undergoes a second switching between C′ and D, leading to a sharp increase of the resistance. The MR effect observed in the cross comes mainly from two sources: (i) the AMR effect due to the switching of the spin block in the junction, and (ii) the possible intrinsic spin-dependent scattering of the electrons propagating through regions of spatially varying magnetization or the so-called DWMR. The resistance change due to the AMR effect between the configurations B and C can be estimated using Equation 8.3

$$(\Delta R / R) = (W / D) \times AMR_{NiFe} \times Cos^2(\theta) \qquad (8.3)$$

In the above equation, $W = 0.5$ μm is the width of the wire and $D = 3.4$ μm is the separation of the voltage pads, $AMR_{NiFe} = 1.4\%$ is the magnitude of the AMR effect

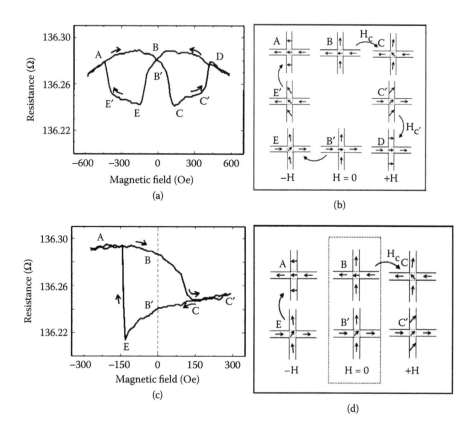

FIGURE 8.11 The longitudinal MR curves of the 0.5 μm straight cross taken with (a) a full loop, (c) maximum applied field reduced to about 300 Oe with the respective CSB switching schematics shown in b and d. (From Y.B. Xu et al., *Phys. Rev. B*, 61, 14901, 2000. With permission.)

of the 0.5 μm cross determined from transverse MR measurements, and θ is the angle between the magnetization and current directions. As shown in Figure 8.11b, the CSB is expected to be aligned close to the diagonal direction in C, that is, $\theta \sim 45°$, which gives a MR value of -1.0×10^{-3}. This is much larger than the observed value of $-3.0 \pm 0.2 \times 10^{-4}$, indicating a significant contribution of a positive intrinsic MR due to spin-dependent scattering of the DW. The AMR contribution may be measured precisely as shown by the MR curve in Figure 8.11c. Here the field is swept between about −300 to 300 Oe starting from the saturation state A, with the switching schematic shown in Figure 8.11d. As the maximum reversal field is smaller than the field at the point C′, the second jump from C′ to D does not occur in this controlled MR measurement. When the reversal field is reduced to zero, a new configuration B′ strikingly different from that of B is created. The CSB in B′ is orientated along the diagonal direction corresponding exactly to the detailed spin configuration established by the results of micromagnetic calculations shown in Figure 8.9, which allows us to calculate quantitatively the AMR contribution. A numerical calculation of the AMR effect is made by further considering the current distribution in the

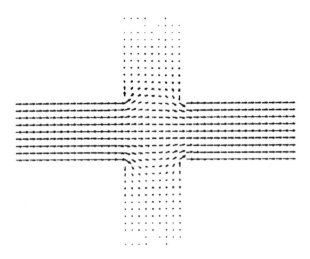

FIGURE 8.12 Current distribution for the 1 μm cross calculated using a finite element method. (From Y.B. Xu et al., *Phys. Rev. B*, 61, 14901, 2000. With permission.)

sample, which is calculated using a finite element method as shown in Figure 8.12. The AMR ratio is then calculated using Equation 8.4

$$(\Delta R / R) = AMR_{NiFe} \times \sum Cos^2(\theta_i) \tag{8.4}$$

where θ_i is the angle between the current and magnetization directions at the unit cell i. The resistance change due to the AMR effect between configurations B and B' is -6.0×10^{-4} integrated over a distance between two voltage pads according to Equation 8.4. This value is smaller than that estimated from Equation 8.3, which may be due to the fact that about 20% of current flows through the vertical arms and that there is a small perpendicular component of the current within the junction. As the AMR contribution of -6.0×10^{-4} is about twice as large as the measured value of $-3.2 \pm 0.2 \times 10^{-4}$, we can thus conclude that the 45° domain wall between the arms and the junction contributes a positive MR with a magnitude of the order of 10^{-4}, that is, the intrinsic spin-dependent scattering for the electrons propagating through regions of spatially varying magnetization leads to an increase of resistance. Solving for the current distribution in the cross-structure geometry and assuming a domain wall thickness of 30 nm,[34] we obtain a resistivity ratio of 2×10^{-2} for the 45° domain wall, comparable to the ratios obtained for the necked $Ni_{80}Fe_{20}$ and Ni wires.

8.7 MAGNETORESISTANCE AND MEASUREMENT CURRENT INTERACTION

It is also of interest to examine the MR response of the samples, as this can reveal details of the magnetization switching and domain wall nucleation. The results from MR measurements of the $Ni_{80}Fe_{20}$ wires in the transverse and longitudinal configurations are shown in Figure 8.13 and Figure 8.14 respectively.[35] The MR measurements

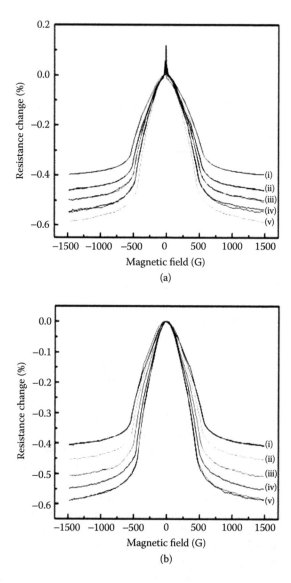

FIGURE 8.13 Transverse MR measurements at (a) 100 μA, (b) 1 mA measurement current for $Ni_{80}Fe_{20}$ samples with (i) 50 nm, (ii) 100 nm, (iii) 150 nm, (iv) 200 nm, (v) 250 nm constriction widths. (From S. Lepadatu, Y.B. Xu, and E. Ahmad, *J. Appl. Phys.*, 97, 10J708, 2005. With permission.)

have been taken at several applied currents. Thus at current densities smaller than 10^7 A/cm^2 an enhancement of MR is observed in both the transverse and longitudinal configurations within ±20 Oe of the zero field for the samples with constriction widths of 50 to 250 nm as shown in Figure 8.13a and Figure 8.14b for a measurement current of 100 μA. For example, this corresponds to a current

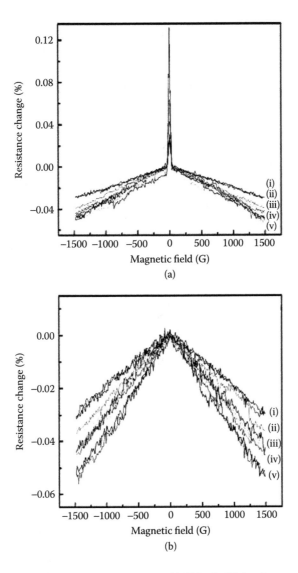

FIGURE 8.14 Longitudinal MR measurements at (a) 100 μA, (b) 1 mA measurement current for $Ni_{80}Fe_{20}$ samples with (i) 50 nm, (ii) 100 nm, (iii) 150 nm, (iv) 200 nm, (v) 250 nm constriction widths. (From S. Lepadatu, Y.B. Xu, and E. Ahmad, *J. Appl. Phys.,* 97, 10J708, 2005. With permission.)

density of 6.7×10^6 A/cm² for the sample with 50 nm constriction width. At current densities larger than 10^7 A/cm², no enhancement of resistance is observed in any of the samples measured as shown in Figure 8.13b and Figure 8.14b for a measurement current of 1 mA. This corresponds to a current density of 6.7×10^7 A/cm² for the sample with 50 nm constriction width. In Figure 8.15, a set of MR measurements for the wire with 50 nm constriction width in the transverse geometry, taken at

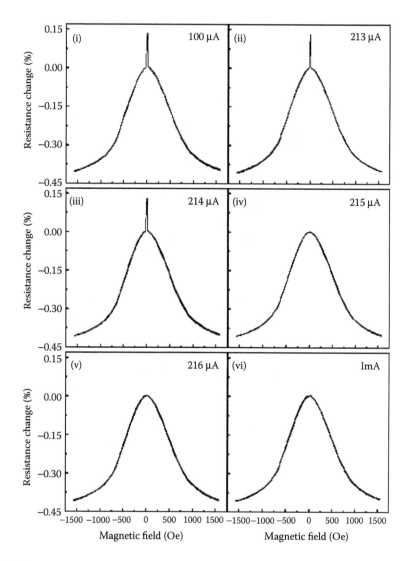

FIGURE 8.15 Transverse MR measurements for a 50 nm constriction $Ni_{80}Fe_{20}$ sample taken at increasing measurement currents, shown for (i) 100 µA, (ii) 213 µA, (iii) 214 µA, (iv) 215 µA, (v) 216 µA, (vi) 1 mA.

measurement currents increasing in 1 µA steps, is shown. These reveal that the transition from the state where an enhancement of MR is observed to the state where no enhancement is observed is abrupt and occurs at a current density of 1.2×10^7 A/cm$^2 \pm 0.6 \times 10^6$ A/cm^2. For the samples with constriction width greater than 300 nm, no difference in MR with varying measurement current is found, which is the same as the straight wire. The enhancement in MR is then due to the nucleation of a domain wall at the constriction at low fields and at applied current densities lower than 10^7 A/cm^2. The positive contribution of the domain wall to the MR gives rise to the

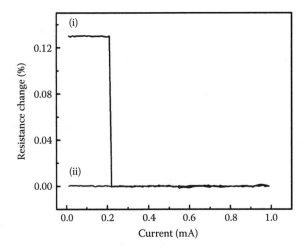

FIGURE 8.16 *I-V* measurements for a 50 nm constriction $Ni_{80}Fe_{20}$ sample taken at (i) 0 Oe and (ii) 20 Oe.

enhancement of resistance observed. At applied current densities greater than 10^7 A/cm^2, no domain wall is nucleated at the constriction due to the spin pressure exerted by the applied current. In this case, no enhancement of resistance is observed. *I-V* measurements have also been performed at varying magnetic fields. It is found that a discontinuous resistance change occurs only for applied magnetic fields within ± 20 Oe of the zero point, with no resistance change for magnetic fields outside this range, as displayed in Figure 8.16 for a 50 nm constriction wire.

There is a very good correlation between the results obtained from *I-V* measurements and those obtained from MR measurements. As seen in Figure 8.13A and Figure 8.14A, an enhancement of resistance at low fields is observed only for the samples with constriction widths in the range of 50 to 250 nm, in both the transverse and longitudinal geometries. Furthermore, the magnitude of this enhancement is found to be equal in both sign and magnitude to the change in resistance observed from *I-V* measurements of necked $Ni_{80}Fe_{20}$ wires for each constriction width, namely from 0.125 to 0.02%. Increasing the applied measurement current above the critical current density, results in removal of the enhancement of resistance for all the constriction widths as shown in Figure 8.13B and Figure 8.14B. To determine the critical current density at which the enhancement of resistance is turned off, several MR measurements have been performed by increasing the measurement current in steps, as shown in Figure 8.15 for a sample with constriction width of 50 nm in the transverse geometry. Starting from a measurement current of 100 µA, the enhancement of resistance is observed up to a measurement current 214 µA (corresponding a current density of $1.2 \times 10^7 A/cm^2$), following which no enhancement of resistance is observed upon increasing the measurement current further. Thus the source of the enhancement of resistance is attributed to the resistance contribution of a domain wall pinned at the constriction at low applied magnetic fields, in both the transverse and longitudinal configurations. Below a current density of 10^7 A/cm^2, a domain wall is formed at the constriction near zero field; wheras above 10^7 A/cm^2, no domain wall is formed due to the spin pressure exerted by the spin-polarized current. Upon reversal from saturation, a domain wall is nucleated at low magnetic fields, within ± 20 Oe of the zero point. Increasing the applied magnetic field on either side results in a removal of the domain wall, as is expected. A consequence of this is observed in the I-V measurements performed at magnetic fields with a magnitude greater than 20 Oe. Thus, as shown in Figure 8.16 for a wire with a 50 nm constriction width, no discontinuous resistance change is observed, as for the I-V measurements performed at zero magnetic fields.

8.8 SUMMARY AND OUTLOOK

In summary, we have studied experimentally the current-induced domain wall motion and domain wall scattering in single layer patterned structures. In particular, a discontinuous resistance change was observed in single layer ferromagnetic wires of $Ni_{80}Fe_{20}$ and Ni patterned with a nanoconstriction of varying width. The resistance change arises due to the removal of the domain wall from the constriction by current-induced domain wall movement, at a critical current density of 10^7 A/cm^2. This shows that the contribution of the domain wall to the resistance is positive in both $Ni_{80}Fe_{20}$ and Ni and was discussed as domain wall scattering. A positive contribution was also observed in $Ni_{80}Fe_{20}$ cross structures, due to a 45° trap at the junction. This was obtained by subtracting the AMR contribution from the overall resistance change observed. For the necked ferromagnetic wires, the MR response revealed an enhancement at low fields in both the transverse and longitudinal configurations when measured at an applied current density less than 10^7 A/cm^2. Increasing the applied current density above 10^7 A/cm^2 removes the enhancement of MR in both the transverse and longitudinal geometries. Furthermore, a discontinuous resistance change in I-V measurements of the necked samples was observed only for low magnetic fields, within ± 20 Oe of the zero point. I-V measurements taken at magnetic fields with a magnitude greater than 20 Oe did not show any discontinuous resistance change with a linear characteristic.

Recently the generation of microwaves in ferromagnetic point contacts by spin-transfer torques has been reported.[36] The possibility that a spin torque can drive steady-state magnetization precession, effectively generating microwaves on a nanometer scale, has also been predicted by Slonczewski[6] and Berger.[7] Interest in this area of research is driven partly by the need to understand and control these mechanisms and by the numerous proposed device applications such as nanometer size microwave oscillators.[37,38] As the device dimensions are reduced to the nanometer scale, the interactions between a spin-polarized current and a thin film can dominate over the effects of an applied magnetic field. The presence of such microwave excitations may have negative consequences for the stability of future hard disk read heads and must be taken into account. However they might also provide for novel techniques of switching in nanomagnetic devices and implementation of MRAM chips.[39] Further developments in research on current-induced switching processes are also expected to yield a new generation of magnetic logic devices. To successfully achieve this aim, several issues must be resolved first. The critical current at which a domain wall is displaced must be reduced for successful commercial devices. One method of achieving this has been shown recently for spin valves using an antiferromagnetic pinning layer.[40] This shows increased spin-transfer efficiency with effective current polarization almost two times higher than normal. On the other hand, larger percentage changes in resistance are also required for easier implementation of magnetic logic devices.

REFERENCES

1. M. Dax, *Semicond. Int.,* 20(10), 84, 1997.
2. D.A. Allwood et al., *Science,* 296, 2003, 2002; C.C. Faulkner et al., *IEEE Trans. Magn.,* 39, 2860, 2003.
3. S.A. Wolf et al., *Science,* 294, 1488, 2001.
4. S. Datta and B.A. Das, *Phys. Lett. A,* 56, 665, 1990.
5. M.N. Baibich et al., *Phys. Rev. Lett.,* 61, 2472, 1988.
6. J.C. Slonczewski, *J. Magn. Magn. Mater.,* 159, 1, 1996.
7. L. Berger, *Phys. Rev. B,* 54, 9353, 1996.
8. F.J. Albert, J.A. Katine, R.A. Buhrman, and D.C. Ralph, *Appl. Phys. Lett.,* 77, 3809, 2000; J.E. Wegrowe et al., *Appl. Phys. Lett.,* 80, 3775, 2002.
9. J. Grollier et al., *J. Appl. Phys.,* 92, 4825, 2002; J. Grollier et al., *Appl. Phys. Lett.,* 83, 509, 2003.
10. L. Berger, *J. Appl. Phys.,* 55, 1954, 1984.
11. P.P. Freitas and L. Berger, *J. Appl. Phys.,* 57, 1266, 1985.
12. L. Gan, S.H. Chung, K.H. Aschenbach, M. Dreyer, and R.D. Gomez, *IEEE Trans. Magn.,* 36, 3047, 2000.
13. M. Tsoi, R.E. Fontana, and S.S.P. Parkin, *Appl. Phys. Lett.,* 83, 2617, 2003.
14. S. Lepadatu and Y.B. Xu, *Phys. Rev. Lett.,* 92, 127201, 2004; S. Lepadatu and Y.B. Xu, *IEEE Trans. Magn.,* 40, 2688, 2004.
15. H. Sato et al., *Phys. Rev. B,* 61, 3227, 2000.
16. T. Tanyiama, I. Nakatani, T. Namikawa, and Y. Yamazaki, *Phys. Rev. Lett.,* 82, 2780, 1999.
17. U. Ruediger, J. Yu, S. Zhang, A.D. Kent, and S.S.P. Parkin, *Phys. Rev. Lett.,* 80, 5639, 1998.
18. J.L. Prieto, M.G. Blamire, and J.E. Evetts, *Phys. Rev. Lett.,* 90, 027201, 2003.
19. Y.B. Xu et al., *Phys. Rev. B,* 61, 14901, 2000.
20. J.L. Tsai et al., *J. Appl. Phys.,* 91, 7983, 2002.
21. M. Viret et al., *Phys. Rev. B,* 53, 8464, 1996.
22. U. Ebels et al., *Phys. Rev. Lett.,* 84, 983, 2000.
23. J.F. Gregg et al., *Phys. Rev. Lett.,* 77, 1580, 1996.
24. P.M. Levy and S. Zhang, *Phys. Rev. Lett.,* 79, 5110, 1997.
25. G. Tatara and H. Fukuyama, *Phys. Rev. Lett.,* 78, 3773, 1997.
26. J.J. Versluijs, M. Bari, and J.M.D. Coey, *Phys. Rev. Lett.,* 87, 026601, 2001.
27. N. García, M. Muñoz, and Y.W. Zhao, *Phys. Rev. Lett.,* 82, 2923, 1999; S.H. Chung et al., *Phys. Rev. Lett.,* 89, 287203, 2002.
28. G. Tatara, Y.W. Zhao, M. Muñoz, and N. García, *Phys. Rev. Lett.,* 83, 2030, 1999.
29. OOMMF package available at http://gams.nist.gov/oommf.
30. P. Bruno, *Phys. Rev. Lett.,* 83, 2425, 1999.

31. A. Hirohata, Y.B. Xu, C.C. Yao, H.T. Leung, W.Y. Lee, S.M. Gardiner, D.G. Hasko, J.A.C. Bland, and S.N. Holmes, *J. Magn. Magn. Mater.,* 226, 1845, 2001.

32. W.J. Carr, *J. Appl. Phys.,* 45, 394, 1974.

33. C.Y. Hung and L. Berger, *J. Appl. Phys.,* 63, 4276, 1988.

34. S. Methfessel, S. Middelhoek, and H. Thomas, *IBM J. Res. Dev.,* 4, 96, 1960.

35. S. Lepadatu, Y.B. Xu, and E. Ahmad, *J. Appl. Phys.,* 97, 10J708, 2005.

36. W.H. Rippard, M.R. Pufall, S. Kaka, S.E. Russek, and T.J. Silva, *Phys. Rev. Lett.,* 92, 027201, 2004.

37. J.C. Slonczewski, U.S. Patent No. 5,695,864, 1997.

38. S.I. Kiselev, J.C. Sankey, I.N. Krivorotov, N.C. Emley, R.J. Schoelkopf, R.A. Buhrman, and D.C. Ralph, *Nature*, 425, 380, 2003.

39. W.H. Rippard, M.R. Pufall, S. Kaka, T.J. Silva, and S.E. Russek, *Phys. Rev. B,* 70, 100406, 2004.

40. T.H.Y. Nguyen, H.J. Yi, S.J. Joo, J.H. Lee, K.H. Shin, and T.W. Kim, *J. Appl. Phys.,* 97, 10C712, 2005.

9 Domain Wall Magnetoresistance in Magnetic Nanowires— Theory

R.L. Stamps, P.E. Falloon, V. Gopar, D. Weinmann, and R.A. Jalabert

CONTENTS

9.1 BACKGROUND AND ISSUES

The subject of electronic transport in magnetic metals is fascinating in terms of fundamental aspects and rich in potential for existing and future applications. The influence of spin on transport has been of great interest for several years because of giant magnetoresistance (GMR).[1,2] The phenomenon is again receiving attention from the point of view of mesoscopic physics in which the phase and the coherence of the electron wave functions are important for transport in geometries with nanometer dimensions. These aspects will be the main concern of the following discussions.

Magnetic domain walls are regions wherein the magnetization of a material changes orientation from one direction to another. Electronic transport through magnetic domain walls brings into play many interesting aspects related to the spin of conduction electrons.[3] For example, a few years ago, experiments on electrodeposited cylindrical Co wires, with diameters down to 35 nm, and polycrystalline Co films 42 nm thick with 150 nm widths suggested a positive contribution to the domain wall resistance in addition to the negative anisotropic magnetoresistance.[4,5]

The positive contribution was attributed to the scattering to of conduction electrons by magnetic domain walls.

In the time since these initial experiments, attention has turned to the possible uses of magnetic domain walls for spin-electronic devices in ferromagnetic nano-structures. A domain wall trapped in a nanostructure can provide a method for storing information,[6] and the devices have already been realized in ferromagnetic nanocon-tacts and semiconductor nanowires.[7,8] Exciting recent developments are based on the idea of current-induced torques that can be created by spin transfer from a spin-polarized current. This phenomenon allows for the reversal of magnetization without application of an external magnetic field, making possible a transport-based mech-anism for switching and controlled motion of domain walls.[9–12]

Spin transport through magnetic nanowires also involves some fundamental problems in mesoscopic physics. Calculations for the ballistic regime have shown that, in the limit of narrow walls, reflection from the wall is an important source of resistance.[13–15] This effect is large for nanocontacts and is very small when many transverse channels contribute, as in the case of nanowires. In the limit of very wide walls, the spin of the electrons follows adiabatically the rotating magnetization direction of the domain wall, and the contribution of the domain wall to the resistance of the wire disappears. Calculations for finite width walls show that deviations from this adiabatic behavior, the so-called mistracking of electron spins upon traversal of the wall enhances the magnetoresistance.[16–18] Another important issue is the question of how impurity scattering and inelastic energy and momentum transfer mechanisms affect the magnetoresistance. Very little has been done in this regard, with existing work primarily concerned with Boltzmann equation approaches in the diffusive regime and some considerations for elastic scattering in random potentials.[17,19,20]

This overview is organized as follows. In Section 9.2, a model for conduction electron transport in a spin-dependent potential is presented using the Landauer–Büttiker formal-ism. Reflection and transmission from the wall are described in Section 9.3, and the channel blocking mechanism for magnetoresistance is presented. Complications due to adiabatic tracking of the spins are discussed in terms of the width of the domain wall. An effective circuit model is presented in Section 9.4 and shown to provide a means of simultaneously describing GMR and ballistic effects. Current-induced torque is discussed in Section 9.5. A summary and outlook are presented in Section 9.6.

9.2 TWO-BAND MODEL

Remarkable progress in the past 20 years has led to very sophisticated and accurate theories for equilibrium band structure. However, a fully consistent description of transport in terms of electronic states calculated from first principles does not yet exist. There is therefore great value in applying suitably simplified models able to capture the essential physics of what is otherwise a complicated many-body problem. A two-band model is the minimum model necessary to describe quantum effects for transport through a magnetic domain wall.

In the two-band model, one makes a distinction between spatially extended electronic states that contribute strongly to conduction and the more localized states that contribute strongly to the formation of local moments. For the transition metal magnets such as

Co, this corresponds to modeling the magnetization produced by exchange correlations in the primarily d-like orbitals as an effective potential for extended s-like conduction states. In the simplest version of this model, the s-like states are treated in a parabolic band approximation with a weakly spin split potential due to s-d hybridization.

The magnetic domain wall is actually a stable configuration of the magnetization with an energy above the ferromagnetic ground state. The exact shape and dimensions of a wall are determined by exchange correlation energies and spin-orbit interactions that contain information about the local atomic environment. One can compute the magnetic structure in a micromagnetic approximation, but from the point of view of the two-band electronic model, the small interaction between d and s states is represented as a simple contact potential. Assuming linear response to an applied voltage, the back action of the s electrons on the magnetization associated with the d electrons is neglected, and the magnetic domain wall appears as a spatially varying spin-dependent potential. In this way, the problem is reduced to a single particle in an effective potential. The micromagnetic description of the magnetic domain wall appears exactly as the effective potential varies spatially in the material. The magnitude of the splitting between the spin-up and spin-down potentials is a parameter related to the exchange correlation energy and the contact potential. Values for this parameter are less than the exchange splitting associated with the magnetization (itself on the order of 1 eV).

An example of an approximate spin-split potential associated with a magnetic domain wall is shown in Figure 9.1 for a domain wall with axis along the z direction. In A, the spin-up and spin-down potentials are shown as functions of position along the wall. The wall region has a width of 2λ, and the wall serves to invert the magnetization by reversing the spin-up and spin-down potentials from one side of the wall to the other.

The local magnetic moment is typically conserved in a ferromagnetic metal, so that a domain wall corresponds simply to a rotation of the local moment. A consequence

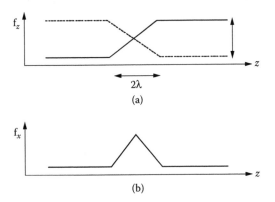

FIGURE 9.1 Spin split effective potential representing a magnetic domain wall in a two band model. In a, the longitudinal component of the potential for a wall with axis along the z direction is sketched. The potential forms a step in each spin component. Conservation of the magnetic moment locally requires a transverse component for the effective potential as shown in b. The transverse potential mixes the up and down spin components of the wave function.

is that the spin-split potential for a domain wall contains a longitudinal component and a transverse component. The Hamiltonian for the conducting s electrons is written in the form

$$H = -\frac{\hbar^2}{2m}\nabla^2 + \frac{\Delta}{2}\mathbf{f}(\mathbf{r})\cdot\boldsymbol{\sigma}\qquad(9.1)$$

Δ represents the magnitude of the exchange split effective potential and $\mathbf{f}(\mathbf{r})$ the functional form of its spatial variation. $\boldsymbol{\sigma}$ is the Pauli spin operator for the conduction electrons. The example longitudinal component sketched in Figure 9.1a has a corresponding transverse component. One example of a transverse component is shown in Figure 9.1b.

It turns out that the detailed shape of the potential has little effect on the qualitative features of the magnetoresistance, and to make the problem mathematically tractable, it can suffice to consider simple geometric shapes as shown in Figure 9.1 that do not actually conserve the magnitude of the magnetic moment. More complicated functional forms that do conserve the moment are discussed in later sections.

The most important features of the effective potential sketched in Figure 9.1 are the inversion of the spin-split potential upon crossing the wall and the presence of a transverse potential. This can be seen by examining the Schrödinger equation for a conduction electron traveling in this potential. The wave function contains spin-up and spin-down components and can be written as

$$\left|\psi(z)\right\rangle = \phi_\uparrow\left|z,\uparrow\right\rangle + \phi_\downarrow\left|z,\downarrow\right\rangle\qquad(9.2)$$

where $\left|z,\uparrow(\downarrow)\right\rangle$ is interpreted as the tensor product of the position eigenvector and the spin-up (-down) state.

Consider a wire geometry with a simple wall described by a potential of the same general form as that shown in Figure 9.1. A sketch of the wire geometry is shown in Figure 9.2. A square wire is assumed for simplicity with sides of height w. It is convenient to define a "longitudinal" energy ε referenced to the Fermi energy E_F according to

$$\varepsilon = E_F - E_{n,m}\qquad(9.3)$$

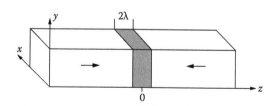

FIGURE 9.2 Geometry for a wall of width 2λ in a square magnetic wire with axis along the z direction. The wire has sides of length w. (After V.A. Gopar, D. Weinmann, R.A. Jalabert, and R.L. Stamps, *Phys. Rev. B*, 69, 014426, 2004. With permission.)

where $E_{n,m}$ is the component of the kinetic energy of a state confined within the wire corresponding to transverse momenta along the x and y directions. In the parabolic band approximation for the conduction states

$$E_{n,m} = \frac{\hbar^2}{2m}\left[\left(\frac{\pi n}{w}\right)^2 + \left(\frac{\pi m}{w}\right)^2\right] \tag{9.4}$$

where n and m are integers. Far from the wall, the potential energy is independent of position. The total energy of a down state, for example, is

$$E_\downarrow = E_{n,m} + \frac{\hbar^2 k_z^2}{2m} \pm \frac{\Delta}{2} \tag{9.5}$$

where the plus sign is chosen for $z \to -\infty$ and the minus sign is chosen for $z \to +\infty$. The second term in Equation 9.5 is the component of the kinetic energy corresponding to the longitudinal momentum and is indexed by the quantum number k_z for the momentum along the z direction. The transverse part of the potential mixes the spin-up and spin-down components, and the Schrödinger equation results in two coupled equations for the position-dependent amplitudes $\phi_{\uparrow(\downarrow)}$ of the longitudinal component of the wave function

$$\frac{d^2}{dz^2}\phi_\uparrow + \frac{2m}{\hbar^2}\left(\varepsilon - \frac{\Delta}{2}f_z\right)\phi_\uparrow = \frac{2m}{\hbar^2}\frac{\Delta}{2}f_x\phi_\downarrow \tag{9.6a}$$

$$\frac{d^2}{dz^2}\phi_\downarrow + \frac{2m}{\hbar^2}\left(\varepsilon + \frac{\Delta}{2}f_z\right)\phi_\downarrow = \frac{2m}{\hbar^2}\frac{\Delta}{2}f_x\phi_\uparrow \tag{9.6b}$$

The picture described by Equation 9.6 is one in which there are waves guided by the wire geometry into momentum "channels" indexed by the integers n and m, with a continuous range of longitudinal momenta available and indexed by k_z. Far from the wall, the spin-up and spin-down states differ in energy by the exchange splitting of the conduction band, and in the wall region the up and down states are mixed by the transverse components of the wall effective potential.

9.3 CHANNEL BLOCKING AND ADIABATICITY

Scattering solutions of Equation 9.6 are used to calculate transmission coefficients from which a conductance is found using Landauer–Büttiker theory by summing over all channels. Before discussing this application of the Landauer–Büttiker theory for the conductance, it is useful to examine the general features of solutions to Equation 9.6.

First consider the extreme case of an abrupt domain wall where $\lambda \sim 0$. The wall is in this case represented by a jump in f_z from -1 to 1 at $z = 0$. The right-hand side of Equation 9.6 vanishes and the spin-up and spin-down electrons are uncoupled. The effect of the discontinuity in f_z is a spin-dependent potential step of height $\pm\Delta$ for the conduction electrons. Incoming spin-up electrons having longitudinal energy $\varepsilon < \Delta/2$ cannot overcome this step and are completely reflected. Because the density of conduction channels is independent of the energy for wires having a two-dimensional cross section, this mechanism blocks a fraction $\Delta/2E_F$ of the conduction channels.[13] Without the domain wall, all of these channels would completely transmit. Neglecting the effect of the potential step on electrons having longitudinal energy ε higher than $\Delta/2$, this channel blocking mechanism leads to a relative change in conductance due to the domain wall given by[13,14]

$$\frac{\delta g}{g} = -\frac{\Delta}{2E_F} \tag{9.7}$$

In the limit $E_F \gg \Delta$ spin-conserving reflections for $\varepsilon > \Delta/2$ increase the channel blocking effect by a factor of 4/3.

Some general effects of a finite width wall can be seen in the short wall limit. The spin-up and spin-down components are weakly mixed if the domain wall is very short and for high longitudinal energy states in which the transmitted electrons spend a short time inside the domain wall region. The coupled Schrödinger equation can be solved perturbatively in this case of weak coupling. Suppose that the wall profile is given by

$$\mathbf{f}(z) = \begin{cases} \hat{\mathbf{x}}\sqrt{1-(z/\lambda)^2} + \hat{\mathbf{z}}z/\lambda & \text{for } |z| < \lambda \\ \hat{\mathbf{z}}\,\text{sgn}(z) & \text{for } |z| \geq \lambda \end{cases} \tag{9.8}$$

Note that this form conserves the magnitude of the magnetic moment because $|\mathbf{f}^2| = 1$. Suppose that an incoming majority electron (spin up) is incident from the left. The solution to the homogenous equations (without the coupling terms on the right-hand-side of Equation 9.6) is

$$\phi_\uparrow^H = e^{ikz} + r_{\uparrow\uparrow}\,e^{-ikz} \tag{9.9}$$

for $z < \lambda$, and

$$\phi_\uparrow^H = t_{\uparrow\uparrow}\,e^{ik'z} \tag{9.10}$$

for $z > \lambda$. The wave numbers are

$$k = \sqrt{\frac{2m}{\hbar}\left(\varepsilon + \frac{\Delta}{2}\right)} \tag{9.11}$$

and

$$k' = \sqrt{\frac{2m}{\hbar}\left(\varepsilon - \frac{\Delta}{2}\right)} \qquad (9.12)$$

Inside the wall region $-\lambda < z < \lambda$ the solution of the homogeneous equations is expressed in terms of Airy functions Ai and Bi as

$$\phi_\uparrow^H = \alpha \; \mathrm{Ai}\left[p^{2/3}\left(\frac{z}{\lambda} - \frac{2\varepsilon}{\Delta}\right)\right] + \beta \; \mathrm{Bi}\left[p^{2/3}\left(\frac{z}{\lambda} - \frac{2\varepsilon}{\Delta}\right)\right] \qquad (9.13)$$

The dimensionless parameter p is defined by

$$p = \frac{\lambda}{\hbar}\sqrt{m\Delta} = (k_F\lambda)\sqrt{\frac{\Delta}{2E_F}} \qquad (9.14)$$

The reflection and transmission coefficients, as well as α and β are obtained by matching solutions of Equation 9.9, Equation 9.10, and Equation 9.13 at the wall boundaries $z = \pm \lambda$. This homogeneous solution describes blocked channels for $\varepsilon < \Delta/2$ (where k' is imaginary, the transmission is zero, and the reflection is unity).

The first-order correction in the coupling of the spin components is spin down and is generated by plugging the homogeneous solution in the coupling terms of Equation 9.6. Because there is no incident spin-down component, outside the wall region the correction is

$$\phi_\downarrow^{(1)} = r_{\uparrow\downarrow}\, e^{-ik'z} \qquad (9.15)$$

for $z < -\lambda$, and

$$\phi_\downarrow^{(1)} = t_{\uparrow\downarrow}\, e^{ikz} \qquad (9.16)$$

for $z > \lambda$. Inside the wall region the solution is a linear combination of Airy functions plus a particular solution

$$\phi_\downarrow^{(1)} = \alpha_1 \; \mathrm{Ai}\left[-p^{2/3}\left(\frac{z}{\lambda} + \frac{2\varepsilon}{\Delta}\right)\right] + \beta_1 \; \mathrm{Bi}\left[-p^{2/3}\left(\frac{z}{\lambda} + \frac{2\varepsilon}{\Delta}\right)\right] + \phi^P \qquad (9.17)$$

The coefficients α_1 and β_1, as well as the transmission and reflection coefficients, are again determined by matching Equation 9.15 through Equation 9.17 at the wall boundaries. The mixing terms generating the first-order correction provide a spin-flip mechanism that projects the up-spin states and down-spin states onto one another.

In the long wavelength limit, the spin-flip transmission probabilities are given by

$$T_{\uparrow\downarrow} = \left|t_{\uparrow\downarrow}\right|^2 = C^2 p^2 (1+x) \qquad (9.18)$$

for $\varepsilon < \Delta/2$, and

$$T_{\uparrow\downarrow} = 4C^2 p^4 \frac{4(x^2-1)+p^2(x+1)\left[\frac{2}{3}(x-1)-\frac{1}{p^2}\right]^2}{\left[\left(\sqrt{x+1}+\sqrt{x-1}\right)^2 + 4p^2(x^2-1)\right]^2} \qquad (9.19)$$

for $\varepsilon > \Delta/2$. In Equation 9.18 and Equation 9.19, $x = 2\varepsilon/\Delta$. The prefactor C is determined by the transverse wall potential via the integral

$$C = \frac{1}{\lambda} \int_{-\infty}^{\infty} f_x(z)dz \qquad (9.20)$$

The spin-flip transmission depends on the quantity Cp and vanishes as Cp goes to zero. The transmission and reflection coefficients depend on Cp in a nontrivial manner that is only weakly dependent on the functional form of the transverse potential, at least for the case of short domain walls.[14]

The spin-flip transmission of Equation 9.18 and Equation 9.19, resulting from this first-order perturbation theory, is plotted as a function of energy in Figure 9.3 for different wall widths (lines). A numerical solution of the coupled Schrödinger equation is shown for comparison and plotted as symbols. Results for different values of p are shown. The parameter p is proportional to the domain wall width λ and the magnitude of the spin-flip transmission is linear in p for $p < 2\varepsilon/\Delta$. The maximum spin-flip transmission occurs for $p = 2\varepsilon/\Delta$.

A physical interpretation of the spin-flip transmission can be obtained by thinking in terms of spin precession about local fields. The exchange splitting responsible for the magnetization corresponds to an energy that can be thought of as related to an effective field associated with the local magnetic moments. A conduction spin experiences the field through the contact potential Δ and will precess when the spin quantization axis is not colinear with the field. The domain wall describes an effective field whose orientation rotates by 180° from one side of the wall to the other. A conduction electron thus enters the wall with a quantization axis parallel to the effective field direction, and attempts to "track" the direction of the effective field direction by precession as the electron travels along the wall axis.

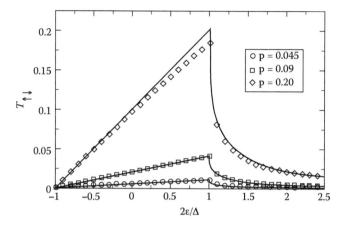

FIGURE 9.3 Spin-flip transmission due to the transverse mixing potential in the long wavelength limit for a linear wall. There is a finite transmission at all energies due to the spin mixing allowed in the wall region by the transverse potential. Spin-flip transmissions for different values of the parameter p are shown. Numerical solutions are shown with symbols and the lines are the results from a perturbative treatment from Equations 9.18 and 9.19. (After V.A. Gopar, D. Weinmann, R.A. Jalabert, and R.L. Stamps, *Phys. Rev. B*, 69, 014426, 2004. With permission.)

Complete tracking results in the reversal of the conduction spin and is called "adiabatic" transport of the electron spin. The amount of precession possible depends on the strength of the field and the speed of the electron along the wall axis. Adiabatic transport of all electronic states is therefore only possible in the limit of very long walls. Adiabaticity is only partial for finite-size walls because the electrons in high longitudinal momentum states do not spend a long time inside the domain wall region, so that the electron spin of these states is not able to completely track the effective field. More precisely, the condition for the local adjustment is that the Larmor precession of the conduction spin about the local magnetization be much faster than the rotation of the local magnetization as viewed by the traveling electron. This gives a condition for adiabaticity in terms of the longitudinal energy and exchange splitting: $\Delta \gg (h/\lambda)\sqrt{\varepsilon/m}$.

Landauer–Büttiker theory provides a simple means of calculating the conductance once the transmission amplitudes are known for all transverse modes (channels) at the longitudinal momentum that corresponds to the Fermi level. The basic relation is

$$g = \sum_{n,m}\sum_{\sigma,\sigma'} T_{n,m}^{\sigma\sigma'}(E_F) \qquad (9.21)$$

where the sums are over all channels and spin states. Each transmission coefficient represents a scattering of electrons in the spin state σ and channel (n,m), with total energy E_F, to spin state σ' in the same momentum channel at the same energy.

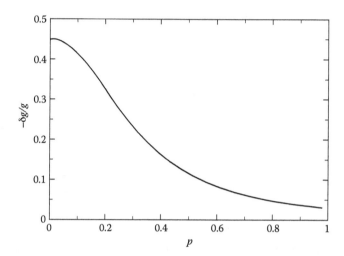

FIGURE 9.4 The relative change in conductance caused by the presence of a domain wall for $E_F = 2\Delta$ shown as a function of the wall width parameter p. The wall is a solitonlike object expected from micromagnetic calculations for an ideal one-dimensional wall. The magnetoconductance is calculated by numerically solving the coupled Schrödinger equation and summing over all conductance channels. (After V.A. Gopar, D. Weinmann, R.A. Jalabert, and R.L. Stamps, *Phys. Rev. B*, 69, 014426, 2004. With permission.)

We assume that the magnetization in the wire does not depend on the transverse coordinates such that the mixing of channels other than spin mixing does not occur.

The conductance of Equation 9.21 describes elastic scattering by the domain wall and is completely determined by the transmission coefficients. The magnetoresistance is therefore entirely due to the reduction of transmission of the channels having low longitudinal energy and the complete reflection of the ones with longitudinal energy $\varepsilon < \Delta/2$. In other words, the channels with small transverse momenta have large longitudinal energies and are able to transmit easily through the wall. States with large transverse momenta have correspondingly small k_z and can be blocked due to their small longitudinal energy.

Adiabaticity suppresses this mechanism by increasing the spin-flip transmission coefficient. In the short wall, long wavelength limit, one can neglect contributions to the spin-flip transmission from states with longitudinal energies much greater than $\Delta/2$. There is a small contribution from adiabatic tracking of states with longitudinal energies below $\Delta/2$, which reduces the magnetoresistance. The magnetoresistance in this approximation is given by

$$\frac{\delta g}{g} = -[1 - C^2 p^2]\frac{\Delta}{2E_F} \qquad (9.22)$$

In the limit of short domain walls, the dependence of the transmission coefficients on the exact shape of the wall appears only in the integral over f_x, which enters the prefactor C. For example, C for the linear wall discussed above has the value $\pi/2$.

The solitonlike profile wall one expects from micromagnetic calculations has the form $f(z) = \hat{x}\,\text{sech}(z/\lambda) + \hat{z}\,\tanh(z/\lambda)$. This wall produces a C with value π. Therefore, the shape enters in the overall magnitude of the magnetoresistance only through the parameter C. In addition, it does not affect the channel blocking mechanism.

As indicated in the discussion surrounding Figure 9.3, it is possible to solve numerically the coupled Schrödinger equation for arbitrary wall profiles. Details of the method are given in Reference 14. Results for the magnetoconductance summed over all channels are shown in Figure 9.4 for the solitonlike wall profile mentioned above. The magnetoconductance is defined as the relative change of the Landauer–Büttiker conductance with and without a domain wall. The total energy in this example is defined by $E_F = 2\Delta$. The conductance is shown as a function of the wall parameter p, which is proportional to the wall width. The magnetoconductance has a maximum for abrupt walls and decreases asymptotically with increasing wall width due to adiabatic tracking of momentum channels that are blocked when $p = 0$.

9.4 CIRCUIT MODEL

The ballistic channel blocking mechanism described above relies on the reflection of conduction spins by the magnetic domain wall. This effect can be large for nanocontacts with very narrow walls, but is a small effect (in relative terms) when many transverse momentum channels contribute or the wall is wide. In the wide wall case, adiabaticity due to tracking of electron spins traversing the wall increases the spin-flip conductance and reduces the relative magnetoconductance.

Another source of magnetoresistance is due to spin-dependent scattering in the wire surrounding the domain wall. This can lead to significant effects depending on the adiabaticity even in cases when the reflection of the electrons by the domain wall itself can be neglected. This is due to the fact that the spin of the electrons remains well-defined outside the domain wall on a length scale that is typically larger than the (temperature-dependent) phase coherence length, but smaller than the distance between the leads. This spin-diffusion length is on the order of 100 nm for transition metals and much smaller than lengths of nanowires typically studied in experiments.

Neglecting the contributions to the magnetoresistance that are due to the reflection by the domain wall, transport through a magnetic wire a few microns long with a single magnetic domain wall should behave like a GMR device. There will be two large domains of antiparallel magnetization separated by a narrow transition region, and spin-dependent scattering (elastic or inelastic) can occur in such a way as to create a difference in the resistance compared to that measured in the single domain state without the domain wall. This is a GMR effect and can be well described by a two-resistor model as done by Valet and Fert in the context of magnetic multilayers.[17] The difference is that in place of a plane interface, which conserves the spin of the conduction electrons, there is a domain wall that can mix the two spin directions.

It is possible to incorporate this difference in a modified two-resistor model by generalizing the effective circuit to include a ballistic conductance representing the domain wall. The key physical assumption in the two-resistor model is that the length scale for scattering events that reverse spin direction, the spin diffusion length l_{sd}, is much larger than the phase coherence length l_ϕ. The transport can be modeled

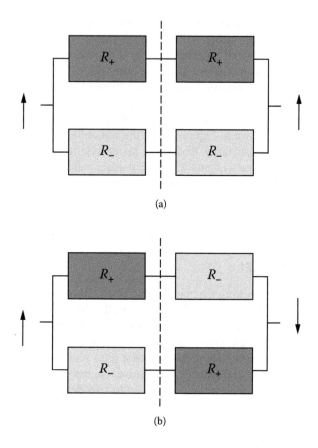

(a)

(b)

FIGURE 9.5 The two-resistor model applied to an interface between two ferromagnetic layers. In a, the magnetizations of the two films are parallel, and in b, the magnetizations are antiparallel. (After P.E. Falloon, R. Jalabert, D. Weinmann, and R.L. Stamps, *Phys. Rev. B,* 70, 174424, 2004. With permission.)

as two resistances in parallel over length scales up to l_{sd}, with the resistors representing the resistivity of each spin state.

In general, resistivity is spin-dependent so that the GMR of an interface between two ferromagnetic layers is described as the difference between the resistances of parallel and antiparallel orientation of the film magnetizations. The corresponding effective circuits are sketched in Figure 9.5. We let R_+ and R_- denote the resistance of the majority and minority spin channels (over a length l_{sd}). In the antiparallel case, the currents in each spin channel are equal, and in the parallel case, the total current has a net polarization

$$\beta = \frac{R_+ - R_-}{R_+ + R_-} \tag{9.23}$$

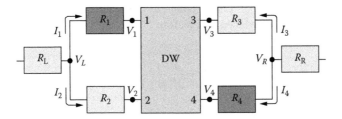

FIGURE 9.6 The circuit model for a domain wall includes an additional resistance element in place of the GMR interface. (After P.E. Falloon, R. Jalabert, D. Weinmann, and R.L. Stamps, *Phys. Rev. B,* 70, 174424, 2004. With permission.)

The main effect of adiabatic tracking of conduction spins traversing the domain wall is to mix the spin states within a momentum channel. This means that the domain wall introduces a new resistance element in place of the interface, as indicated in the sketch of Figure 9.6. The new element has four terminals connecting and mixing the spin-dependent channels. Each terminal connects to resistances representing the diffusive spin-dependent transport occurring over lengths l_{sd} on either side of the domain wall. Electrons in Terminal 1 and Terminal 4 have spin 1/2 (taken with respect to the quantization axis far from the wall), and electrons in Terminal 2 and Terminal 3 have spin −1/2.

The diffusive resistances R_1 and R_4 are equal to the majority resistance R_+, and R_2 and R_3 are equal to the minority resistance R_-. The important feature of the circuit is that the potentials in the spin-up and spin-down channels close to the wall are not necessarily equal even if they are on the same side of the wall. This means that the distribution of current between the spin-up and spin-down channels can differ from that of a homogeneous wire and create a GMR-like enhancement of the resistance. The model is completed by connecting the domain wall region of the wire, which is defined within a spin-diffusion length of the wall, to the remainder of the wire. The spin channels are equilibrated well away from the wall and transport is characterized by spin-independent resistances R_L and R_R. Experimentally fabricated nanowires typically have lengths on the order of micrometers so that in practice one expects that R_R and R_L are each much larger than R_+.

The smallest length scale is assumed to be the phase coherence length. However, domain walls can have lengths on the order of 10 nm in cobalt (and even smaller in the presence of constrictions).[21] It may be then that the domain wall width λ can be comparable to l_ϕ at least at liquid nitrogen temperatures. With this in mind, the magnitudes of $R_\pm$ are determined as follows. Consider incoherent transport over the length l_{sd} as a series of phase coherent segments each of length l_ϕ. In each segment, transport is coherent and diffusive with a spin-dependent elastic mean free path $l_\pm$. The resistance of each segment is

$$R_\pm^\phi = \frac{h}{e^2} \frac{1}{N_\pm} \frac{l_\phi}{l_\pm}$$ (9.24)

where the total number of phase coherent segments in each spin channel is $N = l_{sd} / l_{\phi}$. The total resistance is therefore

$$R_{\pm} = \frac{h}{e^2} \frac{1}{N_{\pm}} \frac{l_{sd}}{l_{\pm}} \tag{9.25}$$

The spin dependence of the resistance therefore arises from differences in $N_{\pm}$ and $l_{\pm}$ between the two spin subbands. It is reasonable to assume that $N_+ > N_-$ but the relationship between l_+ and l_- depends upon details of the material-dependent electronic band structure and scattering processes determined by the available states at the Fermi energy into which electrons can be scattered. These lengths are assumed to be the dominant factor determining spin dependence of the resistance and can be estimated from experimental data taken from GMR measurements.

A multiterminal Landauer–Büttiker formula is used to calculate the currents and potentials at the wall[22]

$$I_a = \sum_{b \neq a} G_{ab}(V_a - V_b) \tag{9.26}$$

The G_{ab} are matrix elements of the domain wall conductance tensor between terminal a and terminal b, whereas I_a and V_a are the current and the voltage corresponding to terminal a, respectively. The domain wall is assumed to be a coherent and disorder-free region connected by perfect leads to the rest of the system. This hypothesis is reasonable for thin wires with elastic and inelastic mean free paths larger than the domain wall width.

The Hamiltonian of Equation 9.1 is symmetric under time reversal, because there are no orbital magnetic fields, so that $G_{ab} = G_{ba}$. Furthermore, if $\mathbf{f}(z)$ has left-right symmetry with respect to interchange of spin direction, then $G_{12} = G_{43}$ and $G_{13} = G_{42}$. These relations together with Kirchhoff's laws allow one to find an expression for the resistance of the domain wall between V_L and V_R.[18]

In the case of an abrupt wall, only G_{13} and G_{24} are nonzero. These can be calculated by summing the transmission function for each conduction channel across the interface.[17] In the adiabatic limit of complete spin tracking, only G_{14} and G_{23} are nonzero. If the transport across the wall is adiabatic, then

$$G_{14} = \frac{e^2}{h} N_{\uparrow} \tag{9.27}$$

and

$$G_{23} = \frac{e^2}{h} N_{\downarrow} \tag{9.28}$$

Because the wall must be very wide to be in the adiabatic limit, it is likely to be larger than the phase coherence length so that an assumption of diffusive transport in each spin channel is reasonable. Using reasoning similar to that leading to Equation 9.25, the corresponding resistances are therefore given by

$$\frac{1}{G_{14}} = \frac{2\lambda}{l_{sd}} R_+$$ (9.29)

and

$$\frac{1}{G_{23}} = \frac{2\lambda}{l_{sd}} R_-$$ (9.30)

In general, transport through the wall is intermediate between the limiting cases of abrupt and adiabatic. As shown above, the exact form of the wall is important for estimating magnitudes of the magnetoresistance. Here a one-dimensional micromagnetic domain wall structure is assumed so that $\mathbf{f(r)} = \mathbf{f}(z)$ as above, with no scattering between different transverse modes (channels). Note that this is a reasonable assumption provided that the wire diameter is small ($w < 40$ nm for cobalt), but for wires with larger diameters, more complicated structures become energetically favorable.[24]

It is possible to obtain analytic expressions for the energy eigenstates of electrons in a potential profile of the form[14,25,26]

$$\mathbf{f}(z) = \begin{cases} \hat{\mathbf{x}} \cos\left(\dfrac{\pi z}{2\lambda}\right) + \hat{\mathbf{z}} \sin\left(\dfrac{\pi z}{2\lambda}\right), & |z| < \lambda \\ \hat{\mathbf{z}} \, \mathrm{sgn}(z), & |z| \geq \lambda \end{cases}$$ (9.31)

The transmission and reflection coefficients calculated from these expressions are used to construct the elements of the conductance tensor. Note that the tensor describes all possible scatterings between the four terminals, and therefore one must sum over the transmission and reflection coefficients as appropriate. Written in terms of the longitudinal energy, the conductance tensor elements are still of the form of Equation 9.21

$$G_{ab} = \frac{e^2}{h} \sum_{n,m} F_{\sigma\sigma'}(\varepsilon)$$ (9.32)

where now $F_{\sigma\sigma'}$ denotes either a transmission or reflection for an electron with spin state σ scattered into spin state σ'. The electrons are either incident from the left or right depending on the tensor element in question.

Symmetries between the different reflection and transmission coefficients can be used to obtain a relatively simple expression for the domain wall resistance in the case of small exchange splitting $\Delta \ll E_F$. The total resistance depends only on the spin-dependent resistances R_+ and R_-, the number of channels N, and a single parameter P characterizing the adiabaticity

$$R_{DW} = \frac{r_o(r_o + R_+ + R_-) + [r_o(R_+ + R_-) + 4R_+R_-]P}{2[r_o + (R_+ + R_-)P]} \qquad (9.33)$$

The ballistic resistance of the nanowire is $r_o = h/(Ne^2)$ and the adiabaticity parameter is defined as

$$P = \frac{1}{\varepsilon_F - 1} \int_1^{\varepsilon_F} T_{\uparrow\downarrow}(\varepsilon) d\varepsilon \qquad (9.34)$$

with $\varepsilon_F = 2E_F/\Delta$. Note that the total number of channels is approximately equal to the number of channels in the energy range $l < \varepsilon < \varepsilon_F$, which gives $N = 8\pi m w^2 E_F/h^2$.

A magnetoresistance can now be defined. The domain wall affects transport in a region of length $2(l_{sd} + \lambda)$. If R_o is the resistance of this region without the domain wall, the relative magnetoresistance due to the wall is

$$MR = \frac{R_{DW} - R_o}{R_{wire}} = \frac{2(l_{sd} + \lambda)}{L_{wire}} \frac{R_{DW} - R_o}{R_o} \qquad (9.35)$$

where R_{wire} is the total resistance of the entire wire and L_{wire} is the total length of the wire. Assuming $r_o \ll R_+$, the use of Equation 9.34 results in a simple expression for the magnetoresistance

$$MR = \left(\frac{l_{sd} + \lambda}{L_{wire}}\right)\left(\frac{2\beta^2}{1-\beta^2}\right)\left(\frac{1-P}{1+\alpha P}\right) \qquad (9.36)$$

where α is defined as the ratio $\alpha = (R_+ + R_-)/r_o$.

The magnetoresistance expressed by Equation 9.36 has a straightforward interpretation. The product of the first two brackets is the GMR of an abrupt interface. The last bracket contains a correction due to tracking of conduction spins in the domain wall. The behavior of this term is shown in Figure 9.7 where it is plotted as a function of P for different values of α. The maximum possible magnetoresistance is determined by the GMR ratio and is realized only for abrupt walls. The magnitude of the reduction for finite-width walls depends on the spin-flip transmission through P, and also on the ratio of diffusive to ballistic conductances specified by α. A large value of α implies a strong sensitivity of the MR reduction to adiabatic effects associated with the wall region.

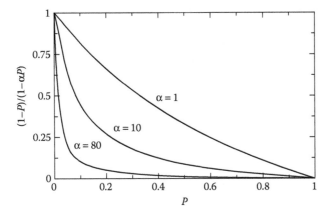

FIGURE 9.7 The magnetoresistance is reduced by a factor related to the spin tracking, or adiabaticity, of conduction electrons while in the wall. This factor is plotted as a function of an adiabaticity parameter P. P is a measure of the spin-flip transmission due to Larmor precession of electrons traversing a magnetic domain wall. (After P.E. Falloon, R. Jalabert, D. Weinmann, and R.L. Stamps, *Phys. Rev. B*, 70, 174424, 2004. With permission.)

9.5 CURRENT-INDUCED TORQUE

The importance of adiabaticity on the magnetoconductance is clear from the expression for *MR* given by Equation 9.36. When viewed in terms of how the angular momentum changes, adiabatic tracking of the conduction spin is particularly intriguing. Suppose that the current through the wall is spin-polarized. Tracking of the local magnetization orientation by rotation requires a change in the angular momentum carried by the current. Conservation of the total angular momentum requires then corresponding changes in the angular momentum associated with the magnetic domain wall. The result is a transfer of angular momentum from the conduction electrons to the domain wall. This transfer of angular momentum creates a torque on the domain wall that can cause it to move. The effect was originally predicted by Berger[27,28] and has been observed in a number of experiments.[9–12]

Angular momentum transfer can be calculated easily within the context of the circuit model presented in the Section 9.4. One needs only to sum the total spin of the incoming and outgoing currents to see that angular momentum is leaving the conduction spin system. As noted earlier, electrons in Terminal 1 and Terminal 3 carry angular momentum $\hbar/2$ and electrons in Terminal 2 and Terminal 4 carry angular momentum $-\hbar/2$. The symmetry of the circuit requires $I_4 = -I_1$ and $I_3 = -I_2$. The rate of angular momentum transfer to the domain wall is therefore proportional to the difference $I_1 - I_2$. This corresponds to a torque acting on the domain wall per unit current given by

$$\frac{\tau}{I} = \frac{\hbar}{e}\frac{I_1 - I_2}{I_1 + I_2} \tag{9.37}$$

In the case of an abrupt wall $I_1 = I_2$ and there is no torque. This is to be expected because the angular momentum transfer, and hence the torque produced, requires some degree of adiabaticity. In the adiabatic limit, $\tau / I = \hbar\beta / e$, a result first obtained by Berger.[28] In the intermediate case for ballistic transport through the wall, Berger's result is generalized to[18]

$$\frac{\tau}{I} = \frac{\hbar\beta}{e} \frac{\alpha P}{1 + \alpha P} \qquad (9.38)$$

This result shows clearly the importance of adiabaticity for the transfer of angular momentum and production of torque. The first ratio is the spin-transfer torque expected for a completely adiabatic wall. The second ratio is a reduction factor due to incomplete spin tracking of electrons through the wall. In other words, the electronic spin channels that are not completely reversed upon traversing the wall do not contribute fully to the transfer of angular momentum. Note that there is always a complete reversal of spin over the length scale of the circuit, $2(l_{sd} + \lambda)$, in the effective circuit model. The induced torque given by Equation 9.38 represents that fraction of the reversal that occurs within the domain wall.

In this context, the role of adiabaticity is thus to increase the torque induced by the current, in contrast to the previous case where its effect was to reduce the magnetoresistance from the larger GMR value. Interestingly, the spin-dependent resistances serve to enhance the difference between up and down spin current components and thereby produce a sizable torque. If this were not the case and the torque were due only to the polarization of spin due to ballistic transmission and reflection at a single domain wall, the magnitude of the torque would be very small, on the order of Δ/E_F.

If there is no pinning of the domain wall within the wire such that it is free to move without dissipation, then the domain wall can respond to the current-induced torque by moving. To be precise, suppose that the torque τ acts on the wall during a time Δt. The transferred angular momentum is $\Delta S = \tau \Delta t$, and the wall can "absorb" ΔS by moving an amount $\Delta z = \Delta S/(w^2\rho)$. ρ is the angular momentum density per unit volume in the wire and determined by the magnetization M of the material according to $\rho = M\hbar / (g\mu_B)$. This argument gives a wall velocity v per unit current of

$$\frac{v}{I/w^2} = \frac{\tau}{\rho I} = \frac{\hbar\beta}{\rho e} \left(\frac{\alpha P}{1 + \alpha P} \right) \qquad (9.39)$$

The wall velocity assumes that motion of the wall completely accounts for all angular momentum lost by the current as it traverses the wall. Implicit in this argument is the assumption that the wall motion can be treated as a negligible perturbation on the reflection and transmission coefficients of the ballistic electron problem. This assumption is likely to be reasonable given that realistic wall velocities are negligible relative to the velocity of conduction electrons at the Fermi energy in most conduction channels.

Note also that because the conduction electrons precess inside the wall, there is a spatially varying torque component that in principle could result in distortions of the wall profile and couple to excitations of the wall.[25] These effects have been neglected in the present model, and the torque has been estimated as an average over a mobile but otherwise "rigid" magnetic structure.

9.6 SUMMARY AND OUTLOOK

The theory of domain wall magnetoresistance has evolved from early work on extended domain walls[29] to modern treatments involving concepts and tools developed in mesoscopic transport theory.[23] Domain wall magnetoresistance can be thought of as being due to two effects. The maximum possible resistance is essentially a GMR effect of transport across an interface between different magnetic domains. If the interface between domains is a continuous wall structure with a finite width, the magnetoresistance is decreased by Larmor precession of the conduction spins as they track a rotating magnetization while traversing the wall. A generalization of the Valet–Fert two-current model is possible in which this reduction is shown to depend upon the ratio of diffusive to ballistic conductances for the two spin channels.

It is useful to compare the effective circuit model predictions to experiments. The experiments reported by Ebels et al.[4] investigated the magnetoresistance due to domain walls in cobalt nanowires with a diameter of 35 nm. They reported contributions of 0.03% due to single walls in wires with a length of 20 μm. The total resistance of the wires was 1.4 kΩ. Using parameters for cobalt of $l_{sd} = 60$ nm and a spin polarization $\beta = 0.4$, an interface GMR of 0.14% is expected. The adiabaticity is calculated to be given by $P \sim 0.25$ with $\alpha = 80$ if $E_F = 10$ eV, $\Delta = 0.1$ eV, $\lambda = 10$ nm, and $p = 5$. A total reduction of the *MR* to 0.0050% is found when these adiabatic effects are included according to Equation 9.36.

A quantitative comparison to experiment is probably unlikely due to the simplicity of the effective circuit model. However the reduction predicted by the model is clearly in the correct direction because the predicted GMR value is an order of magnitude larger than the measured *MR* value. The neglect of realistic band structure effects under the assumption of a spherical Fermi surface is severe,[30] with the result that there is great uncertainty and ambiguity in choosing values for parameters such as Δ, α, β, and *P*. There are additional complications due to experimental difficulties with sample-dependent values and even determining the number of domain walls contributing to the *MR*. Furthermore, it is even possible that walls are pinned at defects or constrictions that change the area and width of the wall. These are sensitive parameters and can easily modify the *MR* predictions by an order of magnitude.

Adiabatic reduction of domain wall magnetoresistance is intrinsic to the physics of how conduction spins traverse a magnetic domain wall. The tracking of a conduction spin due to Larmor precession is responsible not only for adiabatic reduction of *MR*, it is also responsible for the current-induced torque acting on a domain wall. Adiabatic tracking of the domain wall magnetization requires a reversal of the conduction spin, and this in turn involves a change in angular momentum of the current. One way for angular momentum to be conserved is for the domain wall to move with a velocity proportional to the current.

In a recent experiment, Yamaguchi and coworkers measured the displacement of domain walls under the influence of current pulses of varying duration.[12]

The driving voltage was applied to Permalloy™ wires containing domain walls positioned in a clever geometric design. The wire cross section was 240 nm by 10 nm. Assuming that the domain wall moved at constant speed, they found an average wall velocity of 3 m/s for current pulses of density 1.2×10^{12} A/m^2. The wall velocity predicted in the circuit model using Equation 9.39 is two orders of magnitude larger using an adiabaticity $P \sim 1$ as appropriate for Permalloy wall widths on the order of 100 nm. Other relevant parameters for Permalloy are $\beta = 0.5$, $M = 2T/\mu_o$, and $g = 2$. This value, although large, is in agreement with other linear response theories.[31]

There are several ingredients still missing in the wall velocity theory. In addition to the limitations of the theory mentioned within the context of magnetoresistance, domain wall motion involves additional considerations. In particular, the efficiency with which spin-angular momentum is transferred between the conduction electrons and domain wall magnetization may be strongly influenced by other effects such as spin-orbit coupling. Furthermore, motion of the domain wall from the standpoint of micromagnetics involves additional torques describing dissipation that are known to exist from experiments with walls driven by applied magnetic fields.

Another important consideration is the effect of impurity scattering. The theory presented here completely neglects impurity scattering except in the most crude approximation of diffusive transport. Existing work on the problem for domain wall transport is very limited, but the results by Tatara[19] for strong scattering centers illustrates some of the interesting possibilities. Tatara showed that a domain wall may suppress weak localization in the case of very strong scattering, and additional numerical calculations for strong disorder scattering potentials in the coherent diffusive regime support this conclusion.[20] Effects of spin-orbit coupling and spin-dependent scattering within the domain wall may also have large effects and are topics currently under investigation.

ACKNOWLEDGMENT

This work received financial support from the European Union within the Research and Training Network (RTN) program (Contract No. HPRN-CT-2000–00144) and the Australian Research Council Linkages and Discovery Programs. P.E.F. is grateful for the support of an Australian Postgraduate Award and a Jean Rogerson Fellowship from the University of Western Australia and for support from the Université Louis Pasteur in Strasbourg. V.G. thanks the French Ministère délégué à la recherche et aux nouvelles technologies and the Center for Functional Nanostructures of the Deutsche Forschungsgemeinschaft (Project No. B2.10) for their support.

REFERENCES

1. M.N. Baibich, J.M. Broto, A. Fert, F. Nguyen Van Dau, F. Petroff, P. Eitenne, G. Creuzet, A. Friederich, and J. Chazelas, *Phys. Rev. Lett.*. 61, 2472, 1988.
2. G. Binasch, P. Grünberg, F. Saurenbach, and W. Zinn, *Phys. Rev. B*, 39, 4828, 1989.
3. J.F. Gregg, W. Allen, K. Ounadjela, M. Viret, M. Hehn, S.M. Thompson, and J.M.D. Coey, *Phys. Rev. Lett.*, 77, 1580, 1996.

4. U. Ebels, A. Radulescu, Y. Henry, L. Piraux, and K. Ounadjela, *Phys. Rev. Lett.,* 84, 983, 2000.
5. G. Dumpich, T.P. Krome, and B. Hausmanns, *J. Magn. Magn. Mater.,* 248, 241, 2002.
6. M. Kläui, C.A.F. Vaz, J. Rothman, J.A.C. Bland, W. Wernsdorfer, G. Faini, and E. Cambril, *Phys. Rev. Lett.,* 90, 097202, 2003.
7. N. Garcîa, M. Muñoz, and Y.-W. Zhao, *Phys. Rev. Lett.,* 82, 2923, 1999.
8. C. Rüster, T. Borzenko, C. Gould, G. Schmidt, L. Molenkamp, X. Liu, T. J. Wojtowicz, J.K. Furdyna, Z.G. Yu, and M.E. Flatté, *Phys. Rev. Lett.,* 91, 216602, 2003.
9. J. Grollier, D. Lacour, V. Cros, A. Hamzic, A. Vaurés, A. Fert, D. Adam, and G. Faini, *J. Appl. Phys.,* 92, 4825, 2002.
10. N. Vernier, D.A. Allwood, D. Atkinson, M.D. Cooke, and R.P. Cowburn, *Europhys. Lett.,* 65, 526, 2004.
11. S. Lepadatu and Y.B. Xu, *Phys. Rev. Lett.,* 92, 127201, 2004.
12. A. Yamaguchi, T. Ono, S. Nasu, K. Miyake, K. Mibu, and T. Shinjo, *Phys. Rev. Lett.,* 92, 077205, 2004.
13. D. Weinmann, R.L. Stamps, and R.A. Jalabert, in *Electronic Correlations: From Meso- to Nano-physics,* T. Martin, G. Montambaux, and J. Trân Thanh Vân, Eds., EDP Sciences, Les Ulis, France, 2001.
14. V.A. Gopar, D. Weinmann, R.A. Jalabert, and R.L. Stamps, *Phys. Rev. B,* 69, 014426, 2004.
15. P.M. Levy and S. Zhang, *Phys. Rev. Lett.,* 79, 5110, 1997.
16. E. Simanek, *Phys. Rev. B,* 63, 224412, 2001.
17. T. Valet and A. Fert, *Phys. Rev. B,* 48, 7099, 1993.
18. P.E. Falloon, R. Jalabert, D. Weinmann, and R.L. Stamps, *Phys. Rev. B,* 70, 174424, 2004.
19. G. Tatara, *Int. J. Mod. Phys. B,* 15, 321, 2001.
20. P.E. Falloon, D. Weinmann, R.A. Jalabert, and R.L. Stamps, *Phys. Rev. B,* in press.
21. P. Bruno, *Phys. Rev. Lett.,* 83, 2425, 1999.
22. M. Büttiker, *Phys. Rev. Lett.,* 57, 1761, 1986.
23. S. Datta, *Electron Transport in Mesoscopic Systems,* Cambridge University Press, Cambridge, U.K., 1997.
24. H. Forster, T. Schrefl, D. Suess, W. Scholz, V. Tsiantos, R. Dittrich, and J. Fidler, *J. Appl. Phys.,* 91, 6914, 2002.
25. X. Waintal and M. Viret, *Europhys. Lett.,* 65, 427, 2004.
26. A. Brataas, G. Tatara, and G.E.W. Bauer, *Phys. Rev. B,* 60, 3406, 1999.
27. L. Berger, *J. Appl. Phys.,* 55, 1954, 1984.
28. L. Berger, *J. Appl. Phys.,* 71, 2721, 1992.
29. G.G. Cabrera and L.M. Falicov, *Phys. Stat. Sol. B,* 62, 217, 1974.
30. J.B.A.N. van Hoof, K.M. Schep, A. Brataas, G.E.W. Bauer, and P.J. Kelly, *Phys. Rev. B,* 59, 138, 1999.
31. G. Tatara and H. Kohno, *Phys. Rev. Lett.,* 92, 086601, 2004.

10 Introduction to a Theory of Current-Driven Domain Wall Motion

H. Kohno and G. Tatara

CONTENTS

10.1 INTRODUCTION

One of the recent topics in the field of spintronic science and technology is the manipulation of nanoscale magnetization by electric current. Although this phenomenon can be important in technological applications such as the writing process in magnetic random access memory, it will also interest basic scientists (solid state physicists) because it arises from the exchange interaction between conduction electrons and magnetization.

In this chapter, we would like to present a theoretical treatment of the domain wall motion driven by electric current. We first give an elementary introduction to a theoretical description of domain wall dynamics, by taking the (magnetic-) field-driven case as an example. The effect of electric current will then be introduced to study the main subject of current-driven domain wall motion.

10.2 FIELD-DRIVEN DOMAIN WALL MOTION

Suppose we have two nails made of iron. We know that they are ferromagnets at room temperature, but they do not attract each other. The reason is the existence of the magnetic domain structure, which is formed to lower the magnetostatic energy due to long-range dipole-dipole interaction. The typical size of domains is about 1 to 10 μm, and any semimacroscopic sample inevitably has a multidomain structure with vanishing total moment in zero field (if, e.g., it is prepared by cooling from above the Curie temperature). On the other hand, nanoscale magnets have simple domain structures suitable for application and basic studies. The transition region between two neighboring domains, as illustrated in Figure 10.1e, is the magnetic domain wall (DW).

If we apply an external magnetic field, magnetic domains with lowered magnetic energy grow at the expense of other domains (Figure 10.1a to Figure 10.1d). This magnetization process occurs as the movement of DWs. In this section, we shall consider how to describe this field-driven DW motion.[1-4]

10.2.1 PHENOMENOLOGICAL EQUATION OF MOTION

Consider a ferromagnetic wire containing a single DW (Figure 10.2). Let the position of the DW be X. In the presence of external magnetic field B applied in the easy-axis direction, the energy of the whole magnet is linearly dependent on X and B as

$$E_B = -2\gamma\hbar SB\frac{AX}{a^3} + const. \tag{10.1}$$

Here we are considering magnetic moments due to localized spins of magnitude S and gyromagnetic ratio γ.[5] A is the cross-sectional area of the wire, and a^3 is the volume per spin. If we regard the DW as a particle, whose position is X, Equation 10.1 represents the potential energy of this particle, meaning that this particle feels a force,

$$F_B = 2\gamma\hbar SB\frac{A}{a^3} \tag{10.2}$$

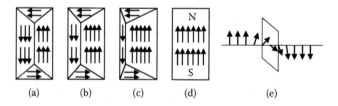

(a)   (b)   (c)   (d)   (e)

FIGURE 10.1 (a): Magnetic domain structure with lowest energy in zero external magnetic field. b → c → d illustrates the magnetization process due to domain wall motion. This occurs under an applied magnetic field, but also under an applied electric current, the latter being the present main subject. (e) An example of magnetic domain wall (Bloch wall).

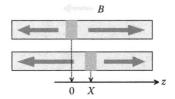

FIGURE 10.2 A ferromagnetic wire containing a single DW. The position of the DW is denoted by X. Thick arrows represent magnetization direction rather than spin direction.

proportional to B. One may then expect, naively, that the motion of this particle is described by an equation of the form

$$M_w \ddot{X} = F_B - \eta \dot{X}. \tag{10.3}$$

Here the dot represents time derivative, and η is some appropriate friction constant. This is indeed a very naive expectation; however it is known that Equation 10.3 holds under certain conditions. In particular, DW has inertial mass M_w known as the Döring mass.[6] To see this, and to derive correct equations of motion for a DW, we shall next look into the dynamics of individual spins in the DW.

10.2.2 EQUATIONS OF MOTION FROM MICROSCOPIC SPIN DYNAMICS

To see how the DW motion occurs at the spin level, let us first set up a microscopic model. We assume that the magnetization is described as localized spins S_i located at each site i with the following spin Hamiltonian

$$H_S = -\tilde{J} \sum_{\langle i,j \rangle} S_i \times S_j + \frac{1}{2} \sum_i (K_\perp S_{i,y}^2 - K S_{i,z}^2). \tag{10.4}$$

Here $\tilde{J}$ is the exchange coupling constant, and K and $K_\perp$ represent easy-and hard-axis anisotropy constants, respectively. We have chosen the easy axis to be the z axis and the hard axis to be the y axis $(K > 0, K_\perp \geq 0)$. A domain wall in its ground state is formed in the xz plane, called the easy plane. See Figure 10.4A.

Each spin obeys a simple law in its dynamics. It continues precessing around an effective field H_{eff} if there is no damping (Figure 10.3a), and such precession will decay and the spin (or, precisely speaking, the associated magnetic moment) eventually points to the direction of H_{eff} in the presence of damping (Figure 10.3b).

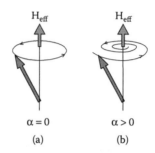

$$\alpha = 0 \qquad\qquad \alpha > 0$$
$$\text{(a)} \qquad\qquad\quad \text{(b)}$$

FIGURE 10.3 Dynamics of a single spin (or precisely, magnetic moment) in an effective field H_{eff}. (a) Without damping. (b) With damping. As for the direction of motion please note Reference 5.

These are, respectively, described by the first and the second terms of (the right-hand side [RHS] of) the following Landau–Lifshits–Gilbert (LLG) equation[1,7]

$$\frac{dS}{dt} = -\gamma S \times H_{\text{eff}} - \alpha \frac{S}{S} \times \frac{dS}{dt} \qquad (10.5)$$

Here, α is a dimensionless damping constant, called the Gilbert damping constant. In this subsection, we consider the case of no damping ($\alpha = 0$) for simplicity.

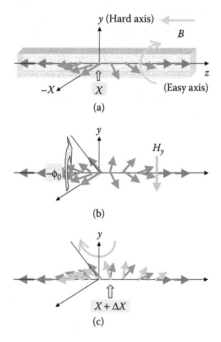

FIGURE 10.4 Dynamics of spins (actually, arrows shown here are magnetic moments rather than spins)[5] in the DW under applied magnetic field $B = (0, 0, -B)$. Spins in the DW rotate around the applied field (a), deviate from the easy plane (b), and rotate around the hard axis (c). As a whole, the DW makes a translational motion.

Back to the DW configuration described by Equation 10.4, we apply a magnetic field $B = (0, 0, -B)$ along the easy axis, as represented by the additional term

$$H_B = -\gamma \hbar B \sum_i S_{i,z} \qquad (10.6)$$

in the Hamiltonian, and consider the resulting motion of the DW. Because the spins in the DW are not parallel to the field B, they will precess around B, that is, around the easy axis (see Figure 10.4b); therefore they deviate from the easy (xz) plane and acquire a hard-axis component S_y. The energy thus increases by

$$E_\perp = \frac{1}{2} K_\perp S_y^2 \qquad (10.7)$$

per spin, and the spins feel an effective field[8]

$$H_y = \frac{1}{\gamma \hbar} \frac{\partial E_\perp}{\partial S_y} = \frac{1}{\gamma \hbar} K_\perp S_y \qquad (10.8)$$

in the hard-axis direction and rotate around this field (Figure 10.4c). Namely, spins in the DW rotate in the easy plane, and this is nothing but the translational motion of the DW. The direction of the translation is, as you see, in the correct direction, so that the magnetization occurs in the correct way.

Let us describe this process a little more quantitatively. First, the applied B induces a rotation around the easy axis. If we denote the rotation angle ϕ_0, it obeys the equation

$$\dot{\phi}_0 = -\gamma B. \qquad (10.9)$$

Now we have finite ϕ_0 (hence finite $S_y = S \sin \phi_0$), and thus have an increase in energy, $E_\perp = 1/2 K_\perp S^2 \sin^2 \phi_0$, due to hard-axis anisotropy. This causes a torque

$$T_z = -\frac{\partial E_\perp}{\partial \phi_0} = -K_\perp S^2 \sin \phi_0 \cos \phi_0 \qquad (10.10)$$

in the easy-axis direction, and this torque induces a change of $S_{tot}^z = \sum_i S_i^z$, which is nothing but the DW motion. This process is quantified as

$$\frac{A\dot{X}}{a^3} 2\hbar S = -N \left(\frac{1}{2} K_\perp S^2 \sin 2\phi_0 \right) \qquad (10.11)$$

or

$$\frac{\dot{X}}{\lambda} = -\frac{K_\perp S}{2\hbar}\sin 2\phi_0,\qquad(10.12)$$

where $N = 2A\lambda/a^3$ is the number of spins in the wall. In Equation 10.11, the left-hand side (LHS) represents the change of angular momentum S_{tot}^z per unit time, and the RHS represents a torque acting on N spins.[9] Equation 10.9 corresponds to the x, y components of the LLG equation (Equation 10.5), and Equation 10.11 corresponds to the z component, with[10]

$$\boldsymbol{H}_{eff} = \boldsymbol{B} + \frac{1}{\gamma\hbar}\frac{\partial E_\perp}{\partial \boldsymbol{S}}.\qquad(10.13)$$

In Equation 10.9, the magnetic field B appears on the RHS. In the particle picture, we have seen that it is the force acting on the DW, which, according to Newton's equation of motion, should be equal to the time-derivative of "momentum." In this sense, ϕ_0 may be regarded as a momentum of the DW.[11] If $|\phi_0| \ll 1$, we may linearize the sine term of Equation 10.12 and see that ϕ_0 is indeed proportional to $\dot{X}$, the DW velocity. From the coefficient, we can read the DW mass as[12]

$$M_w = \frac{\hbar^2 N}{\lambda^2}\frac{1}{K_\perp}.\qquad(10.14)$$

This is known as the Döring mass.[6,13] It is inversely proportional to $K_\perp$; if $K_\perp = 0$, the mass of the DW is infinite, and the DW cannot move even if it feels a force. This is a consequence of the conservation of angular momentum S_{tot}^z.[14,15] Thus $K_\perp$ works as a source of angular momentum; in this sense, $K_\perp$ assists the DW motion. This feature is characteristic to the force-driven case; later we shall see that $K_\perp$ tends to prevent the DW motion if it is driven by spin current (by the spin-transfer effect).

10.2.3 LAGRANGIAN FORMULATION

In this section, we proceed to a little more theoretical treatment and put the analysis into Lagrangian formalism. The Lagrangian formalism is a very convenient method to focus on a few degrees of freedom (such as X and ϕ_0) out of many degrees of freedom. But let us start with a single spin.

The equation of motion of a single spin is given by the LLG equation (Equation 10.5). Apart from the damping term, it is derived from the Lagrangian[5]

$$L = \hbar S\dot{\phi}(\cos\theta - 1) - \gamma\hbar\boldsymbol{S}\times\boldsymbol{B} - V_{ani}(\theta,\phi)\qquad(10.15)$$

with $H_{eff} = B + (1/\gamma\hbar)(\partial V_{ani}/\partial S)$ in Equation 10.5. We have parametrized the spin orientation as

$$S(t) = S(\sin\theta\cos\phi, \sin\theta\sin\phi, \cos\theta) \qquad (10.16)$$

where $\theta = \theta(t)$ and $\phi = \phi(t)$. In Equation 10.15, the first term is the kinetic term; this form is known as the spin Berry phase in quantum mechanics;[16,17] it is important also in classical mechanics in that it leads to the equation of motion characteristic to angular momentum. (In the present study, we treat the magnetization and DW as classical objects.) One can check that the Euler–Lagrange equations

$$\frac{d}{dt}\frac{\partial L}{\partial \dot{q}} - \frac{\partial L}{\partial q} = 0 \qquad (10.17)$$

for $q = \theta$ and $q = \phi$, lead to the LLG equation (Equation 10.5) with $\alpha = 0$.

The Euler–Lagrange formalism does not include damping effects: it conserves energy (if L is not explicitly time-dependent). The damping can be included by the Rayleigh's method.[18] With a so-called dissipation function

$$W = \frac{\alpha}{2}\frac{\hbar}{S}\dot{S}^2 = \frac{\alpha}{2}\hbar S(\dot{\theta}^2 + \dot{\phi}^2\sin^2\theta), \qquad (10.18)$$

the Euler–Lagrange equation is now modified as

$$\frac{d}{dt}\frac{\partial L}{\partial \dot{q}} - \frac{\partial L}{\partial q} = -\frac{\partial W}{\partial \dot{q}}. \qquad (10.19)$$

One can show that $dH/dt = -2W$, where $H = \dot{q}(\partial L/\partial\dot{q}) - L$ is the Hamiltonian; therefore, the quantity $2W$ has a physical meaning of the energy dissipation rate.[19] Equation 10.19, with Equation 10.15 and Equation 10.18, then reproduces the full LLG equation (Equation 10.5).

It is straightforward to extend the above description for a single spin to many-spin systems. Here we restrict ourselves to spin configurations that are slowly varying in space and time and adopt a continuum description, $S_i(t) \rightarrow S(x, t)$, and write it as Equation 10.16 with $\theta = \theta(x, t)$ and $\phi = \phi(x, t)$ now describing spin configurations in a continuum space x and time t. (We consider an effectively one-dimensional magnet and neglect transverse variations.) The Lagrangian is then given by

$$L_S = \int \frac{d^3x}{a^3}[\hbar S\dot{\phi}(\cos\theta - 1) + \gamma\hbar BS\cos\theta$$
$$\qquad (10.20)$$
$$-\frac{S^2}{2}\{J[(\nabla\theta)^2 + \sin^2\theta(\nabla\phi)^2] + \sin^2\theta(K + K_\perp\sin^2\phi)\}]$$

where $J = \tilde{J}a^2$.

We now restrict ourselves to fixed magnetization profile representing a single DW.[20,21] The Euler–Lagrange equation derived from (Equation 10.20) has a static DW solution $\theta = \theta_w(x-X), \phi = 0$, where

$$\cos\theta_w(x) = \pm\tanh(x/\lambda) \qquad (10.21)$$

or $\sin\theta_w(x) = [\cosh(x/\lambda)]^{-1}$. We then regard the DW position X as a dynamical variable and allow its time dependence, $X = X(t)$. Using this DW solution in (Equation 10.20), we obtain the Lagrangian for a DW as

$$L_{dw} = \pm\frac{\hbar NS}{\lambda}\dot{X}\phi_0 - \frac{1}{2}K_\perp NS^2\sin^2\phi_0 \mp F_B X - V_{pin}(X) \qquad (10.22)$$

up to total time-derivative, where

$$\phi_0(t) \equiv \int\frac{dx}{2\lambda}\phi(x,t)\sin^2\theta_w[x-X(t)] \qquad (10.23)$$

is essentially the angle $\phi(x,t)$ at the DW center. We have introduced a pinning potential $V_{pin}(X)$ for the DW coming from spatial irregularities. The first term of Equation 10.22 means that, except for a proportionality constant, X and ϕ_0 are canonical conjugate to each other; namely, ϕ_0 is the canonical momentum conjugate to X. We have noted this fact before by an intuitive argument, but now have shown it mathematically.[21] Similarly, the dissipation function becomes

$$W_{dw} = \alpha\hbar S\frac{A}{a^3}\left(\frac{\dot{X}^2}{\lambda} + \lambda\dot{\phi}_0^2\right) \qquad (10.24)$$

Taking variations with respect to X and ϕ_0, respectively, we obtain the equations of motion for a DW as follows

$$\frac{\hbar NS}{\lambda}\left(\pm\dot{\phi}_0 + \alpha\frac{\dot{X}}{\lambda}\right) = \mp F_B + F_{pin}, \qquad (10.25)$$

$$\frac{\hbar NS}{\lambda}(\pm\dot{X} - \alpha\lambda\dot{\phi}_0) = \frac{NS^2 K_\perp}{2}\sin 2\phi_0. \qquad (10.26)$$

These equations (especially, with the lower sign[22]) are equivalent to Equation 10.9 and Equation 10.11, but now include damping effects. The magnetic field acts as a force, $F_B = \gamma\hbar(NS/\lambda)B$, and promotes ϕ_0 (i.e., spin rotation around the easy axis),

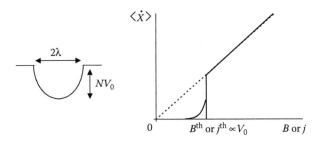

FIGURE 10.5 (Left) Model pinning potential. (Right) Time-averaged DW velocity $\langle \dot{X} \rangle$ as a function of applied field B for the field-driven case. At absolute zero (thick line) and at finite temperature (thin line). By replacing B by electric (charge) current j (except for some factor), this graph also applies to the current-driven DW motion if the momentum-transfer mechanism dominates over the spin-transfer mechanism (See Section 10.3).

and the hard-axis anisotropy $K_\perp$ acts as a torque and drives X (i.e., translation of the DW). If the system is not spatially uniform, we have a pinning force $F_{pin} = -(\partial V_{pin}/\partial X)$. Note that X and ϕ_0 have different time-reversal properties, and their mixing (α terms) means breaking of time-reversal symmetry and leads to damping. For $|\phi_0| \ll 1$, Equation 10.25 and Equation 10.26 reduce to Equation 10.3 with $\eta = \alpha(\hbar NS / \lambda^2)$.[23]

10.2.4 DYNAMICS DRIVEN BY FORCE

Here we give a brief study on the dynamics of a DW based on Equation 10.25 and Equation 10.26.

In the absence of a pinning potential ($V_{pin} = 0$), Equation 10.25 and Equation 10.26 have a stationary solution describing a moving DW with constant velocity $\dot{X} = \mp \gamma B \lambda / \alpha$ and constant ϕ_0 (i.e., $\dot{\phi}_0 = 0$). The value of ϕ_0 is determined by Equation 10.26 This stationary solution exists only if $|\dot{X}| < K_\perp S \lambda / 2\hbar$ or $|B| < \alpha K_\perp S / 2\gamma\hbar$. Above this field, ϕ_0 also becomes time dependent, and the DW motion acquires oscillatory components.

To study the effect of a pinning potential, we model it as (see Figure 10.5, left)

$$V_{pin} = \frac{NV_0}{\xi^2}(X^2 - \xi^2)\Theta(\xi - |X|), \tag{10.27}$$

where $\Theta(x)$ is the Heaviside step function, V_0 the pinning strength, and ξ is the pinning range. In this case, the DW is pinned below the threshold field $B^{th} = 2V_0\lambda/(\gamma\hbar S\xi)$. The DW velocity vs. B relation is shown in Figure 10.5 (right). Here and hereafter, we take $\xi = \lambda$ for simplicity.

10.3 CURRENT-DRIVEN DOMAIN WALL MOTION

In the previous section, we have reviewed the DW motion driven by a magnetic field. The materials presented there will be well-known results. In particular, the theoretical descriptions given in Section 10.2.2 and Section 10.2.3 are equivalent to the LLG equation with a constraint that the magnetization profile is fixed to that of a single DW. In this section, we introduce the effect of electric current (due to conduction electrons) into the same formalism. We treat conduction-electron and magnetization degrees of freedom as independent.[24] In the following, "magnetization" means that of localized spins. Also, we refer to conduction electrons simply as "electrons."

10.3.1 INTERACTION OF DOMAIN WALL AND ELECTRON FLOW: AN INTUITIVE PICTURE

How does the electric current affect the DW motion? What is the relevant interaction between electrons and the magnetization? It is the *s-d* exchange interaction

$$H_{sd} = -J_{sd} \int d^3x S(x) \cdot (x) \cdot \vec{\sigma}(x) \tag{10.28}$$

between localized spins $S(x)$ and electron spins. Here $\vec{\sigma}(x) = c^\dagger(x)\vec{\sigma}c(x)$ represents the spin density of conduction electrons, where $c(x)$ is the electron annihilation operator. There are certainly other possible effects, such as hydromagnetic or electromagnetic, but they are ineffective in small or thin systems,[25] or can be excluded experimentally.

The *s-d* interaction affects the DW motion in two different ways. One is the momentum transfer, or force, and the other is the spin transfer, or torque.

Consider a metallic ferromagnet containing a single DW, and suppose there is an electron flow from left to right (Figure 10.6). If an electron is reflected by the DW, its momentum is changed (Figure 10.6a). This process acts as a force on the DW by transferring the linear momentum from the electron to the DW. (Note that the *s-d* interaction (Equation 10.28) is translationally invariant and conserves the total momentum of electrons and magnetization.) This force is proportional to the charge current j and DW resistivity ρ_w as will be shown later.

On the other hand, if the electron is transmitted through the DW *adiabatically,* namely, by keeping its spin direction closely parallel to the local magnetization, the spin angular momentum of the electron is changed (Figure 10.6b). This process acts as a torque on the DW by transferring the spin angular momentum from the electron to the DW. (Note that the *s-d* interaction [Equation 10.28] also conserves the total spin angular momentum of the electrons and magnetization.) In other words, this change of the electron spin should be absorbed by the magnetization, leading to a

(a)   (b)

FIGURE 10.6 Two effects of electric current on the DW via the *s-d* exchange interaction. (a) Reflected electron has transferred linear momentum to the DW. (b) Adiabatically transmitted electron has transferred spin angular momentum to the DW.

translational motion of the DW. This torque can be shown to be proportional to the spin current j_s in the adiabatic limit. This is the spin-transfer effect; a conduction electron exerts a spin torque on the DW. These two effects were originally found by Berger long ago.[25-27] The spin-transfer effect is now familiar through recent studies on multilayer of pillar systems.[28,29]

These two effects enter the equations of motion as follows[22,30]

$$\frac{\hbar NS}{\lambda}\left(\dot{\phi}_0 + \alpha\frac{\dot{X}}{\lambda}\right) = F_{el} + F_B + F_{pin} \qquad (10.29)$$

$$\frac{\hbar NS}{\lambda}(\dot{X} - \alpha\lambda\dot{\phi}_0) = \frac{NS^2 K_\perp}{2}\sin 2\phi_0 + T_{el,z} \qquad (10.30)$$

where F_{el} is the force and $T_{el,z}$ is the torque, both from electrons. One might easily convince oneself that the force F_{el} from electrons appears in the equations in exactly the same way as the other forces (F_B and F_{pin}) already considered. As for the torque $T_{el,z}$ we may understand it as follows.

Suppose a spin current I_s (in unit of $\hbar/2$, the dimension being, e.g., an ampere) is flowing in the left far region to the DW. In the adiabatic case, the spin current flowing in the right far region to the DW is given by $-I_s$. Because the change of angular momentum of a single electron is $\hbar$ after a passage through the DW, the total angular momentum of electrons is changing at a rate $\hbar I_s/e$, which is transferred to the DW, and acts as a torque

$$T_{el,z} = \frac{\hbar}{e}I_s \qquad (10.31)$$

on the DW. This torque is now added to the RHS of Equation 10.11, leading to Equation 10.30 above.

For thick walls as realized in metallic wires,[31-35] the reflection probability (or DW resistivity) will be vanishingly small, and the spin-transfer effect will be the dominant driving mechanism. (It turns out that this is not the case for AC current.[36]) The momentum-transfer effect is considered to be effective only for thin walls, as in nanocontacts and possibly in magnetic semiconductors.[37]

10.3.2 MATHEMATICAL DERIVATION

Mathematically, what is new compared to the field-driven case in the s-d exchange term H_{sd}. Supplementing the electron part by, for example, that of a free-electron system

$$L_{el} = \sum_k c_k^\dagger(i\hbar\partial_t - \varepsilon_k)c_k \qquad (10.32)$$

with $\varepsilon_k = \hbar^2 k^2/2m$, we consider the total Lagrangian $L_{tot} = L_{dw} + L_{el} - H_{sd}$ and derive the equations of motion by taking variations with respect to X and ϕ_0.

The s-d coupling introduces two new terms into the equations of motion.[30] One comes from the X derivative, which is force

$$F_{el} \equiv -\left\langle \frac{\partial H_{sd}}{\partial X} \right\rangle = -\int d^3x \, \nabla_x M_w(x) \cdot n(x) \qquad (10.33)$$

where $M_w(x) = J_{sd} S_{dw}(x - X)$, with S_{dw} being the DW spin configuration, and $n(x) = \langle \bar{\sigma}(x) \rangle$ represents spin polarization of electrons. The other comes from the ϕ_0 derivative, which is torque (force on $-\phi_0$)

$$T_{el,z} \equiv +\left\langle \frac{\partial H_{sd}}{\partial \phi_0} \right\rangle = -\int d^3x \, [M_w(x) \times n(x)]_z \qquad (10.34)$$

As seen in the previous Section 10.3.1, the torque $T_{el,z}$ is proportional to the spin current flowing in the bulk (far from the DW) in the adiabatic case.

The force comes from the gradient of magnetization when electrons are scattered by this spatial nonuniformity. There is an exchange of linear momentum between electrons and the DW. The torque comes from the mismatch in direction between DW magnetization and electron spin polarization. They precess around each other and exchange spin angular momentum with each other.

The equations of motion are obtained as Equation 10.29 and Equation 10.30 above. These equations are directly obtained from the effective Lagrangian

$$L_{eff} = -\frac{\hbar NS}{\lambda} X \dot{\phi}_0 - \frac{1}{2} K_\perp NS^2 \sin^2 \phi_0 + (F_B + F_{el})X - T_{el,z}\phi_0 - V_{pin}(X) \qquad (10.35)$$

together with the dissipation function (Equation 10.24). Among others, there is a direct linear coupling between ϕ_0 and spin current (or spin torque $T_{el,z}$).

10.3.3 MICROSCOPIC CALCULATION OF FORCE AND TORQUE

Force and torque both vanish if there is no current (and if the DW is at rest). They can be finite in the presence of current. In this sense, the calculation of force and torque resembles that of transport coefficients, and we can use the techniques developed for the calculation of transport coefficients, such as those named after the great physicists, Boltzmann, Kubo, Landauer, Keldysh, and others. Such serious calculations will be presented elsewhere. The result is that the torque is, as seen above, proportional to the spin current in the adiabatic limit.

As for the force, consider the situation that the electrons are accelerated by an applied electric field E, and a steady current state is maintained by the scattering from a DW. We assume that only the DW scatters electrons. Then, the electrons acquire momentum $eN_{el}E$ from E per unit time (this is nothing but the force), where N_{el} is the total number of electrons, and this whole momentum is released to the DW under the stationarity assumption. Thus the DW feels a force $F_{el} = eN_{el}E$ from the electrons. The current density j is related to E as $E = \rho_w j$ via the resistivity ρ_w due to a single DW. Eliminating E from these two relations, we obtain

$$F_{el} = eN_{el}\rho_w j \tag{10.36}$$

10.3.4 Dynamics Driven by Spin Transfer

To study the dynamics, we first rewrite Equation 10.29 and Equation 10.30 in the form

$$\dot{\phi}_0 + \alpha\frac{\dot{X}}{\lambda} = f - v_{pin}\frac{X}{\lambda} \tag{10.37}$$

$$\frac{\dot{X}}{\lambda} - \alpha\dot{\phi}_0 = \kappa_\perp \sin 2\phi_0 + v_s \tag{10.38}$$

where all parameters, except for α, have dimensions of frequency; $f = \lambda/\hbar NS(F_{el} + F_B)$, $v_{pin} = 2V_0/\hbar S$, $\kappa_\perp = SK_\perp/2\hbar$, and $v_s = T_{el,z}/(\hbar NS) = I_s/(eNS)$. We have adopted Equation 10.27 for the pinning potential. In this section, we focus on the DW dynamics driven by the spin-transfer effect and set $f = 0$. (The momentum-transfer effect, or force, is formally the same as the field-driven case; both enter the equation through f defined above.) We consider only the case of DC current and assume v_s is time-independent. (This also includes the case of a pulsed current.)

We first consider the case without pinning potential, $v_{pin} = 0$. In Equation 10.30, the spin transfer, or spin torque, enters as a source to the DW velocity $\dot{X}$; it tries to drive $\dot{X}$ directly. However, there is also a hard-axis anisotropy ($\kappa_\perp$) term, and this term tends to absorb the transferred spin. In fact, if $|v_s|$ is smaller than $\kappa_\perp$, spin transfer v_s is completely absorbed by the $\kappa_\perp$ term, that is, transferred to the lattice, and is not used for the translational motion ($\dot{X}$) of the DW: DW is apparently pinned and not driven to a stream motion even in the absence of a pinning potential.[30]

The time dependence of X and ϕ_0 are shown in Figure 10.7a. The DW approaches a static state with finite displacements of X and ϕ_0, which is maintained by dissipating the spin transfer from electrons to the lattice via $\kappa_\perp$. If $|v_s|$ exceeds $\kappa_\perp$, DW moves with constant average velocity $\langle\dot{X}\rangle$ with oscillating components superposed (Figure 10.7b). This oscillation is due to the alternating exchange of angular momentum between X and ϕ_0. (ϕ_0 also varies with some constant average angular velocity $\langle\dot{\phi}_0\rangle$, which is smaller than $\langle\dot{X}\rangle$ by a factor of α.) Therefore, for the DW motion driven by the spin-transfer effect, there is a finite threshold spin current

$$j_s^{th(1)} = \frac{eS^2}{a^3\hbar}K_\perp\lambda \tag{10.39}$$

essentially given by $K_\perp$. This threshold is finite even if there is no pinning potential. This is an intrinsic pinning due to $K_\perp$. Here, $K_\perp$ tends to prevent the DW motion by absorbing the transferred spin angular momentum, in contrast to the field-driven

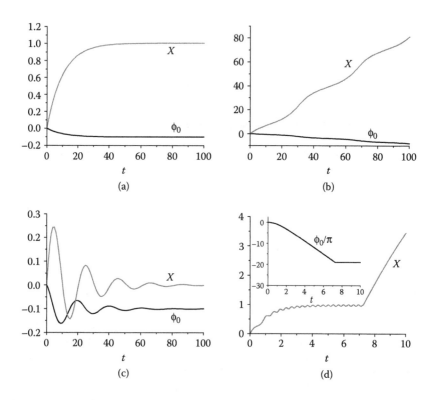

FIGURE 10.7 Time dependence of X and ϕ_0 under spin current v_s, without pinning potential $v_{pin} = 0$ (a, b) and with pinning potential $v_{pin} > 0$ (c,d), for $\kappa_\perp = 0.5$ and $\alpha = 0.1$ (Reference 38). (a) Below threshold $j_s^{th(1)}$ ($v_s < \kappa_\perp$, $v_s = 0.1$). (b) Above threshold $j_s^{th(1)}$ ($v_s > \kappa_\perp$, $v_s = 1$). (c) Below $j_s^{th(1)}$, pinned regime ($v_s < \kappa_\perp$; $v_s = 0.1$, $v_{pin} = 0.1$). (d) Above $j_s^{th(2)}$ ($v_s > \kappa_\perp$, $v_s = 0.977$, $v_{pin} = 10$), the depinning is seen.

or force-driven case, where, as we have seen before, $K_\perp$ helps the DW motion by supplying angular momentum. Above the threshold, the time-averaged DW velocity is given by

$$\langle \dot{X} \rangle = \frac{1}{1+\alpha^2} \frac{1}{2S} \frac{a^3}{e} \sqrt{j_s^2 - (j_s^{th(1)})^2} \qquad (10.40)$$

We next consider the case with extrinsic pinning potential, $v_{pin} > 0$. For $|v_s| < \kappa_\perp$, after the initial transient period with oscillation, the DW is eventually pulled back to the pinning center ($X = 0$) and ϕ_0 approaches a constant value (Figure 10.7c). For $|v_s| > \kappa_\perp$, DW makes a finite displacement ΔX and oscillates around the mean position (Figure 10.7d.) This is the pinned state. In this pinned state, transferred spin is solely used to drive ϕ_0, and ϕ_0 continues to vary rapidly. In this pinned state, AC noise or electromagnetic radiation may be expected due to this ϕ_0 motion. If X happens to go beyond ("step over") the pinning range, the DW is depinned. After depinning, the ϕ_0-motion slows down, and the transferred spin is mainly used for the translational motion of the DW.

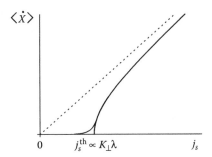

FIGURE 10.8 Time-averaged DW velocity $\langle \dot{X} \rangle$ as a function of spin current j_s at absolute zero (thick line)[30] and at finite temperature (thin line).[41] The dotted line represents $\langle \dot{X} \rangle$ for the case of complete spin transfer ($K_\perp = 0$).

Let us examine the depinning condition for the latter case (Figure 10.7D). In the pinned state ($\dot{X} \sim 0$), the average DW position may be estimated from Equation 10.37 as $\Delta X \sim -\lambda \dot{\phi}_0 / v_{\text{pin}}$, whereas $\dot{\phi}_0 \sim -v_s / \alpha$ from Equation 10.38. The condition for depinning, $|\Delta X| > \lambda$, thus leads to the second threshold

$$j_s^{\text{th}(2)} = \frac{4e}{a^3 \hbar} \alpha V_0 \lambda \tag{10.41}$$

for the spin current.

The actual threshold spin current j_s^{th} is given by the larger of $j_s^{\text{th}(1)}$ and $j_s^{\text{th}(2)}$. Because $j_s^{\text{th}(2)}$ contains a factor of α, which is considered to be very small (~ 0.01), j_s^{th} is determined by the hard-axis anisotropy,[39] if the pinning potential is not extremely strong.[30,40] For this case, the time-averaged DW velocity $\langle \dot{X} \rangle$ is plotted as a function of spin current j_s in Figure 10.8.

So far, experimental studies on the DW motion have been focused on the X motion. It will be interesting to detect the motion of ϕ_0 as well.

10.4 SUMMARY

We have attempted to present an elementary explanation of the domain wall dynamics, for both field-driven and current-driven cases. An important observation is that the angle ϕ_0 should be treated on an equal footing with the DW position X,[4] because, mathematically, they are canonically conjugate to each other.[21]

For the current-driven case, there are two mechanisms to drive the DW; one is force, or momentum transfer, which is proportional to charge current and DW resistivity, and the other is torque, or spin transfer, which is proportional to spin current. The spin-transfer mechanism will be dominant for thick walls as in metals, whereas the momentum-transfer mechanism may be effective for thin walls as in nanocontacts and magnetic semiconductors. In the spin-transfer mechanism, even in the absence of a pinning potential, there is a finite threshold spin current to drive the DW, which is essentially given by the hard-axis anisotropy constant.

Extensive experimental studies have recently been carried out on current-driven DW motion. For details, please see Reference 31 through Reference 37 and Chapter 8 of this book.

ACKNOWLEDGMENTS

We would like to thank H. Ohno, T. Ono, E. Saitoh, J. Shibata, A. Yamaguchi, and M. Yamanouchi for valuable discussions. H.K. is indebted to K. Miyake for his continual encouragement.

REFERENCES AND NOTES

1. S. Chikazumi, *Physics of Ferromagnetism*, 2nd ed. Oxford University Press, Oxford 1997.
2. A. Hubert and R. Schäfer, *Magnetic Domains*, Springer-Verlag, New York, 1998.
3. F. H. de Leeuw, R. van den Doel, and U. Enz, *Rep. Prog. Phys.*, 43, 659, 1980.
4. J. C. Slonczewski, *Int. J. Magn.*, 2, 85, 1972; *J. Appl. Phys.*, 45, 2705, 1974.
5. We consider the case that the spin S and the associated magnetic moment μ are opposite in direction, $\mu = -\gamma\hbar S$ with $\gamma > 0$, as is the case for a free electron.
6. W. Döring, *Z. Naturforsch*, 3A, 373, 1948.
7. The equation for magnetization M, instead of spin S (Reference 5), has a plus sign (instead of minus sign) in front of the damping ($\alpha-$) term in Equation 10.5.
8. This may include the demagnetizing field from surface magnetic charges (as in nanowires) or from volume magnetic charges (as in moving Bloch walls[1,20] as well as the effective field due to crystalline magnetic anisotropy.
9. We may also estimate the torque as $T_z = -\gamma\hbar(S \times H_{eff})_z = -\gamma\hbar S_x H_y = -K_\perp S_x S_y$ (from Equation 10.8), with $S_x = S\cos\phi_0$ and $S_y = S\sin\phi_0$.
10. Effective fields coming from $\tilde{J}$ and K in the Hamilton (Equation 10.4) cancel each other thanks to the stationary property of DW solution.
11. If we rewrite Equation 10.9 as $-(\hbar S N/\lambda)\dot{\phi}_0 = F_B$, with F_B given by (Equation 10.2), we see that $P = -(\hbar S N/\lambda)\phi_0$ is the corresponding momentum. In Section 10.2.3, we will see that this is the canonical momentum rather than the dynamical momentum.
12. From Equation 10.12 and Reference 11, we have $P = (\hbar^2 N/\lambda^2 K_\perp)\dot{X}$ for "momentum," which may be identified with $M_w\dot{X}$.
13. For a Bloch wall, the demagnetizing field from volume magnetic charges due to nonuniform $\phi[x,t]$) contributes to $K_\perp$ by $2\pi M_0^2$, where $M_0 = \gamma\hbar S/a^3$ is the magnetization. Retaining only this contribution leads to the original expression by Döring (Reference 20).
14. The $K_\perp$ breaks spin rotational symmetry around the z axis, and hence breaks the conservation of S_{tot}^z. Note that the applied field B conserves S_{tot}^z.
15. Introduction of finite damping α also leads to nonconservation of S_{tot}^z. In this case, it is possible to exchange angular momentum between X and ϕ_0, as seen from Equation 10.25 and Equation 10.26. Then, even if $K_\perp = 0$, there is a solution of a moving DW, where the angular momentum necessary for the DW translation (X) is provided by ϕ_0.
16. A. Auerbach, *Interacting Electrons and Quantum Magnetism*, Springer Verlag, New York, 1994, Chap. 10.
17. M. V. Berry, *Proc. Roy. Soc. London* A, 392, 45, 1984.
18. H. Goldstein, C. Poole, and J. Safko, *Classical Mechanics*, 3rd ed. Addison Wesley, Boston, MA, 2002, Chap. 1, Sec. 5.
19. Therefore, W is independent of the choice (parametrization) of dynamical variables, and one can check that Equation 10.19 is covariant under general variable changes.

20. H.-B. Braun and D. Loss, *Phys. Rev. B*, 54, 3237, 1996; H.-B. Braun, J. Kyriakidis, and D. Loss, *Phys. Rev. B*, 56, 8129, 1997.

21. S. Takagi and G. Tatara, *Phys. Rev B*, 54, 9920, 1996.

22. If we take the upper sign in Equation 10.25 and Equation 10.26 and assume that the external field is applied in the positive z-direction, $B = (0, 0, +B)$, they coincide with those of Reference 30. Note that Equation 10.25 and Equation 10.26 (and Equation 10.9 and Equation 10.11) have been derived for $B = (0, 0, -B)$.

23. Precisely speaking, because of the special form of the damping (α) terms, the mass and force are renormalized as $M_w = (\hbar^2 N / \lambda^2 K_\perp)(1 + \alpha^2)$ and $F = f + \alpha(\hbar/SK_\perp)\dot{f}$, where $f = F_B + F_{\text{pin}}$.

24. This is not an essential assumption in the following part, but just for simplicity. In the case where the exchange interaction between itinerant electrons is not negligible (or the magnetization is solely due to itinerant electrons), we need to consider a self-consistency condition, which, however, does not seem to change the essential results at the present level of the theory.

25. L. Berger, *J. Appl. Phys.,* 49, 2156, 1978.

26. L. Berger, *J. Appl. Phys.,* 55, 1954, 1984.

27. L. Berger, *J. Appl. Phys.,* 71, 2721, 1992; E. Salhi and L. Berger, *ibid.,* 73, 6405, 1993.

28. J. C. Slonczewski, *J. Magn. Magn. Mater.,* 159, L1, 1996.

29. L. Berger, *Phys. Rev. B,* 54, 9353, 1996.

30. G. Tatara and H. Kohno, *Phys. Rev. Lett.,* 92, 086601, 2004.

31. A. Yamaguchi, T. Ono, S. Nasu, K. Miyake, K. Mibu, and T. Shinjo, *Phys. Rev. Lett.,* 92, 077205, 2004.

32. J. Grollier, D. Lacour, V. Cros, A. Hamzic, A. Vaurés, A. Fert, D. Adam, and G. Faini, *J. Appl. Phys.,* 92, 4825, 2002; J. Grollier, P. Boulenc, V. Cros, A. Hamzic, A. Vaurés, A. Fert, and G. Faini, *Appl. Phys. Lett.,* 83, 509, 2003.

33. M. Tsoi, R.E. Fontana, and S.S.P. Parkin, *Appl. Phys. Lett.,* 83, 2617, 2003.

34. M. Kläui, C.A.F. Vaz, J.A.C. Bland, W. Wernsdorfer, G. Faini, E. Cambril, and L.J. Heyderman, *Appl. Phys. Lett.,* 83, 105, 2003.

35. N. Vernier, D.A. Allwood, D. Atkinson, M.D. Cooke, and R.P. Cowburn, *Europhys. Lett.,* 65, 526, 2004.

36. E. Saitoh, H. Miyajima, T. Yamaoka, and G. Tatara, *Nature,* 432, 203, 2004.

37. M. Yamanouchi, D. Chiba, F. Matsukura, and H. Ohno, *Nature,* 428, 539, 2004.

38. For a more realistic value of $\alpha = 0.01$, characteristic features of the DW motion are modified, compared with those shown in Figure 10.7, as follows: A, B Time scale for both X and ϕ_0 becomes longer in proportion to α^{-1}. Magnitude of X is enhanced ($\propto \alpha^{-1}$) according to $\dot{X}/\lambda = -\alpha^{-1}\dot{\phi}_0$. (C) Decay time becomes longer ($\sim \alpha^{-1}$) whereas the oscillation period ($\Delta t \sim 20$) is unchanged, therefore, X and ϕ_0 oscillate many times ($\sim$30 to 40) before the decay. (D) The oscillation amplitude of X in the pinned state is much reduced.

39. Some preliminary experimental results seem to be consistent with this theoretical result. S.S.P. Parkin, private communication; A. Yamaguchi, H. Tanigawa, T. Ono, and S. Nasu, private communication.

40. Note that $j_s^{\text{th}(2)}/j_s^{\text{th}(1)} = \alpha(2\gamma\hbar B^{\text{th}})/(SK_\perp)$, where B^{th} is the depinning field defined in Section 10.2.4.

41. G. Tatara, N. Vernier, and J. Ferre, *Appl. Phys. Lett.,* 86, 252509, 2005.

Part III

Spin Injection and Spin Devices

Part III

Spin Injection and Spin Devices

11 Silicon-Based Spin Electronic Devices: Toward a Spin Transistor

Sarah M. Thompson, David Pugh, Duncan Loraine, Cindi L. Dennis, John F. Gregg, Chitnarong Sirisathitkul, and William Allen

CONTENTS

11.1 INTRODUCTION

A burgeoning trend in spin electronics[1-3] is the development of spin-sensitive semiconductor devices. Implementing spin-polarized carriers in semiconductors expands potential device functionality by differentiating between up and down spin carriers in both the conduction and valence bands. This spin selectivity may be realized by using magnetic semiconductors or by employing external spin-selective ferromagnets to inject spin-polarized current into nonmagnetic semiconductors. As the length scales are reduced, ferromagnetic single electron devices become feasible.

245

Crucial to the operation of all these devices are the fundamental processes of spin injection[4] into the semiconductor, the spin-diffusion length within the semiconductor, and spin-dependent detection.

There have been several attempts to fabricate spin transistors that exploit the spin-dependent scattering of charge carriers to yield a device with high current gain and high magnetic sensitivity. The first such attempt was Johnson's all-metal three-terminal device,[5] which added a third terminal to the middle paramagnetic layer of a giant magnetoresistive (GMR) multilayer. The electrical characteristics of this purely ohmic device are magnetically tuneable, but, due to its all-metal construction, its operation yields only small voltage output changes and no power gain. Subsequent versions of the spin transistor have attempted to integrate semiconductors with spin electronics to generate novel functionality. There are two major variants. In the first, the metallic components retain their spin selectivity while the semiconductor is used only to control the distribution of applied potentials across the device. Most versions[6-8] fall into this category, including the Monsma hot electron spin-valve transistor.[6] This device sandwiched a GMR multilayer between two pieces of silicon, forming an emitter Schottky barrier that injected electrons into the metallic base and a collector Schottky barrier whose height determined which carriers were collected. The magnetic configuration of the metallic GMR multilayer controlled the energy of the electrons reaching the collector Schottky barrier. Although the electrical characteristics of this transistor were magnetically sensitive, all of the manipulation of the electrons' spin occurred in the GMR multilayer. Furthermore, the thickness of the metal base layer, which must accommodate at least two (preferably more) ferromagnetic layers plus spacer layers, defined not only the degree of magnetic sensitivity, but also the magnitude of the current gain; the more layers present, the greater the magnetic sensitivity, but the lower the current gain. The best value for the ratio of collector current to emitter current in the hot electron spin-valve transistor is currently 1×10^{-3} (see Chapter 16 of this book).

The second variant, the experimental realization of a spin field effect transistor based on the proposal by Datta and Das,[9] was being developed concurrently. This transistor is a modification of a field effect transistor (FET) in which conventionally an applied electric field changes the width of the depletion region and hence the output current magnitude. In a spin FET, spin-polarized electrons are injected from a magnetic source into a semiconductor channel. During passage through the channel, these electrons undergo Rashba precession, the frequency of which depends upon the gate voltage. Finally, the electrons are analyzed by spin-selective scattering in the magnetic drain. The electrical characteristics are therefore dependent upon not only the magnetic orientation of the source and drain, but also on the gate voltage. Gardelis and coworkers[10] have made progress toward realizing this device, but the gate functionality remains to be convincingly demonstrated.

The spin transistor[11-13] described here is markedly different from those described above in that it exploits spin transport in the silicon as well as using minority carriers to achieve high current gain and magnetic sensitivity. It functions by using a ferromagnet to inject spin-polarized electrons via a tunnel barrier into a silicon base. These spin-polarized minority carriers traverse the base diffusively and are harvested by another ferromagnet via another tunnel barrier. The spin selectivity derives from the back-biased detector presenting a different density of final spin states to the spin-polarized minority carriers in the base. Furthermore, as in the bipolar junction transistor, the carrier scattering and recombination rates in the base may be very low, so that the current gain (given by the differential ratio of the detector current to the base current: $\beta = \Delta I_D/\Delta I_B$) is potentially very large compared to the Johnson and Monsma metal-based devices discussed above. Unlike the other three terminal spin devices, this spin-transistor design exhibits a maximum current gain greater than unity (1.4) and an average current gain (over all applied base currents) slightly less than unity (0.9), in addition to being magnetically sensitive (magnetocurrent changes by up to 140% in −110 Oe).

Before the full spin transistor is described, we first make the case for using silicon as our base material on which to build spin-electronic devices and consider some simpler devices designed to test some of the different device components as well as design and constructional issues. This enables us to explore the feasibility of such spin-electronic devices before examining the silicon spin transistor itself.

11.2 THE CASE FOR SILICON

Silicon remains industry's most important semiconductor, holding 98% of the world market. This is in part due to its well-known excellent material properties such as the ability to grow large, low defect, single crystals that have low brittleness and high thermal conductivity. The vast complexity and sophistication of silicon-based devices and chip fabrication processes has resulted in a breadth and depth of studies concerning the properties of silicon. Indeed, silicon was the first semiconductor to be studied using optical pumping,[14] but the mechanism is inefficient and yields a polarized spin population less than 0.01%. Spin resonance absorption in silicon and on impurities in silicon were first performed in the 1950s[15,16] the most comprehensive being by Feher concerning silicon.[17] However, the majority of work on the applications of spin electronics have been based on n-type GaAs, primarily due to the convenience of optical pumping to observe the spin lifetime and diffusion lengths of the carriers.[18] Surprisingly, only a handful of studies have considered silicon as the semiconductor in a spin-electronic device,[12,19,20] but if spin-electronic devices are to be integrated with existing silicon-based electronics, then it would be more practical for these devices themselves to be based on silicon.

To consider silicon as a material for spin-electronic devices, we need to consider its properties relating to spin transport, in particular its spin-diffusion length, and make the comparison with other commonly studied materials such as GaAs. Here we present a simple model of the spin-diffusion length of electrons in n-type silicon, which allows it to be calculated as a function of doping and temperature. We use phenomenological device equations used extensively in silicon device design that yield a simple usable equation that illustrates not only that silicon is a suitable semiconductor with a long spin-diffusion length, but also that the spin diffusion length can be engineered with temperature and doping. Further details of these calculations can be found in Reference 21.

11.2.1 Spin Relaxation in Semiconductors

In semiconductors, the most important interaction that causes spin relaxation is the spin-orbit (SO) interaction. The SO interaction couples the spin degree of freedom to the spatial motion of the electron, which significantly influences the transport properties. Three main mechanisms for spin dephasing in semiconductors have been proposed: the Elliot–Yafet[22] (EY); D'Yakonov Perel' (DP); and Bir, Aronov, and Pikus[23] (BAP) mechanisms, which we will briefly outline for those not familiar with semiconductors; for a recent review see Fabian and Sharma.[24] In the EY mechanism, the SO interaction leads to the mixing of wave functions of opposite spins. This results in spin flip due to impurity and phonon scattering. The DP mechanism is due to the SO interaction in crystals without an inversion symmetry, which results in a spin state splitting of the conduction band at $k \neq 0$. This is equivalent to an effective magnetic field acting on the spin with its magnitude and orientation depending on k. Between two scattering events, the spin precesses around this internal field. The BAP mechanism originates from the mixing of heavy hole and light hole bands induced by SO coupling. Spin-flip scattering of electrons by holes due to coulombic interaction is therefore permitted, which gives rise to spin dephasing.

The BAP mechanism becomes important in p-doped or intrinsic samples. However in n-type samples, the holes rapidly recombine with the high numbers of electrons, and the regular BAP mechanism is blocked. Hence in III-V GaAs, the DP mechanism is the main mechanism for spin dephasing. Silicon, however, has an inversion symmetry and as a result, the DP mechanism is absent. Hence, in lightly doped and n-type silicon, the EY mechanism dominates. This allows calculation of the spin-diffusion length in n-type silicon. The EY mechanism is described by the equation[25]

$$\frac{1}{\tau_{sf}} = A \left(\frac{\Delta}{\Delta + E_g} \right)^2 \left(\frac{E_g}{\varepsilon_e} \right)^2 \frac{1}{\tau_e} \tag{11.1}$$

where τ_{sf} is the spin relaxation time, τ_e is the momentum scattering time, Δ is the spin splitting of the valence band, ε is the kinetic energy of the electron, and E_g is the energy gap. A is a parameter dependent upon the scattering mechanism and is close to unity in all cases considered. Using this equation, the spin-relaxation time and spin-diffusion length of silicon can be calculated.

11.2.2 Calculation of the Spin-Diffusion Length in *N*-Type Silicon

The spin diffusion length is a critical device design parameter, as it dictates the physical dimensions of any spin devices.[11] The phenomenological model of transport in the spacer layer of current perpendicular-to-the-plane (CPP) geometry structures defines the spin-diffusion length as

$$l_{sf} = \sqrt{D\tau_{sf}} \tag{11.2}$$

The transport of free carriers in nondegenerate silicon is characterized by the fundamental Boltzmann transport relationship. The effects of applying an external field and the scattering processes involved leads to the concept of an electron with mobility μ. The response of the electrons to a concentration gradient is defined in terms of a diffusion coefficient D, which can be related to the mobility μ at thermal equilibrium by the Einstein relation $D = \mu kT/e$.

$$l_{sf} = \sqrt{\frac{\mu kT}{e} \tau_{sf}} \tag{11.3}$$

The mean free time, τ_{sf}, which is central to the Elliot-Yafet (EY) effect, can also be written in terms of the carrier mobility $\mu = e\tau_e/m_e^*$. This leads to a revised equation for the spin-diffusion length, written in terms of mobility

$$l_{sf} = \frac{\mu}{e}\left(\frac{\Delta + E_g}{\Delta}\right)\left(\frac{E_g}{\varepsilon_e}\right)\sqrt{\frac{kTm_e^*}{A}} \tag{11.4}$$

The mobility of carrier electrons in bulk silicon μ is a well-explored parameter in terms of doping and temperature variability. It can be modeled with an accuracy greater than 5% over a temperature range T between 200 and 400 K and for doping levels below 10^{19} cm^{-3} using the approach described in Reference 26. The method is summarized as follows. The intrinsic mobility of silicon is limited by scattering of the electrons by phonons. The simplest model for intrinsic mobility is described by the equation

$$\mu_L = \mu_0\left(\frac{T}{T_0}\right)^{-\alpha} \tag{11.5}$$

This simple equation belies the complexity of the phonon scattering processes that contribute to the carrier scattering, but is more than adequate for the analysis of silicon devices. The deliberate introduction of electrically active impurities into silicon allows the control of silicon devices. Over the temperature range of interest, it will be assumed that these impurities are fully active and therefore neutral impurity scattering will be ignored. The contribution due to impurity scattering over temperature can be described by the equation

$$\mu_I = \frac{AT^{3/2}}{N_{scat}}\left[\ln\left(1 + \frac{BT^2}{N_{carr}}\right) - \left(\frac{BT^2}{N_{carr} + BT^2}\right)\right]^{-1} \tag{11.6}$$

where N_{carr} is the number of free carriers available for screening, and N_{scat} is the number of scattering sites introduced by carriers. The contribution to the mobility from impurity scattering can then be combined with that from phonons

$$\mu = \mu_I f(x) \tag{11.7}$$

where the $f(x)$ can be approximated by

$$f(x) = \frac{1.025}{1+\left(\dfrac{x}{1.68}\right)^{1.43}} - 0.025 \tag{11.8}$$

and

$$x = \left(\frac{6\mu_L}{\mu_I}\right)^{\frac{1}{2}} \tag{11.9}$$

This model of mobility is used widely to analyze silicon-based devices. The models assume moderate doping (nondegenerate, $N_d < 10^{19}$ cm^{-3}), low applied fields, and thermal equilibrium. In these limits the effective mass, diffusion coefficient, and mobility can be considered scalar and the Einstein relations hold. The mobility of bulk silicon produced by this model is shown in Figure 11.1 using the parameters shown in Table 11.1. These values can then be used in Equation 11.4 to calculate the spin-diffusion length in silicon with temperature and doping.

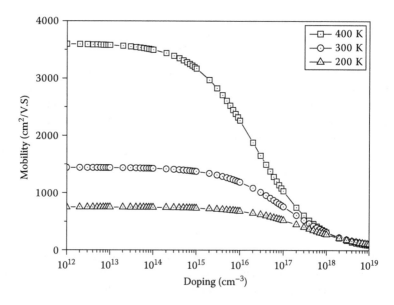

FIGURE 11.1 Electron mobility vs. doping density for various temperatures from 200 to 400 K.

TABLE 11.1

Parameters Used for Calculating Mobility and Spin-Diffusion Length in n-Type Silicon

Parameter	Value
A	4.61×10^{17}/cm/V/s/K$^{3/2}$
B	1.52×10^{15} cm^{-3}K^{-2}
α	2.26
N_{carr}	$N_d - N_a$ free carriers available for screening
N_{scat}	$N_d + N_a$ number of scattering sites introduced by carriers
μ_0	1439 cm^{2}V^{-1}s^{-1}
T_0	300 K
K	1.38×10^{-23} JK^{-1}
M_0	9.11×10^{-31} kg

The following parameters were also included in the analysis. The temperature dependence of the direct bandgap of silicon was modeled using the following relationship, which leads to a greater sensitivity to temperature[27,28]

$$E_g(T) = 1.17 - \frac{4.73 \times 10^{-4} T^2}{T + 636} \qquad (11.10)$$

This gives a slight decrease in the band gap as temperature increases, thereby reducing the spin-diffusion length. The band gap is also dependent on doping levels; however, a consensus on the doping dependence does not exist. The doping does not seem to have a significant impact on the band gap until doping levels $N_d >$ 10^{19} cm^{-3}, and so are neglected in this analysis.[29] The kinetic energy of the electrons in silicon is taken to approximate the thermal energy kT/e for bulk silicon. The effective mass of the electron is assumed to stay constant with doping and temperature. In reality it would increase with both temperature and doping, leading to a small increase in the spin-diffusion length. However, this change is small and considered negligible.

The results are shown in Figure 11.2 and Figure 11.3. Figure 11.2 shows the spin-diffusion length variation for temperatures between 200 and 400 K, and Figure 11.3 shows the variation for doping levels between 10^{12} to 10^{17} cm^{-3}. The spin-diffusion length at 300 K is of the order of 30 to 70 μm for doping levels between 10^{12} to 10^{17} cm^{-3} and is shown to be dependent on both doping and temperature, decreasing as temperature is increased. The effect is more pronounced at lower doping levels where the spin-diffusion length is much higher at 200 K. Doping decreases the spin-diffusion length, and this is more pronounced at lower temperatures. For doping levels above 10^{15} cm^{-3}, the spin-diffusion length drops significantly for all temperatures considered. Up to this level, the spin-diffusion length remains approximately constant with doping.

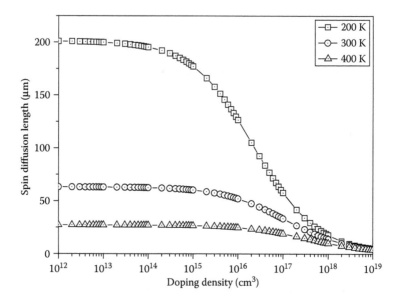

FIGURE 11.2 The spin-diffusion length in silicon (μm) against doping density for temperatures ranging from 200 to 400 K.

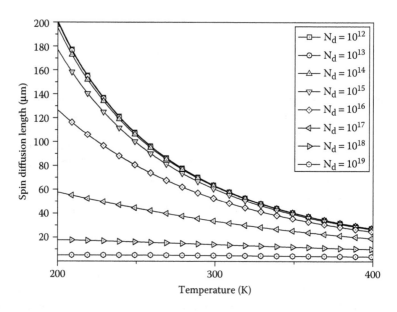

FIGURE 11.3 The spin-diffusion length of electrons in silicon with temperature for doping levels ranging from $N_d = 1 \times 10^{12}$ to 1×10^{17} cm^{-3}. The general trend of the spin-diffusion length follows that of mobility.

We note that the mobility dictates the temperature and concentration dependence. The ionized impurity scattering dominates at lower temperatures, and lattice scattering increases and becomes more significant at higher temperatures. Increasing the doping concentration at a given temperature results in the reduction of carrier mobility and hence the mean free time and spin-diffusion length. The calculated spin-diffusion lengths are long, e.g., 62.5 μm for 10^{14} cm^{-3} ($\tau_{sf} = 1.1$ μs) at room temperature. This length scale is large compared to conventional complementary metal-oxide semiconductor (CMOS) devices and is promising for a room temperature silicon-based spin-electronics device.

Comparison with work by Feher[17] is difficult as he only made one measurement above 5 K. This was made using a microwave technique at 20 K and a relaxation rate of 10^6 s^{-1} was measured, indicating separate processes than those observed below 5 K. Even so, the relaxation times observed indicated a long spin-diffusion length. Taking 1000 s, for example, leads to a spin-diffusion length of 0.2 m at 1.5 K. This is expected to fall rapidly with temperature as the mobility decreases, but still supports the large spin-diffusion lengths calculated in this chapter.

11.2.3 COMPARISON WITH THE SPIN-DIFFUSION LENGTH IN GaAs

It has been well established experimentally that the DP mechanism dominates in GaAs at lower temperatures and higher doping levels.[16,18] The spin-relaxation rate due to the DP mechanism is given by

$$\frac{1}{\tau_{sf}} = BQ^2 \frac{\varepsilon^3}{\hbar^2 E_g} \tau_m \qquad (11.11)$$

where Q is a splitting parameter, which is approximated as

$$Q = \frac{4\Delta}{[(E_g + \Delta)(3E_g + 2\Delta)]^{\frac{1}{2}}} \cdot \frac{m_e}{m_{ev}} \qquad (11.12)$$

m_{ev} is a constant close in magnitude to the mass of the free electron.

Again using the Einstein relation, and assuming $\mu = e\tau_m/m_e^*$, the spin-diffusion length can be estimated for the DP mechanism. As the spin time τ_s is now inversely proportional to τ_p, the final expression is independent of the mobility and so the l_{sd} is dictated by the direct variation of temperature and energy gap with temperature and doping. The experimental comparison between the predicted spin time for GaAs and experimental results has also been discussed in depth by Kikkawa[18] and Aronov.[25] In summary it has been seen that in the doping and temperature regimes considered in this chapter, the DP mechanism gives a good estimation of the measured spin lifetime in GaAs, and that these and the resulting spin-diffusion lengths are smaller than those predicted for silicon using the EY mechanism. For example, at 300 K, a spin time of 80 ps is predicted, which leads to a spin length of 0.2 mm

---

TABLE 11.2
Comparison of the Main Factors Contributing
to Spin Relaxation in Silicon and GaAs

	GaAs	Si
Spin splitting Δ (eV)	0.35	0.044
Energy gap Eg (eV)	1.43	1.12
Mobility of electrons at 300 K ($cm^2V^{-1}s^{-1}$)	3900	1430
Inversion symmetry	No	Yes
Dominant spin relaxation mechanism for n-type at 300 K	DP	EY

---

assuming a mobility of 8500 $cm^2V^{-1}s^{-1}$ and effective mass of 0.065 m_o. Table 11.2 summarizes the main differences between silicon and GaAs in terms of spin relaxation.

11.2.4 SPIN-DIFFUSION LENGTH IN P-TYPE SILICON

A full discussion of the spin-diffusion length of conduction electrons injected into p-type silicon is beyond the scope of this chapter. It is known without doubt that the lifetime of electrons in p-type silicon is long; in fact this is central to the operation of the bipolar transistor. The lifetime is sensitive to impurities such as O, Fe, and C and also the degree of crystalline perfection, but lifetimes of up to 20 ms have been observed in unprocessed float-zone p-type silicon.[26] However, the presence of excess holes within the system will cause interaction between the conduction electrons and holes and may cause accelerated spin relaxation via the BAP mechanism. It is unclear at what doping levels and temperatures the EY or BAP mechanism will dominate in p-type silicon. Studies on p-type GaAs show that the BAP mechanism becomes significant at lower temperatures and higher doping levels.[25] It is envisaged that the BAP mechanism may begin to dominate at lower doping levels and higher temperatures in silicon compared with GaAs, due to the longer spin times expected from the EY mechanism. Lower spin diffusion lengths may be expected in p-type silicon, but they will still be similar to those observed in p-type GaAs.

11.2.5 SPIN ENGINEERING IN SILICON

It has been shown clearly that the spin-diffusion length in silicon can be manipulated and controlled via temperature and doping. The use of doping is particularly attractive as this also allows a gain within a silicon device and is a well-established and controlled fabrication process. Also, the ability to engineer the spin by any of the parameters that affect the momentum relaxation time allows the concept of different spin-based devices to be envisaged; for example, strain devices based on the piezoelectric effect on mobility. Note also that the EY mechanism is independent of the applied magnetic field, which is in contrast to the DP mechanism in GaAs. This may also be of use in magnetic-based devices where the application of an external magnetic field may influence device behavior.

11.3 TEST DEVICES

To evaluate the performance of the constituent parts of a silicon-based spin-electronics device, various lateral and test devices were fabricated. Lateral devices such as the type shown in Figure 11.4 are relatively easy to fabricate, but suffer from the additional complications of ill-defined surface regions, surface effects, and spurious current paths.

Vertical devices are easier to analyze as the current path is more direct, but they are harder to fabricate due to the difficulty of depositing device-quality silicon on top of a metallic ferromagnet. One solution, as used in the Monsma transistor,[6] is to cold bond the two parts of the device. The alternative method, used here and illustrated in Figure 11.5, is to use a thin silicon membrane on both sides of which high-quality materials can be deposited. This has the advantage that device-quality silicon, with known doping, is guaranteed and such silicon-on-insulator (SOI) wafers are commercially available with silicon membranes as thin as 1 μm. In our devices, the silicon SOI wafer was etched to form a thin 1 μm Si membrane with a 5 to 100 μm^2 area, this membrane area was then selectively doped to form specific contact areas. Isolation moats were created around the device down to the buried oxide layer.

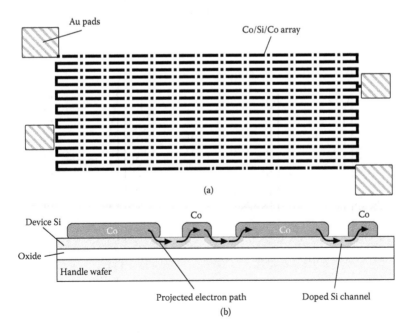

(a)

(b)

FIGURE 11.4 Examples of lateral test device shown schematically to emphasize the projected electron path. The separation of the Co pads is varied from 2 to 50 μm.

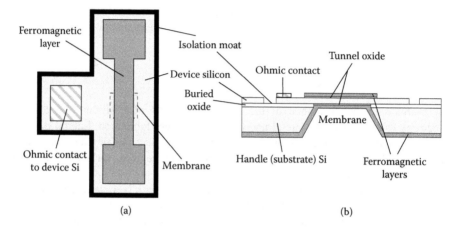

FIGURE 11.5 Vertical devices fabricated on a SOI wafer at the NASA Goddard Space Center.

Additional contacts were made to the silicon layer for testing and optimization of device performance, and additional test structures were also added to independently characterize each interface. Co layers were deposited on either side of the wafer by DC sputtering in an Ar atmosphere and the top Co layer patterned to introduce shape anisotropy and different switching fields for the magnetic layers. For convenience, the thermal oxide SiO_2 was used as a tunnel barrier and formed by heating in dry oxygen. A schematic diagram of the operation of the devices can be seen in Figure 11.6.

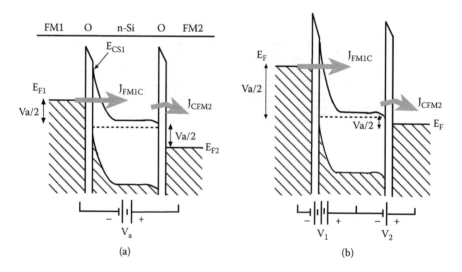

FIGURE 11.6 Schematic diagram of the reverse-biased band structure of the test devices illustrating the principle of operation. The bottom diagram shows the addition of the third contact.

In the reverse bias configuration shown in Figure 11.6, the dominant current density across the FM1/SiO/Si interface is from the spin-polarized FM into the silicon by tunneling. This spin-polarized current is swept to the second interface where it then traverses across the second Si/SiO/FM2 barrier, again by spin-dependent tunneling. The current density should increase with positive bias and be controlled by the relative magnetic orientation of the two ferromagnetic layers. The third contact to the Si allows independent control of each interface bias to optimize performance. In addition to providing a high-resistance source of spin-polarized electrons, the SiO also serves as a diffusion barrier.

The results for a set of lateral test devices are presented here, these had Co-Si-Co gaps of 5 to 50 μm. Both n- and p-type silicon was used; although we will concentrate here on the p-type devices for which the injection of minority spin-polarized electrons proved most effective. The current-voltage characteristics of the device were measured at different magnetic fields and the results are shown in Figure 11.7 and Figure 11.8. The variation of magnetoresistance shown in Figure 11.8 corresponds to the three voltage regions of the I-V curve shown in Figure 11.7. The linear region around $V = 0$ corresponds to conduction over all barriers due to the presence of a parallel resistance channel. For $V > \pm 0.2$V, the transport is dominated by the voltage-dependent leakage current through the reverse-biased barrier. In this region, the current is dominated by the minority carriers which are spin-polarized electrons. Gradual breakdown starts at approximately ± 0.75V.

The magnetoresistance observed in these devices, and similar ones,[21,30–32] demonstrate that the spin-diffusion length in silicon is indeed many microns, which is in agreement with our calculations, and that it is possible to design and fabricate spin-electronic devices using a thin silicon membrane etched from a SOI wafer. They lead the way to the development of the first silicon-based spin transistor described in Section 11.4.

11.4 A SILICON-BASED SPIN TRANSISTOR

11.4.1 THEORETICAL BASIS

Recent work[33] has analyzed the spin-injection efficiencies of different spin-electronic devices. For ease in analyzing the spin-transistor fabricated and measured here, we make several assumptions. The modeled transistor has a direct-injected base with a half-metallic ferromagnetic injector separated from the base by a tunnel barrier. We assume that the base width is less than both the recombination length and the minority carrier spin-diffusion length. In addition, the detector has spin channels with a common diffusion constant D, but different densities of states $\rho_\uparrow$ and $\rho_\downarrow$. The detector spin-diffusion length l_F is large compared with the detector-base depletion layer thickness so that variations in the electrochemical potentials $\mu_\uparrow$ and $\mu_\downarrow$ across the depletion layer can be ignored. Because the spin-up current is smaller than the spin-down current by a factor of b/c, it may be neglected. Therefore, the figure of merit for this device, which is the collected current ratio for oppositely magnetized

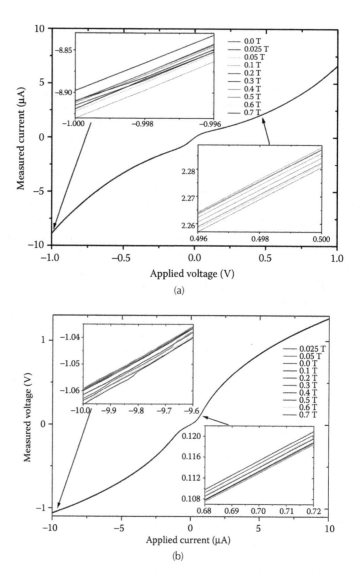

FIGURE 11.7 *I-V* and *V-I* characteristics for the devices in a magnetic field varying from 0 to 0.7 T.

collector configurations, may be readily shown to be

$$
\frac{J_{T\downarrow}}{J_{T\uparrow}} = \frac{2b\left(\rho_\uparrow + \rho_\downarrow\right)kT + \dfrac{D_S}{D_F}l_F n_0 e^{\frac{qV}{kT}}\left(\dfrac{\rho_\downarrow}{\rho_\uparrow}\right)}{2b\left(\rho_\uparrow + \rho_\downarrow\right)kT + \dfrac{D_S}{D_F}l_F n_0 e^{\frac{qV}{kT}}\left(\dfrac{\rho_\uparrow}{\rho_\downarrow}\right)} \tag{11.13}
$$

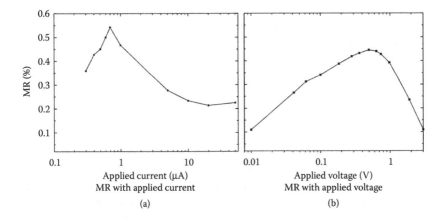

FIGURE 11.8 The variation of the magnetoresistance across the device as function of applied current and applied current.

where D_S and D_F are the control and detector diffusion coefficients, respectively; n_0 is the equilibrium minority carrier density in the base; and V is the detector-control bias voltage. Under high-voltage conditions ($qV \gg kT$), this reduces to $(\rho_\downarrow/\rho_\uparrow)^2$ for a direct injected device. For a tunnel-injected device, the corresponding ratio is $(\rho_\downarrow/\rho_\uparrow)$.

11.4.2 DESIGN AND FABRICATION

The device structure is shown in Figure 11.9. Charge carriers are injected into a thin silicon layer from a ferromagnet via a tunnel barrier by applying a voltage between the ferromagnet and the silicon. This spin-polarized current traverses the silicon (by diffusion and drift) to a second tunnel barrier separating the silicon from another ferromagnet, which is back biased to extract carriers. The existence of

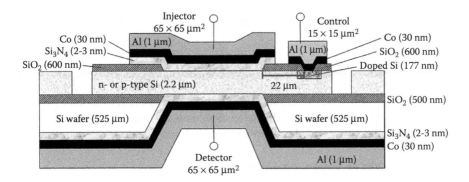

FIGURE 11.9 Structural schematic of the fabricated silicon device. The injector-detector separation is 2.2 μm and the injector-control separation is 22 μm. The injector and detector contacts are metal-insulator-semiconductor junctions, whereas the control contact is ohmic.

a spin-polarized current in the silicon is indicated by the output current dependence on the relative orientation of the two ferromagnets, whose magnetizations may be differentially switched.

The devices were fabricated using standard photolithography on n- and p-type SOI wafers with an active silicon layer resistivity of 17 to 33 Ω-cm. All of the devices were fabricated according to the following process:

1. An insulating layer of 600 nm of SiO_2 was grown on the front (active silicon) side of the SOI wafer.
2. The SiO_2 was removed in selected areas to create the control contacts.
3. The silicon in the control contacts was ion implanted with As^+ or BF_2^+ for active silicon doped n- or p-type, respectively. This yielded a surface concentration of ~1×10^{20} atoms/cm^3, forming a good ohmic contact.
4. The SiO_2 was removed in selected areas to create the injector contacts.
5. On the back (handle silicon) of the wafers, a layer of Si_3N_4 was deposited with low-pressure chemical vapor deposition.
6. The Si_3N_4 was removed in selected areas and a pit was wet-etched down to the buried oxide layer.
7. The remaining back-surface Si_3N_4 layer and the buried oxide layer (in the selected region) were removed. This contacted the active silicon layer where the detector will be formed.
8. Tunneling barriers[34,35] of Si_3N_4 were formed on both the front and back of the wafers by low-pressure epitaxy using a self-limiting nitride process. However, on the front of the wafers, the Si_3N_4 was removed by reactive ion etching (RIE) to make the control contacts ohmic. The detector and injector contacts are metal-insulator-semiconductor junctions and the control contact is a metal-semiconductor (ohmic) junction.
9. A total of 30 nm of Co and 1 μm of Al (for the electrical contacts) were deposited on both sides of the wafers.
10. The front side was then etched (to remove the Al) and ion milled (to remove the Co) to isolate the control and injector contacts. The detector remains as a common contact for all of the transistors.
11. An isolation trench was milled by RIE to isolate each transistor from its neighbors.

The transistors were fabricated on two 4 in. SOI wafers — one with the active silicon layer doped p-type and the other with the active silicon layer doped n-type. From a single chip of each wafer containing 18 devices, 6 were chosen (based upon the quality of their two-terminal I-V characteristics), mounted into a chip package with silver epoxy, and wire bonded for further electrical and magnetic characterization.

11.4.3 ELECTRICAL CHARACTERIZATION

Typical room temperature I-V characteristics of the control-detector barrier with a floating injector are shown in Figure 11.10. (The results are similar for the control-injector barrier.) A single conduction mechanism cannot account for the behavior

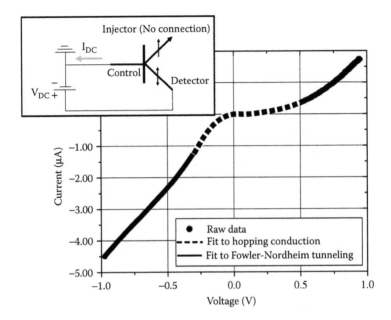

FIGURE 11.10 Typical example of the detector-control characteristics with the injector contact floating and the control at ground for the *p*-type Si. The dots are measured data points and the dashed and solid lines are fits to Equation 11.14 or Equation 11.15, respectively, according to the region the data are in. (Inset: Shows the circuit diagram for these measurements where the single-headed, vertical arrow indicates the "fixed" magnetic layer and the double-headed, vertical arrow indicates the "free" magnetic layer.)

over the whole voltage range. At low bias voltages, the functional form of the *I-V* curves is given by

$$I = Ae^{B|V|^{-1/4}} + CV + D \qquad (11.14)$$

where A and D have units of amperes, B has units of $V^{1/4}$, and C is an admittance with siemens units. (The linear term accounts for parasitic currents.) The form of this curve implies that the dominant conduction mechanism in this regime is Mott's variable range hopping conduction (HC)[36,37] where the barrier is composed of the Si_3N_4 low-transmission barrier and a Schottky barrier in the silicon.[31] This is in agreement with previous work,[38–40] which found that Si_3N_4 low-transmission barriers conduct primarily by electron hopping near the Fermi level due to defects or dangling bonds.

At voltages greater than ~|0.5| V (as determined from numerical differentiation), it can be seen that the functional form of the data changes. This change is similar to that observed previously[34] and is ascribed to Fowler–Nordheim tunneling (FNT) conduction[41] overtaking the hopping conduction. In our structures, this change is due to injection of the carriers above the Schottky barrier and through the Si_3N_4

low-transmission barrier only. The functional form for FNT (Equation 11.15) accu-
rately fits this data for voltages greater than ~|0.5| V.

$$I = AV^2 e^{B/V} + CV + D \qquad (11.15)$$

where A has ampere per volts squared units, B has volts units, C is an admittance
with siemens units, and D has ampere units. The doping type and concentration of
the silicon influence only the detailed form of the I-V characteristics.

It is well-established[42] that HC destroys the spin polarization of carriers due to
their extended time at each localized state for an average time that generally exceeds
their spin lifetime. Therefore, little magnetic sensitivity of spin-dependent origin
should be exhibited by the device in the regime where HC dominates. This observation can be exploited as a means to help separate genuine spin-injection effects
in the device from the multitude of spurious magnetic artifacts (anisotropic magnetoresistance, Lorentz magnetoresistance [LMR], Hall effect, etc.) that plague these
types of magnetotransport measurements.

The three-terminal I-V characteristics with fixed control current, presented in
Figure 11.11 and Figure 11.12 were measured at room temperature with a magnetic
field applied in the plane of the device and perpendicular to the current direction.
Owing to the substantial additional noise associated with magnetic field sweeping
(due to magnetocaloric effects combined with the temperature dependence of silicon), the data were measured by sweeping the voltage-current characteristics at a
selection of fixed magnetic fields. For the data presented, these fixed field values
were established by approaching from the negative magnetic field direction.

Due to shape anisotropy, the magnetic moments of the injector and detector pads
are parallel for applied magnetic fields lower than +2.4 kA/m (+30 Oe) and greater
than +9.2 kA/m (+115 Oe), and antiparallel for positive applied magnetic fields
between 2.4 kA/m (30 Oe) and 9.2 kA/m (115 Oe), as shown by the triangles in
Figure 11.11. There are three significant results from these measurements. First, the
I-V characteristics are a function of applied magnetic field. Second, disregarding
LMR (see below), little magnetic sensitivity is observed for voltages below the onset
of FNT (see Figure 11.11). Third, the data of Figure 11.12 (which are taken at an
applied voltage above the HC/FNT threshold and are corrected for LMR) show that
when the magnetic moments of the injector and detector pads are parallel, the injector
current is larger than in the antiparallel configuration. Furthermore, the data in the
HC regime changes little after correction for LMR.

The raw data of Figure 11.11 was corrected for LMR by the following: The I-V
characteristics of the output current at fixed control current and voltage were measured
as a function of magnetic field. These data were then fitted to a polynomial of order 2,
and the fit was used to calculate the output current purely from LMR at each of the
measured fields. This value was then subtracted from the raw data (with an added
constant background value so that the current magnitude remains the same) to yield
the processed data in Figure 11.12. (It should be noted that, due to the noise associated
with the magnetic field sweeping, the fit takes an average which yields only the LMR
signal; this is the same regardless of the applied control current or voltage.)

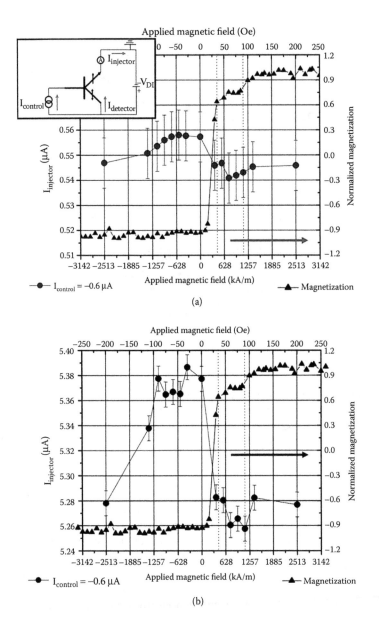

FIGURE 11.11 The raw injector current (dots) as a function of applied magnetic field for the p-type Si in (top) the dominant HC regime at V_{ID} = 0.15 V and $I_{control}$ = 0.6 µA and (bottom) the dominant FNT regime at V_{ID} = 1 V and $I_{control}$ = −0.6 µA. Half of a hysteresis loop as measured on a vibrating sample magnetometer (VSM) is shown by the triangles. The horizontal arrow indicates the direction of the magnetic field sweep of the measurements, following saturation at fields <−80 kA/m (−1 kOe) and the dotted vertical lines show the region of antiparallel alignment. (Inset for A shows the circuit diagram for these measurements where the encircled "A" represents an ammeter. The single-headed, vertical arrow indicates the "fixed" magnetic layer and the double-headed, vertical arrow indicates the "free" magnetic layer.)

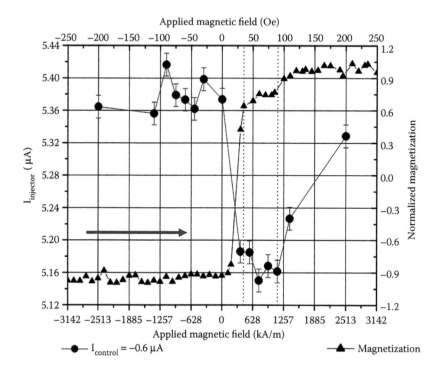

FIGURE 11.12 The injector current (dots) as a function of applied magnetic field for the *p*-type Si after correction for LMR at $V_{ID} = 1$ V and $I_{control} = -0.6$ µA (in the dominantly FNT regime). Half of a hysteresis loop as measured on a VSM is shown by the triangles. The arrow indicates the direction of the magnetic field sweep of the measurements, following saturation at fields < -80 kA/m (-1kOe) and the dotted vertical lines show the region of antiparallel alignment.

To confirm the origins of the observed magnetotransport effects, we must consider not just the spin-dependent tunneling effects with which we are primarily concerned, but also a large family of other potential artifacts with which the spin-dependent transport might be confused. These include anisotropic magnetoresistance (AMR), LMR, and the first-order and second-order Hall effects. The magnetic fields driving these possible artifacts are twofold in origin: the actual magnetic fields that we apply to switch the electrode magnetizations (these do not exceed 80 kA/m) and also the fringing fields at the extremities of the magnetic electrodes whose values may locally approach 1 T over very small volumes.

The AMR effect is typically of the order of 1%. Cobalt has a nominal resistivity of 6 µΩ-cm, which yields a resistance of less than 450 nΩ for the cobalt contacts and approximately 4 nΩ for any AMR contribution to the magnetotransport of the device. AMR effects can thus be discounted in our measurements.

Any first-order Hall effect voltages due to the applied or fringing magnetic fields should have odd symmetry (i.e., would change sign as the applied field is reversed). From the known Hall coefficient of our SOI wafer and the device geometry, we would not expect to see a measurable Hall effect at any magnetic field in Figure 11.11. This is confirmed: there is no such linear term in the magnetotransport. Even more importantly, there is no significant component with odd symmetry in the signal as a function of magnetic field that follows the magnetization of the magnetic contacts. This latter observation allows us to conclude that the Hall effect due to fringing fields may also be neglected: although these fringing fields may be large over a restricted volume, they are apparently sufficiently far from the main current path not to influence the transport. This observation also enables us to put an upper limit on the effect of the fringing fields in causing both the second-order Hall effect and LMR (both of these signals are symmetric and hence potentially confusable with tunneling magnetoresistance [TMR]). Consequently, we can also discount these fringe field effects as possible origins of the observed symmetric magnetotransport.

The final potential artifact is LMR due to the applied magnetic field. This is definitely present and clearly observable (data not shown). Because any LMR in the ferromagnet, which is only 30 nm thick (compared to 2.2 μm for the Si), is negligibly small, the effect originates solely in the silicon. Its magnitude agrees with predictions — approximately 4% (~40 kΩ) of the total resistance over the applied field range. This is also a symmetric (parabolic) effect, and we have subtracted it from the data of Figure 11.11 to yield Figure 11.12 (where its primary effect is to alter the injector currents of the $H = \pm 16$ kA/m [± 200 Oe] data) to leave a 4% signal variation that we believe arises from spin-dependent transport.

Our confidence in this result arises not only from having accounted for the potential competing artifacts, but also from the correlation between the observed spin-dependent effect and the voltage regime of the barriers. If spin-dependent transport was absent, then the two different methods of conduction (FNT and HC) should have no influence on the magnetic signal. The type of magnetic sensitivity present at low voltages should also be present at high voltages and vice versa. In practice, the onset of strong asymmetric magnetic sensitivity is voltage-dependent. This is consistent with the HC substantially destroying the spin-polarization of the carriers in the low-bias regime; then when bias is increased and FNT begins to dominate, the spin-polarization is retained, rendering the current sensitive to the relative magnetic orientations of the magnetic-detector and injector-contact pads. Using this mechanism, the spin-polarization can be estimated from Julliere's model[43] for spin tunneling to be $2.5 \pm 0.5\%$ for the p-type Si. (Assuming holes are the carrier in the n-type Si, the same arguments apply and yield a spin-polarization of $10 \pm 1\%$.) Note that this is similar to the 12% MR observed by Wong and Xu in GaAs devices and described in Chapter 15 of this book. This value for the spin-polarization is significantly lower than the theoretical value of 38%, as well as being lower than typical values in the literature. This is to be expected because HC, though no longer dominant, is still active at the high-barrier bias and it destroys part of the injected spin polarization.

The results in Figure 11.12 show a variation in the collector current as a function of magnetic field indicating that the transistor behaves as a magnetically tunable device with a field-dependent gain. The maximum variation of the average (taken over all applied control currents) current gain was $2.2 \pm 0.3\%$ which occurred at −110 Oe. This percentage change was independent of device type. The currents calculated by removing the LMR contribution represent the injector current in parallel or antiparallel configuration, depending upon the applied field. Therefore, the magnetocurrent (defined as) $MC = ((I_P - I_{AP})/I_{AP})100$ can be calculated to be $98 \pm 7\%$ ($140 \pm 13\%$) for p-type (n-type).

Finally, the figure of merit is calculated using the theory developed in Section 11.4.1. For Co, $\rho_\uparrow = 0.1740$ and $\rho_\downarrow = 0.7349$ states/eV, yielding an ideal figure of merit of 4.2 for tunnel injection and 17.8 for direct injection (calculated from Equation 11.13 assuming high bias). The experimental figure of merit is 1.05 (1.1) for p-type (n-type) silicon. At this point, it should be noted that one of the assumptions in deriving Equation 11.13 is that the detector is a half-metallic ferromagnet. This is clearly not the case in the actual device. A normal ferromagnet would be expected to permit some spins aligned antiparallel to its magnetization direction to tunnel into the collector. This would essentially dilute the purity of the tunneling spins, thereby decreasing the figure of merit. However, this is not expected to be sufficient to explain the full factor of 4 difference between the theoretical and experimental figure of merit. Full details about the silicon-spin transistor can be found in Reference 44.

11.5 CONCLUSIONS

We have demonstrated that silicon is a viable material to use as a base for constructing spin-electronic devices. Our calculations do not claim to be as accurate as say nonperturbative calculations of spin times; however, they do provide an insight into the order of magnitude of expected spin-diffusion lengths and the effect of doping and temperature expected for silicon. The prediction that the spin-diffusion lengths in n-type silicon are an order of magnitude longer than those predicted and observed in GaAs makes silicon a promising material for spin electronics as long as the problems of spin injection are overcome.

The lateral and vertical test devices demonstrated that the spin-diffusion length in silicon is indeed many microns. With symmetry considerations from the device geometry, the voltage-dependent magnetic sensitivity strongly supported the existence of spin injection into silicon, thus demonstrating the feasibility of exploiting rectifying junctions and tunnel barriers as a means of introducing spin-dependent behavior into semiconductor devices.

Proof of concept has been established by a high current gain (greater than 1), magnetically sensitive, silicon-based spin transistor. In zero magnetic field and at room temperature, the detector I-V characteristics are similar to those of conventional transistors and offer a current gain of 0.97 (1.4) for p-type (n-type) silicon. The current gain of the device can be magnetically tuned (up to a $2.2 \pm 0.3\%$ change in -110 Oe in the average current gain, $3.5 \pm 1\%$ [$12 \pm 1\%$] for p-type [n-type] silicon for the TMR contribution alone, and $98 \pm 7\%$ [$140 \pm 13\%$] for p-type [n-type] silicon for the calculated magnetocurrent). Moreover, the control current and injector-detector voltage control this field-dependent gain. However, several improvements can be made to future generations for better device characteristics, including (i) improved tunnel barriers, (ii) improved differential magnetic switching,

(iii) optimized electrode geometry as determined from recent modeling[29] and (iv) uniform doping profile.

In addition to improving the particular device described here, the future lies in using our imagination to create new spin-electronic devices that exploit spin-dependent transport within a semiconductor-based device. This should lead the way to devices with new functionality and also to the realization of power and transistor efficient logic circuits as envisaged by Sugahara and Tanaka[45] using spin metal-oxide-semiconductor field-effect transistors.

ACKNOWLEDGMENTS

The test devices were made at the NASA Goddard Space Center in collaboration with Shahid Aslam and Nilesh Tralshawala from Cornell University and with Jim Eckert and Patti Sparks from Harvey Mudd College. The silicon spin transistors were made in collaboration with Graham Ensell at the microelectronics facility at the University of Southampton (United Kingdom). We are also grateful to Randall Kirschman for useful discussions. Funding was provided by the U.K. Engineering and Physical Sciences Research Council and E.U. projects SPINOSA, Dynaspin, MagNoise, and HotSEAMS.

REFERENCES

1. J.F. Gregg, I. Petj, E. Jouguelet, and C.L. Dennis, *J. Phys. D*, 35(18), R121, 2002.
2. I. Žutić et al., *Rev. Mod. Phys.*, 76(2), 323, 2004.
3. G. Schmidt, *J. Phys. D: Appl. Phys.*, 38, R107, 2005.
4. R.P. Borges, C.L. Dennis, J.F. Gregg, E. Jouguelet, K. Ounadjela, I. Petj, S.M. Thompson, and M.J. Thornton, *J. Phys. D: Appl. Phys.*, 35, 186, 2002.
5. M. Johnson, *Science*, 260, 320, 1993.
6. D.J. Monsma et al., *Phys. Rev. Lett.*, 74, 5260, 1995.
7. R. Sato and K. Mizushima, *Appl. Phys. Lett.*, 79, 1157, 2001.
8. S. van Dijken, X. Jiang, and S.S.P. Parkin, *Appl. Phys. Lett.*, 80, 3364, 2002.
9. S. Datta and B. Das, *Appl. Phys. Lett.*, 56, 665, 1990.
10. S. Gardelis et al., *Phys. Rev. B*, 60, 7764, 1999.
11. J. Gregg et al., *J. Magn. Magn. Mater.*, 175, 1, 1997.
12. J.F. Gregg and P.D. Sparks, U.S. Patent 6,218,718, 2001.
13. C.L. Dennis, C. Sirisathitkul, G.J. Ensell, J.F. Gregg, and S.M. Thompson, *J. Phys. D: Appl. Phys.*, 36, 81, 2003.
14. G. Lampel, *Phys. Rev. Lett.*, 65, 1643, 1968.
15. A.M. Portis, H. van Kempen, H. van Leken, R.A. de Groot, W. van Roy, and J.D. Boeck, *J. Phys. Cond. Matt.*, 7, 9447, 1995.
16. F.K. Willenbrock and N. Bloembergen, *Phys. Rev.*, 91, 1281, 1953.
17. G. Feher and E.A. Gere, *Phys. Rev.*, 114(5), 1245, 1961.
18. J.M. Kikkawa and D.D. Awschalom, *Phys. Rev. Lett.*, 80, 4313, 1998.
19. Y.Q. Jia, R.C. Shu, and S.Y. Chou, *IEEE Trans. Magn.*, 32, 4707, 1996.
20. D.R. Lorraine, D.I. Pugh, H. Jenniches, R. Kirschman, S.M. Thompson, W. Allen, C. Sirisathikul, and J.F. Gregg, *J. Appl. Phys.*, 87, 5161, 2000.
21. D.I. Pugh, Ph.D. thesis, University of York, 2001.

22. R.J. Elliot, *Phys. Rev.,* 96, 266, 1954.
23. G.L. Bir, A.G. Aronov, and G.E. Pikus, *Sov. Phys. JETP,* 42, 705, 1975.
24. J. Fabian and S. Das Sarma, *J. Vac. Sci. Technol. B,* 17, 1708, 1999.
25. A.G. Aronov, G.E. Pikus, and A.N. Titkov, *Sov. Phys. JETP,* 57(3), 680, 1983.
26. INSPEC, *Properties of Silicon,* EMIS Data Review Series No. 4., INSPEC, 1988.
27. S.M. Sze, *Physics of Semiconductor Devices,* 2nd ed., Wiley Interscience, Hoboken, NJ, 1981.
28. C.D. Thurmound, *J. Electrochem. Soc.,* 122, 1133, 1975.
29. R.J. Van Overstraeten and R.P. Mertens, *Sol. State Electron.,* 30(11), 1077, 1987.
30. C. Sirisathitkul, Ph.D. thesis, University of Oxford, 2000.
31. D. Loraine, Ph.D. thesis, University of York, 2000.
32. W.A. Allen, Ph.D. thesis, University of Oxford, 2000.
33. J.F. Gregg, R.P. Borges, E. Jouguelet, C.L. Dennis, I. Petj, S.M. Thompson, and K. Ounadjela, *J. Magn. Magn. Mater.,* 265, 274, 2003.
34. T.E. Hartman, J.C. Blair, and R. Bauer, *J. Appl. Phys.,* 37, 2468, 1966.
35. C.L. Dennis, C.V. Tiusan, R.A. Ferreira, J.F. Gregg, G.J. Ensell, S.M. Thompson, and P.P. Freitas, *J. Magn. Magn. Mater.,* 290–291, 1383, 2005.
36. N.F. Mott and E.A. Davis, *Electronic Processes in Non-Crystalline Materials,* Oxford University Press, NewYork, 1979.
37. Y. Xu, D. Ephron, and M.R. Beasley, *Phys. Rev. B,* 52, 2843, 1995.
38. C. Chaneliere et al., *J. Non-Cryst. Sol.,* 245, 73, 1999.
39. Y. Manabe and T. Mitsuyu, *J. Appl. Phys.,* 66, 2475, 1989.
40. T. Goto and T. Hirai, *J. Mater. Sci.,* 24, 821, 1989.
41. R.H. Fowler and L.W. Nordheim, *Proc. R. Soc. London A,* 119, 173, 1928.
42. S.S. Manoharan et al., *Appl. Phys. Lett.,* 72, 984, 1998.
43. M. Julliere, *Phys. Lett. A,* 54, 225, 1975.
44. C.L. Dennis, Ph.D. thesis, University of Oxford, 2004.
45. S. Sugahara and M. Tanaka, *Appl. Phys. Lett.,* 84(13), 2307, 2004; private communication, 2004.

12 Spin LEDs: Fundamentals and Applications

Willem Van Roy, Pol Van Dorpe, Vasyl F. Motsnyi, Gustaaf Borghs, and Jo De Boeck

CONTENTS

12.1 INTRODUCTION

Semiconductor spintronics is the field where the spin degree of freedom of charge carriers in a semiconductor is actively controlled and manipulated, resulting in improved device functionality compared to charge-only devices. The intended improvement can come in various ways: improved performance, lower power, increased functionality, built-in memory capability, and so on. A recent review of the state of the art, including a collection of proposed device concepts and a number of remaining challenges, has been given by Žutić et al.[1] Spin transistors are being considered as one of the possible concepts to extend the scaling of traditional complementary metal-oxide semiconductor (CMOS) technology or to provide a successor beyond CMOS.[2]

The realization of spin-based device operation requires a succession of steps: generation (injection) of spin-polarized carriers in a semiconductor, transport, manipulation, storage, retrieval, and detection. This review focuses mainly on the injection and to some extent also on the detection steps. Most of the successful experiments reported in the literature have been done on GaAs or other direct band-gap semiconductors where optical and electrical methods can be combined. Once the electrical control of spins is firmly established, it is expected that many of the techniques can be transferred to indirect gap semiconductors such as Si.

Spin orientation can be obtained in a variety of ways, including optical, resonance, and transport (i.e., electrical) techniques. Only the electrical injection- detection technique is well suited for large-scale device integration. This technique is based on the fact that transport in ferromagnetic metals is spin polarized, and the idea is to use ferromagnetic contacts to inject or detect preferentially one spin orientation into or from a semiconductor. Because many ferromagnetic metals have Curie temperatures well above room temperature (RT), this technique should be suitable for RT operation.

Early all-electrical injection-detection experiments using *ohmic* ferromagnetic contacts only showed effects smaller than ~1% that could not be isolated from parasitic effects due to stray magnetic fields (Hall effects, magnetoresistance effects).[3] A breakthrough came with the realization that conductance mismatch suppresses the injected spin polarization when ohmic contacts are used.[4] (This amounts essentially to the fact that the spin-independent resistance of the semiconductor channel is much larger than the spin-dependent resistance of the ferromagnetic contacts, such that the currents of minority and majority spins are nearly equal.) This problem can be circumvented either by using half-metallic contacts (i.e., contacts where the conduction is strictly 100% polarized), by using ferromagnets (e.g., magnetic semiconductors) whose resistance is better matched to the semiconductor channel, or by inserting a tunnel barrier to introduce a large spin-dependent resistance.[5,6]

12.2 SPIN LEDs AND FARADAY VS. OBLIQUE HANLE GEOMETRIES

The early experiences with parasitic effects in all-electrical injection-detection devices have also highlighted the value of breaking up the spin-dependent transport experiments and clearly understanding the individual steps before attempting a fully integrated device. In a large class of III-V and II-VI zinc-blende semiconductors with a direct band gap, optical techniques can be used to generate and probe the carrier spins inside the semiconductor. Optical selection rules link the circular polarization of the light to the spin polarization of the carriers involved in the process.[1] A detailed discussion can be found in Reference 7. These techniques can be used to study the spin lifetimes, transport, and dynamics inside the semiconductor using all-optical injection and detection, or they can be used to study the efficiency of electrical injection (a spin light-emitting diode [LED]) or electrical detection (a spin-dependent photodiode).

In a spin LED (Figure 12.1), polarized electrons are injected from a magnetic contact. They recombine with unpolarized holes supplied by the *p*-type substrate, and the circular polarization of the emitted light reflects the spin polarization of the injected electrons.

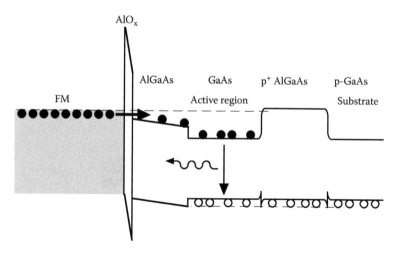

FIGURE 12.1 Band profile of an AlO_x-based LED under bias. Polarized electrons are injected from a magnetic contact and recombined with unpolarized holes supplied by the p-type substrate. The circular polarization of the emitted light reflects the spin polarization of the injected electrons (see Reference 26 for details).

Some spin manipulation is required to obtain reliable results. The circular polarization of the light reflects only the electron spin component parallel to the propagation direction of the light. Most ferromagnetic thin films have an in-plane magnetization due to shape anisotropy (Figure 12.2a) suggesting that a side-emitting LED is best suited. However, it has been demonstrated that this configuration may give unreliable results, even when the mesa width is shrunk down to 1 μm and waveguide effects and resonant pumping (photon recycling) can be excluded.[8] This was tentatively attributed to the selection rules for side emission in quantum wells, and the situation might be different for bulk active regions.

The most reliable configuration, both for quantum wells and bulk active regions, is to use top-emission LEDs where the top contact is sufficiently thin (5 to 15 nm) to become semitransparent. The most common way to obtain a perpendicular spin component is by applying a sufficiently strong perpendicular magnetic field that pulls the magnetization of the ferromagnetic thin film out of the plane (Faraday configuration, Figure 12.2b). A drawback of this configuration is that relatively large fields are needed (1 to 2 T), and this can result in parasitic effects due in Zeeman splitting in the semiconductor. This is especially true in narrow gap semiconductors (InAs, InSb, etc.) with large g factors, where the thermodynamic spin polarization due to Zeeman effects can largely dominate the nonequilibrium spin polarization due to injection from the ferromagnetic contact, even leading to sign changes of the optical polarization.[9,10]

The oblique Hanle geometry (Figure 12.2c) uses spin manipulation inside the semiconductor to obtain a circular spin component. A small field $\mathbf{B}_{45} = 0.1$ to 0.5 T is applied at an angle (ideally 45°) with the surface. This field is insufficient to pull the magnetization of the contact completely out of plane, but it causes the injected electron spins to precess around $\mathbf{B}$ with the Larmor frequency $\Omega = g^*\mu_B/\hbar B$ (Figure 12.3).

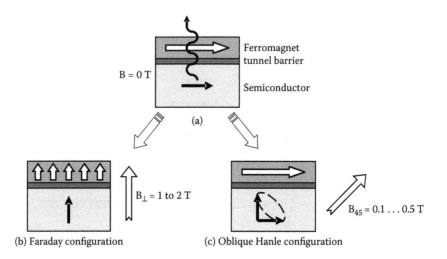

(a)

(b) Faraday configuration   (c) Oblique Hanle configuration

FIGURE 12.2 (a) Most ferromagnetic thin films have in-plane magnetization, whereas the circular polarization of a surface-emitting LED carries only information about the perpendicular spin component. (b) In the Faraday configuration, a large perpendicular field is applied to pull the magnetization out of the plane (spin manipulation inside the ferromagnet). (c) In the oblique Hanle geometry, a small oblique magnetic field creates spin precession of the electrons after they are injected into the semiconductor (spin manipulation inside the semiconductor).

Here g^* is the effective g factor ($g^* = -0.44$ for electrons in GaAs); $\hbar$ is Planck's constant; and $\mu_B = e\hbar/2m_e = 9.27 \times 10^{-24}\,\text{Am}^2$ is the Bohr magneton. The average out-of-plane spin component depends on the product of the precession frequency Ω and the spin lifetime T_s, where T_s includes both the spin lifetime τ_s and the electron lifetime (or

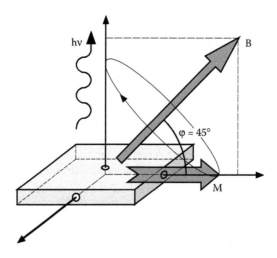

FIGURE 12.3 Oblique Hanle measurement geometry (see Reference 26 for details).

recombination time) τ : $T_s^{-1} = \tau_s^{-1} + \tau^{-1}$. At small fields ($\Omega T_s \ll 1$), the average precession angle is very small, and the spins are hardly disturbed. At large fields ($\Omega T_s \gg 1$), the spins go through many precession cycles before they recombine, resulting effectively in the projection of the injected spins on the oblique field.

The technique can be used both with electrically injected spins (in-plane injection) and with optical spin injection using circularly polarized light (out-of-plane injection) and results in the following dependence of the optical circular polarization (normal emission) on the oblique magnetic field

$$P(B) = \frac{\Pi_{inj}}{2} \cdot \frac{T_s}{\tau} \cdot \frac{\frac{1}{2}(\Omega T_s)^2}{1+(\Omega T_s)^2} = \frac{\Pi_{inj}}{2} \cdot \frac{T_s}{\tau} \cdot \frac{\frac{1}{2}(B/\Delta B)^2}{1+(B/\Delta B)^2}, \quad \text{in-plane injection}$$

(12.1)

$$P(B) = \frac{\Pi_{inj}}{2} \cdot \frac{T_s}{\tau} \cdot \frac{1+\frac{1}{2}(\Omega T_s)^2}{1+(\Omega T_s)^2} = \frac{\Pi_{inj}}{2} \cdot \frac{T_s}{\tau} \cdot \frac{1+\frac{1}{2}(B/\Delta B)^2}{1+(B/\Delta B)^2}, \quad \text{out-of-plane injection}$$

(12.2)

where $\Pi_{ini} = (n^\uparrow - n^\downarrow)/(n^\uparrow + n^\downarrow)$, the degree of spin polarization with which the electrons are injected, and the prefactor 1/2 reflects the optical selection rules in a bulk active region (this factor would be absent in a quantum well where the degeneracy between light and heavy holes is lifted). The half-width

$$\Delta B = \left(\frac{g^* \mu_B}{\hbar} \cdot T_s \right)^{-1}$$

(12.3)

corresponds to the condition $\Omega T_s = 1$. The resulting functional shapes are shown in Figure 12.4.

An additional benefit of the oblique Hanle technique is that reference experiments with circular optical excitation are easily possible in the same configuration. The injected spin polarization by optical excitation is known, and this allows the full characterization of electron recombination times and spin lifetimes inside the semiconductor. With this information, it becomes possible to extract the loss in polarization during the interval between injection and recombination, and hence to determine the spin polarization of the electrons at the time they are injected. The Faraday geometry cannot easily disentangle these effects and yields only the spin polarization at the time of recombination.[11]

Because of the characteristic functional shape of the Hanle curves, it is easy to separate the true spin-injection signal from possible parasitic effects. In reality, the oblique field will slightly tilt the magnetization of the film out of the plane. The angle θ with the film plane is easily obtained from a magnetostatic energy

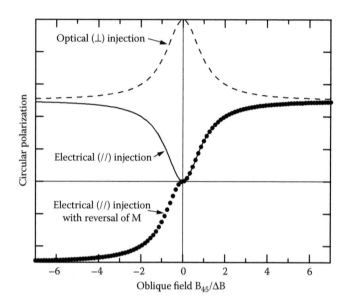

FIGURE 12.4 Expected dependence of the (perpendicular) circular polarization: (dashed line) optical injection of spins perpendicular to the surface, (full line) electrical injection of in-plane spins, (dotted line) electrical injection, taking into account that the in-plane component of the oblique field will reverse the magnetization direction of a soft magnetic film.

calculation,[12] and the resulting correction on the optical circular polarization is given by

$$P(B) = \frac{\Pi_{inj} \cos(\theta)}{2} \cdot \frac{T_s}{\tau} \cdot \frac{\frac{1}{2}(\Omega T_s)^2}{1+(\Omega T_s)^2} + \frac{\Pi_{inj} \sin(\theta)}{2} \cdot \frac{T_s}{\tau} \cdot \frac{1+\frac{1}{2}(\Omega T_s)^2}{1+(\Omega T_s)^2} \quad (12.4)$$

When the emitted light passes through the ferromagnetic film, any perpendicular component of the magnetization may result in magnetic circular dichroism (MCD), which adds to the circular polarization of the light. This effect is less severe on the oblique Hanle configuration as compared to the Faraday configuration due to the smaller out-of-plane magnetization component and can be easily taken into account by performing a reference experiment using optical excitation with linearly polarized light at sufficiently large photon energies to generate unpolarized electrons.[7]

A more comprehensive description of the oblique Hanle technique in spin LEDs can be found in Reference 12.

Figure 12.5 gives an example of the oblique Hanle measurements on AlO_x-based spin injection into a LED with a p-type active region. The panels (a) and (c) show the optical reference experiments. The upper curves show the detected circular polarization as a function of the oblique field, when spins are injected optically with a circularly polarized laser beam at 1.58 eV. These measurements allow a full

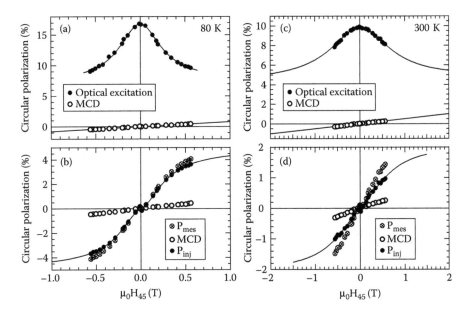

FIGURE 12.5 Experimental spin-injection data for an AlO_x-based spin LED with a p-type active region, at 80 and 300 K, as a function of the oblique magnetic field $\mu_0 H_{45}$. The top panels show the measured circular polarization under optical excitation with circularly polarized light at 1.58 eV (●) with a fit according to Equation 12.2 and reference measurements of the MCD under linearly polarized excitation at 1.96 eV (O). The bottom panels show the measured circular polarization under electrical spin injection, the MCD data from the top panel, and the net spin-injection signal (●) with a fit according to Equation 12.4. The fit results for these measurements are (a) $T_s/\tau = 0.67 \pm 0.08$, $\Delta B = (0.28 \pm 0.03)$ T, and thus $T_s = (92 \pm 10)$ ps, (b) $\Pi_{inj,el} = (21 \pm 3)\%$, (c) $T_s/\tau = 0.39 \pm 0.05$, $\Delta B = (0.8 \pm 0.1)$ T, and thus $T_s = (32 \pm 3)$ ps, (d) $\Pi_{inj,el} = (16 \pm 2)\%$ (see Reference 12 for details).

characterization of the active region of the spin LED: from the width ΔB of the curve, the spin lifetime T_s can be determined according to Equation 12.3. From the measured circular polarization at $B = 0$ T, and the known injected spin polarization Π_{inj} ($= \Pi_{inj,opt} = 50\%$ in the case of a bulk active region), the value of T_s/τ can be calculated through Equation 12.4. Also τ_s can then be calculated from $T_s^{-1} = \tau_s^{-1} + \tau^{-1}$.

With this information, the curves for electrical spin injection on Panel b and Panel d can then be analyzed, where the only remaining fit coefficient in Equation 12.1 is $\Pi_{inj} = \Pi_{inj,el}$. This procedure works even when the applied oblique field is too small to saturate the circular polarization, as is the case at 300 K (Panel d) where the reduced spin lifetime T_s results in a broadened Hanle curve.

Panel a and Panel c also show the MCD contribution measured under linearly polarized optical excitation at 1.96 eV, which is superimposed on the measurements for both optical and electrical spin injection. This contribution is (much) smaller than the actual spin-injection signal and can easily be corrected for in the analysis.

12.3　SPIN-INJECTION CONTACTS: RESULTS AND BIAS DEPENDENCE

Some popular spin-injection contacts include oxide tunnel barriers, Schottky tunnel barriers, and (ferro)magnetic semiconductors.

12.3.1　Oxide-Based Tunnel Injectors

A typical oxide-based tunnel spin LED is sketched in Figure 12.1 with the following layer structure: p^+ GaAs substrate/200 nm p-$Al_{0.3}Ga_{0.7}As$ (2×10^{18} cm^{-3})/100 nm GaAs (active region)/15 nm $Al_{0.2}Ga_{0.8}As$ (undoped)/1.3 nm AlO_x/2 nm CoFe/8 nm NiFe/5 nm Cu. We will discuss two different doping levels in the active region: not intentionally doped and p-type doped at 2×10^{18} cm^{-3}. The semiconductor heterostructure was grown by molecular beam epitaxy. Immediately after the growth, the sample was transferred through air into the sputtering machine for the fabrication of the tunnel barrier and the ferromagnetic top electrode. The oxide layer was deposited in a two-step process, which facilitates a full oxidation of the Al, reduces the chance on pinholes, and enables the fabrication of a thicker barrier. First, a thin Al layer is sputtered and naturally oxidized in a controlled oxygen atmosphere at 140 torr. After sputtering and oxidation of the first Al layer, a second Al layer was sputtered and oxidized, followed by the deposition the ferromagnetic stack. Typical injected spin polarizations that are obtained with this configuration are 25 to 30% at low temperature (4.2 to 77 K),[12–15] and up to 16% at room temperature.[12]

Recently, epitaxial CoFe/MgO(001) tunnel injectors on GaAs have been demonstrated. Thanks to the high tunnel spin polarization of this epitaxial junction,[16,17] an increased spin injection of 52% at 100 K and 32% at 300 K can be achieved.[18]

The bias dependence of the injected spin polarization depends on the type of active layer as shown in Figure 12.6. For a p-type active layer, the injected spin

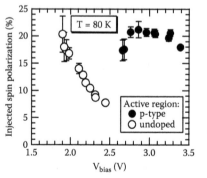 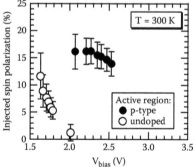

FIGURE 12.6 Experimental bias dependence of the injected spin polarization for two spin LEDs with an AlO_x tunnel injector, at 80 and 300 K; whereas a p-type active region results in an injected polarization that is nearly independent of the applied bias, the LED with an undoped active region results in a rapid decrease of the injected spin polarization with applied bias. See the text for a discussion of the origin of this difference (see also Reference 12 and Reference 19 for details).

polarization is roughly independent of the applied bias, at both measurement temperatures, whereas the injected spin polarization (i.e., the polarization of the electrons that reach the bottom of the conduction band in the active region) decreases with increasing bias.[12,19] This bias dependence is not related to the polarization of the magnetic tunnel contact, but is the result of the increased spin relaxation in the semiconductor when electrons are injected at higher energies than the conduction band minimum. The responsible mechanism is the D'yakaonov–Perel (DP) spin-relaxation mechanism,[7,20] with an efficiency that strongly increases with energy. Due to the lack of inversion symmetry in GaAs, the conduction band is spin split. The splitting depends on the energy: $\sim(E - E_C)^{3/2}$. This splitting is seen by the spins as an effective magnetic field that depends on their quasimomentum and will cause the spins to precess around an axis that depends on their momentum. The relevant time scale of this precession is the relaxation time to the bottom of the conduction band. This is the longitudinal-optic (LO)–phonon time τ_{LO} for undoped GaAs. This time is long enough to cause a random precession of the spins before they reach the bottom of the conduction band and leads to a reduction of the spin polarization at high applied bias. In highly doped GaAs, the momentum scattering time for ionized impurity scattering τ_{ip} is smaller than τ_{LO}. The electron undergoes several momentum scattering events before it thermalizes. This results in a dynamic narrowing of the total precession angle such that the spin is conserved, and the spin polarization remains largely independent of the applied bias.

12.3.2 SCHOTTKY TUNNEL INJECTORS

An alternative way of creating a tunnel barrier on GaAs is by inserting a thin, highly n^{++} doped surface layer to create a very thin Schottky barrier that allows electrons to tunnel through (Figure 12.7).[19,21–23] Both δ-doping methods and degenerate doping into the high 10^{18} or low 10^{19} cm^{-3} ranges are possible. The highly doped layer needs to be designed carefully such that it is fully depleted: it should bring the conduction

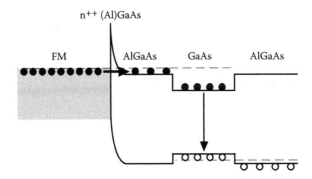

FIGURE 12.7 Schematic diagram of a spin LED with a Schottky tunnel injector under bias. The doping of the n^{++} region is so high that the current is dominated by electrons tunneling through the narrow triangular tunnel barrier, instead of thermionic emission over the barrier as in a low-doped Schottky diode.

band of the semiconductor down to the Fermi level of the magnetic contact without leaving an undepleted n^{++} part. If the layer is too thin or the doping level too low, then part of the Schottky barrier will remain and electron injection will become impossible. If the doping level is too high or the layer too thick, then the remaining undepleted n^{++} region will dilute the injected electron spins. An optimized structure that results in a flat conduction band in the electron transport region is, for example,[19] p^+ GaAs substrate/180 nm p-$Al_{0.3}Ga_{0.7}As$ (10^{18} cm^{-3})/20 nm i-$Al_{0.3}Ga_{0.7}As$/100 nm undoped GaAs/20 nm n-$Al_{0.1}Ga_{0.9}As$ (5×10^{16} cm^{-3})/20 nm n-$Al_{0.1}Ga_{0.9}As$ (3×10^{17} cm^{-3})/ 10 nm n-$Al_{0.1}Ga_{0.9}As$ (3×10^{18} cm^{-3})/5 nm n-$Al_{0.1}Ga_{0.9}As$ (7.5×10^{18} cm^{-3})/8 nm n-$Al_{0.1}Ga_{0.9}As$ (1.5×10^{19} cm^{-3}). The 100 nm wide undoped GaAs active region in this LED is the same as in the AlO_x-injector with undoped active region.

The ferromagnetic contact has to be deposited *in situ*, because oxidation of the top part of the Schottky tunnel barrier would modify the carefully designed n^{++} layer. One advantage of the tunnel Schottky barrier is that it allows the growth of epitaxial, single crystalline magnetic contacts, with better control over the magnetic properties than layers deposited on amorphous AlO_x. A possible disadvantage is a reduced chemical stability and an increased risk of interdiffusion. The injected spin polarizations demonstrated with epitaxial Fe contacts are very similar to those obtained with Fe/AlO_x or CoFe/AlO_x injectors.[15] Epitaxial MnSb injectors can be grown with twofold magnetic anisotropy, square hysteresis loops, and 100% remanence. This provides a possible memory function and also allows larger parts of the Hanle curves to be traced (reversal of the magnetization can be suppressed). In combination with the higher electrical efficiency (see below), this allows the generation and detailed observation of nuclear polarization effects by electrical spin injection.[19, 23–25]

The bias dependence of the injected spin polarization may show a behavior that differs from the AlO_x-based injectors. Figure 12.8 shows the bias dependence for

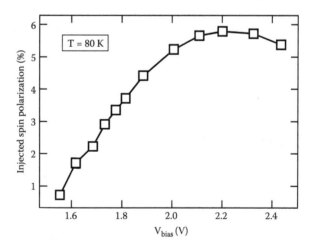

FIGURE 12.8 Experimental bias dependence of the injected electron spin polarization for a spin LED with a Schottky tunnel injector, MnSb magnetic contact, and an undoped active region at 80 K (see Reference 19 for details). The bias dependence is different from that of a spin LED with the same undoped active region but an AlO_x injector, see Figure 12.6.

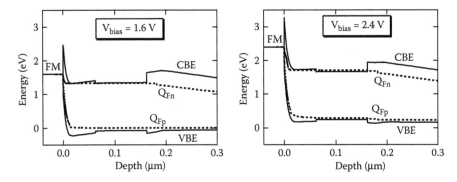

FIGURE 12.9 Self-consistent band profile simulations for a tunnel Schottky LED at an applied bias of 1.6 and 2.4 V. CBE and VBE: conduction and valence band edges, Q_{Fn} and Q_{Fp}: quasi-Fermi levels for electrons and holes (see Reference 26 for details).

the layer structure that was described above and with a MnSb epitaxial contact.[19] The threshold for light emission was around 1.47 V. However, at this low bias, no signal was detected that originated from spin injection: the circular polarization at V_{bias} = 1.47 V showed a weak linear dependence on the applied magnetic field with no hysteresis in the low field range. It had a linear slope that originates from MCD in the MnSb film. This has been used to correct the MCD contribution in measurements at higher bias, which have subsequently been analyzed using Equation 12.4. The circular polarization increases sharply up to an applied bias of 2.3 V and decreases again above 2.3 V. This bias dependence is quite different from the AlO_x injector with undoped active region (Figure 12.6).

The behavior can be explained by a slight overdoping of the injection region during the MBE growth.[26] At low bias (Figure 12.9) the (Al,Ga)As spin-transport layer is not fully depleted. As a result, the transit time for the injected electrons to pass through this layer to the active region is by far longer than the expected spin lifetime. This means that the injected electron ensemble has lost its spin before it reaches the GaAs active region. (An alternative view is that the injected electron spins are diluted by unpolarized electrons residing in this region.) With increasing bias, the transit time through the (Al,Ga)As region is reduced until the conduction band becomes flat at V_{bias} ~ 2.4 V, which is in good agreement with the experiment. As a consequence, the injected electrons arrive in the GaAs active region with a higher polarization. The reduction at bias voltages above flat band has the same origin as the bias dependence in AlO_x spin LEDs with an undoped active region, namely enhanced spin relaxation by the DP mechanism.

12.3.3 MAGNETIC SEMICONDUCTOR INJECTORS

The first successful demonstration of spin injection into a nonmagnetic semiconductor was obtained with magnetic semiconductors. The record for the highest injected spin polarization ($\geq$ 90%)[27] is still held by the diluted magnetic semiconductor (Zn,Be,Mn)Se, however this material operates only at low temperatures and

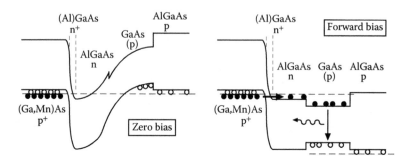

FIGURE 12.10 Band profile of a spin LED with a (Ga,Mn)As Zener tunnel injector. The depletion region of the GaMnAs p^+–GaAs n^+ junction is sufficiently narrow to allow tunneling of spin-polarized carriers out of the exchange-split (Ga,Mn)As valence band into the conduction band of GaAs.

needs a large applied magnetic field. Early results using the p-type ferromagnetic semiconductor (Ga,Mn)As yielded only small injected spin polarizations due to the short spin lifetime of holes.[28] However, by using a Zener tunnel junction the spin-polarized holes in (Ga,Mn)As can be converted in spin-polarized electrons injected into GaAs.[29,30] With an optimized doping design and high-quality (Ga,Mn)As films, we have demonstrated injected electron spin polarizations of 80% at 4.2 K and 25% at 80 K.[31]

The band diagram of a spin LED with a Zener tunnel injector is shown in Figure 12.10. A negative voltage is applied to the (Ga,Mn)As electrode to bias the Zener diode in the reverse direction and the p-n junction of the LED in the forward direction. From about 1.6 V onward, spin-polarized electrons are injected from the valence band of (Ga,Mn)As into the conduction band of the AlGaAs spin-transport layer and further into the GaAs active region of the LED. A second current component consisting of holes flowing from the substrate toward the (Ga,Mn)As film becomes dominating above 2.5 V as shown by both simulations and experiment.[31]

As for the highly doped Schottky injectors, a careful design of the doping profile is essential to avoid long transit times and spin relaxation in the central AlGaAs transport layer. Our optimized design has the following layer structure: p^+ GaAs substrate/200 nm p-Al$_{0.3}$Ga$_{0.7}$As (2×10^{18} cm^{-3})/100 nm p-GaAs (2×10^{18} cm^{-3})/60 nm n-Al$_x$Ga$_{1-x}$As (1×10^{17} cm^{-3})/30 nm n-Al$_x$Ga$_{1-x}$As (1×10^{18} cm^{-3})/9 nm n-GaAs (9×10^{18} cm^{-3}).

The injected spin polarization of this (Ga,Mn)As Zener injector shows a very steep drop-off with increasing bias or injection current (Figure 12.11). In contrast to the oxide and Schottky tunnel injectors, this drop is *now* related to the injector and not to the spin-transport region or the active region of the LED detector. The crucial factor is the thin interfacial (Ga,Mn)As/GaAs region from which the electrons are injected. Under different bias conditions, the width of the Zener junction, the overlap between the valence band of (Ga,Mn)As, and the conduction band of GaAs, and thus the exact region from which electrons tunnel, are all different. At higher bias conditions, the region from which most electrons tunnel shifts away from the

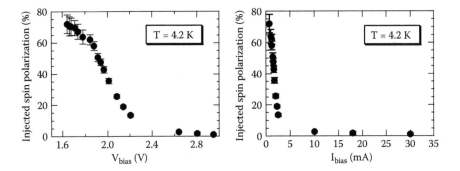

FIGURE 12.11 The injected electron spin polarization in a (Ga,Mn)As Zener spin LED at 4.2 K shows a strong bias dependence (see Reference 31). The left panel shows the injected spin polarization as a function of the total bias applied over the injector plus the LED junction; the right panel shows the same data as a function of the current through both junctions.

bulk like (Ga,Mn)As toward more depleted (Ga,Mn)As and even GaAs, where the spin polarization is expected to be lower. A full treatment can be found in Reference 31 and Reference 32.

12.4 ELECTRICAL CONSIDERATIONS

12.4.1 BIPOLAR SPIN LEDS AND PARASITIC HOLE CURRENT

Apart from the spin polarization achievable with the different contact configurations, their electrical behavior is also important: the total number of carriers that can be injected, the presence or absence of parasitic currents in the spin LED, and, crucially, their ability to operate in an all-electrical, unipolar device.

In most oxide-based tunnel barriers, the Fermi level of the metal will not spontaneously line up with the conduction band of the semiconductor. In most III-V semiconductors (with the notable exception of InAs), surface states pin the surface Fermi level around the center of the forbidden gap, Figure 12.12a. At unpinned interfaces, such as the Si/SiO_2 interface that is at the heart of the very successful CMOS industry, the band lineup is determined by the difference in work function between metal and semiconductor.

Consequently, a sizable voltage has to be dropped over the oxide to align the Fermi level of the magnetic contact with the conduction band of the semiconductor, Figure 12.12b. Part of the associated electric field leaks into the semiconductor, attracting holes from the substrate toward the tunnel barrier, and resulting in a large and undesired hole leakage current flowing in parallel to the desired electron injection current. The threshold current for luminescence in our AlO_x-based spin LEDs is often in the mA range and consists mostly of holes.

Self-consistent simulations (Figure 12.13) confirm this intuitive picture. An electric field is present over the entire AlGaAs transport layer, with the correct polarity to sweep holes from the active region toward the surface. Close to the AlO_x

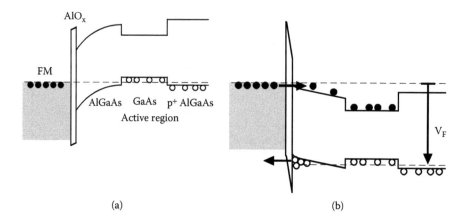

(a)   (b)

FIGURE 12.12 Band lineup in an oxide-based spin LED: (a) at zero bias, and (b) under forward bias with an electric field leaking into the AlGaAs transport layer and sweeping holes from the substrate to the metal contact.

tunnel barrier, the quasi-Fermi level of the holes (Q_{Fp}) crosses the valence band edge, indicating a large accumulation of holes that can tunnel into the contact. The electron quasi-Fermi level (Q_{Fn}) in the AlGaAs transport region is much deeper below the valence conduction band edge, indicating a much lower carrier concentration compared to the holes and thus also a small electron current compared to the hole current.

Tunnel Schottky injectors behave in a completely different way. Intuitively, one might expect an even larger hole-leakage current because there is no longer a tunnel barrier for the holes. However, the highly doped n^{++} surface layer creates a flat-band situation inside the semiconductor, even without applied bias, and there is no electric field that sweeps the holes from the active region toward the surface. Experiments show

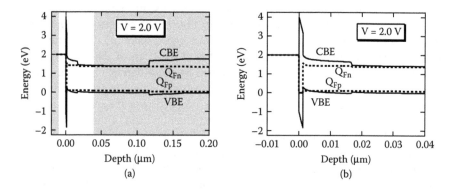

FIGURE 12.13 Self-consistent simulations of a spin LED with an oxide tunnel injector and an undoped active region under bias (2 V). The two panels show the full active layer and an expanded view of the injection contact. CBE and VBE: conduction and valence band edges, Q_{Fn} and Q_{Fp}: quasi-Fermi levels for electrons and holes (see Reference 24 for details).

threshold currents for luminescence as small as 10 μA in tunnel Schottky spin LEDs, which is much smaller than those in AlO_x-based LEDs. Because the electron current at the luminescence threshold is expected to be nearly the same, most of this reduction is due to the elimination of the parasitic hole current. The simulations of a Schottky spin LED under bias (Figure 12.9) confirm that the depletion region screens the surface field from the holes. Around the threshold (~1.6 V), a strong built-in field repels the holes from the surface, and even at 2.4 V bias, the holes are not attracted to the surface. Also the positions of the quasi-Fermi levels Q_{Fp} and Q_{Fn} indicate that the hole concentration in the AlGaAs transport layer is much smaller than the electron concentration, even at 2.4 V bias.

The electrical efficiency of (Ga,Mn)As Zener injectors is similar to that of Schottky injectors: here also a well-designed doping results in a band profile without built-in electric fields that can sweep a parasitic hole current through the device.

12.4.2 PERFORMANCE IN UNIPOLAR DEVICES

A more fundamental issue is the behavior of the contacts in all-electrical devices. Such devices are expected to be unipolar, with the substrate or the channel between the injection and detection contacts undoped or slightly n-doped to obtain the longest possible spin-diffusion length. A p-type layer would remove the injected electrons by electron-hole recombination, thereby preventing any injected electrons from reaching the detection contact. A naïve view of the behavior of an oxide-based tunnel contact in a unipolar device is shown in Figure 12.14. Panel a shows the band lineup at zero bias. It is clear that for both forward bias (spin injection, Panel b), and for reverse bias (spin detection, Panel c) a large voltage needs to be applied over the contact to allow the current to flow. This results in an asymmetry in the behavior of the contacts in injection and detection modes, and a large threshold voltage before the device turns on.

A well-designed Schottky tunnel injector shows much better behavior: the built-in voltage drop over the highly doped depletion layer causes the metal Fermi level to line up with the semiconductor conduction band at zero bias, and the electron current can start flowing in both forward and reverse directions without any threshold voltage.

Accurate modeling of oxide-based tunnel contacts shows that the situation is even more serious. In the potential profile shown in Figure 12.14b, a large electric field is present over the oxide tunnel barrier. To generate such a field, strong dipole charges need to be present at both sides of the oxide: negative charges on the metal side, positive charges on the semiconductor side. In a spin LED, which is a bipolar device, the positive charges on the semiconductor side are provided by the accumulation of holes supplied by the substrate. This accumulation is clearly visible in Figure 12.13 where the hole–quasi-Fermi level Q_{Fp} close to the interface is located well within the valence band. In unipolar devices, holes are absent, and the buildup of the dipole charge is impossible. As a result, the voltage is not dropped over the tunnel barrier, but mostly over the semiconductor (Figure 12.15a), and the contact behaves as a regular Schottky contact with current flow in only in one direction (Figure 12.15b): the contact can only extract carriers from the semiconductor, but

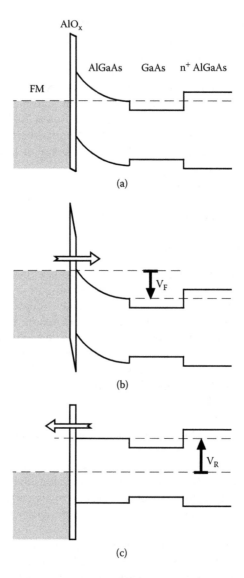

FIGURE 12.14 Naïve (and incorrect) view of an oxide-based injection-detection contact in a unipolar device. (a) Zero bias — a band offset exists at the metal-insulator-semiconductor interface, similar to that in a direct metal-semiconductor contact. (b) Injection geometry ("forward" bias, equivalent to the reverse direction of the underlying metal-semiconductor Schottky contact) — a threshold voltage V_F is dropped over the oxide tunnel barrier to raise the metal work function above the semiconductor conduction band. (c) Detection geometry ("reverse" bias, corresponding to the forward direction of the Schottky diode) — a threshold voltage V_R is required before current can flow (V_R is equivalent to the turn-on voltage of the underlying Schottky diode).

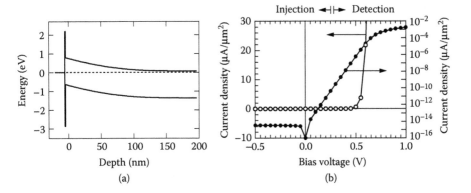

FIGURE 12.15 (a) Simulated band structure for a CoFe/AlO$_x$/n-GaAs (5×10^{16} cm^{-3}) contact (RT, zero bias). (b) Simulated *I-V* characteristics of this structure at RT (see Reference 24 for details).

not inject into the semiconductor. As a consequence, the simple oxide tunnel injector, which was reasonably successful in spin LEDs, cannot be used in unipolar injection-detection devices. Even in the detection (extraction) polarity, which corresponds to the forward direction of the Schottky diode, thermionic emission is needed to send the carriers over the Schottky barrier, and the electrical efficiency remains small until a sufficiently large voltage is applied to reach the flat band.

The solution is to replace the hole accumulation of the spin LED by built-in positive charges in the form of ionized (depleted) donors, Figure 12.16a. This is essentially the combination of a tunnel Schottky contact in a series with an oxide tunnel barrier. The resulting contact has a linear *I-V* characteristic, and charges can

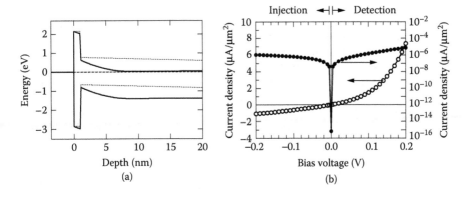

FIGURE 12.16 (a) Same as Figure 12.15, but now with a n^{++} GaAs region just to the right of the AlO$_x$ barrier (full line). For comparison the result of Figure 12.15 is also shown (dashed line). (b) Simulated *I-V* characteristics of the structure with n^{++} layer at RT (see Reference 24 for details).

start flowing at small bias in both injection and detection polarities, as in the single Schottky tunnel barrier.

12.5 NOTE ON NUCLEAR POLARIZATION EFFECTS

The reader should be aware of the hyperfine interaction between the injected spin-polarized electrons and the nuclei. In the presence of a nonzero magnetic field (actually, an external field larger than the nuclear dipole-dipole interactions), this can result in a net nuclear spin polarization, especially when a large amount of electron spins are pumped into the semiconductor. The nuclear polarization can act back on the electron spins and influence their dynamics. These effects are not discussed here, more information can be found in Reference 19, Reference 23 to Reference 25, and Reference 33 to Reference 35.

REFERENCES AND NOTES

1. I. Žutić , J. Fabian, and S. Das Sarma, Spintronics: fundamentals and applications, *Rev. Mod. Phys.*, 76, 323–410, 2004.
2. See the International Technology Roadmap for Semiconductors (ITRS), Emerging Research Devices section (ERD), http://www.itrs.net/Common/2005ITRS/ERD2005. pdf.
3. F.G. Monzon and M.L. Roukes, Spin injection and the local Hall effect in InAs quantum wells, *J. Magn. Magn. Mater.*, 198–199, 632–635, 1999.
4. G. Schmidt, D. Ferrand, L.W. Molenkamp, A.T. Filip, and B.J. van Wees, Fundamental obstacle for electrical spin injection from a ferromagnetic metal into a diffusive semiconductor, *Phys. Rev. B*, 62, R4790–R4793, 2000.
5. E.I. Rashba, Theory of electrical spin injection: tunnel contacts as a solution of the conductivity mismatch problem, *Phys. Rev. B,* 62, R16267–R16270, 2000.
6. A. Fert and H. Jaffres, Conditions for efficient spin injection from a ferromagnetic metal into a semiconductor, *Phys. Rev. B*, 64, 184420, 2001.
7. M.I. D'Yakonov and V.I. Perel, Theory of optical spin orientation of electrons and nuclei in semiconductors, in *Optical Orientation,* Modern Problems in Condensed Matter Sciences, Vol. 8, F. Meier and B. P. Zakharchenya, Eds., North-Holland, Amsterdam, 1984, pp. 11–71.
8. R. Fiederling, P. Grabs, W. Ossau, G. Schmidt, and L.W. Molenkamp, Detection of electrical spin injection by light-emitting diodes in top- and side-emission configurations, *Appl. Phys. Lett.*, 82, 2160–2162, 2003.
9. K. Yoh, H. Ohno, Y. Katano, K. Sueoka, K. Mukasa, and M.E. Ramsteiner, Spin polarization in photo- and electroluminescence of InAs and metal/InAs hybrid structures, *Semicond. Sci. Technol.*, 19, S386–S389, 2004.
10. C.J. Meining, K.A. Korolev, B.D. McCombe, P. Grabs, I. Chado, G. Schmidt, and L.W. Molenkamp, Spin polarization measurements of InAs-based LEDs, in Proceedings of the APS March Meeting, Montreal, Canada, 2004, *J. Supercond.*, 18, 391–397, 2005.
11. Reference experiments with optical excitation based on the regular Hanle effect may be possible in the Voigt geometry (in-plane field, perpendicular optical excitation, and perpendicular detection). However, these yield the spin lifetime at small magnetic fields (same as the oblique Hanle configuration), which may be different from the spin lifetime at the much higher magnetic fields that are used in the electrical experiments in the Faraday configuration.

12. V.F. Motsnyi, P. Van Dorpe, W. Van Roy, E. Goovaerts, V.I. Safarov, G. Borghs, and J. De Boeck, Optical investigation of electrical spin injection into semiconductors, *Phys. Rev. B*, 68, 245319, 2003.

13. V.F. Motsnyi, J. De Boeck, J. Das, W. Van Roy, G. Borghs, E. Goovaerts, and V.I. Safarov, Electrical spin injection in a ferromagnet/tunnel barrier/semiconductor heterostructure, *Appl. Phys. Lett.*, 81, 265–267, 2002.

14. P. Van Dorpe, V.F. Motsnyi, M. Nijboer, E. Goovaerts, V.I. Safarov, J. Das, W. Van Roy, G. Borghs, and J. De Boeck, Highly efficient room temperature spin injection in a metal-insulator-semiconductor light-emitting diode, *Jpn. J. Appl. Phys.*, 42, L502–L504, 2003.

15. O.M.J. van't Erve, G. Kioseoglou, A.T. Hanbicki, C.H. Li, B.T. Jonker, R. Mallory, M. Yasar, and A. Petrou, Comparison of Fe/Schottky and Fe/Al$_2$O$_3$ tunnel barrier contacts for electrical spin injection into GaAs, *Appl. Phys. Lett.*, 84, 4334–4336, 2004.

16. W.H. Butler, X.-G. Zhang, T.C. Schulthess, and J.M. MacLaren, Spin-dependent tunneling conductance of Fe/MgO/Fe sandwiches, *Phys. Rev. B*, 63, 054416, 2001.

17. J. Mathon and A. Umerski, Theory of tunneling magnetoresistance of an epitaxial Fe/MgO/Fe(001) junction, *Phys. Rev. B*, 63, 220403(R), 2001.

18. X. Jiang, R. Wang, R.M. Shelby, R.M. Macfarlane, S.R. Bank, J.S. Harris, and S.S.P. Parkin, Highly spin-polarized room-temperature tunnel injector for semiconductor spintronics using MgO(100), *Phys. Rev. Lett.*, 94, 056601, 2005.

19. P. Van Dorpe, W. Van Roy, V.F. Motsnyi, G. Borghs, and J. De Boeck, Efficient electrical spin injection in GaAs: a comparison between AlO$_x$ and Schottky injectors, *J. Vac. Sci. Technol. A*, 22, 1862–1867, 2004.

20. G. Fishman and G. Lampel, Spin relaxation of photoelectrons in *p*-type gallium arsenide, *Phys. Rev. B*, 16, 820–831, 1977.

21. A.T. Hanbicki, B.T. Jonker, G. Itskos, G. Kioseoglou, and A. Petrou, Efficient electrical spin injection from a magnetic metal/tunnel barrier contact into a semiconductor, *Appl. Phys. Lett.*, 80, 1240–1242, 2002.

22. A.T. Hanbicki, O.M.J. van 't Erve, R. Magno, G. Kioseoglou, C.H. Li, B.T. Jonker, G. Itskos, R. Mallory, M. Yasar, and A. Petrou, Analysis of the transport process providing spin injection through an Fe/AlGaAs Schottky barrier, *Appl. Phys. Lett.*, 82, 4092–4094, 2003.

23. J. Strand, B.D. Schultz, A.F. Isakovic, C.J. Palmstrøm, and P.A. Crowell, Dynamic nuclear polarization by electrical spin injection in ferromagnet-semiconductor heterostructures, *Phys. Rev. Lett.*, 91, 036602, 2003.

24. P. Van Dorpe, W. Van Roy, J. De Boeck, and G. Borghs, Spin dependent transport properties in spin-LED's, a survey, *Proc. SPIE*, 5732, 426–437, 2005.

25. P. Van Dorpe, W. Van Roy, J. De Boeck, and G. Borghs, Nuclear spin pumping by electrical spin injection, submitted.

26. W. Van Roy, P. Van Dorpe, V. Motsnyi, Z. Liu, G. Borghs, and J. De Boeck, Spin-injection in semiconductors: materials challenges and device aspects, *Phys. Stat. Sol. B*, 241, 1470–1476, 2004.

27. R. Fiederling, M. Keim, G. Reuscher, W. Ossau, G. Schmidt, A. Waag, and L.W. Molenkamp, Injection and detection of a spin-polarized current in a light-emitting diode, *Nature*, 402, 787–790, 1999.

28. Y. Ohno, D.K. Young, B. Beschoten, F. Matsukura, H. Ohno, and D.D. Awschalom, Electrical spin-injection in a ferromagnetic semiconductor heterostructure, *Nature*, 402, 790–792, 1999.

29. E. Johnston-Halperin, D. Lofgreen, R.K. Kawakami, D.K. Young, L. Coldren, A.C. Gossard, and D.D. Awschalom, Spin-polarized Zener tunneling in (Ga,Mn)As, *Phys. Rev. B*, 65, 041306, 2002.

30. M. Kohda, Y. Ohno, K. Takamura, F. Matsukura, and H. Ohno, A spin Esaki diode, *Jpn. J. Appl. Phys.*, 40, L1274–L1276, 2001.

31. P. Van Dorpe, Z. Liu, W. Van Roy, V.F. Motsnyi, M. Sawicki, G. Borghs, and J. De Boeck, Very high spin polarization in GaAs by injection from a (Ga,Mn)As Zener diode, *Appl. Phys. Lett.*, 84, 3495, 2004.

32. P. Van Dorpe, W. Van Roy, J. De Boeck, G. Borghs, P. Sankowski, P. Kacman, J.A. Majewski, and T. Dietl, Voltage controlled spin injection in a (Ga,Mn)As/(Al,Ga)As Zener diode, *Phys. Rev. B*, 72, 205322, 2005.

33. J. Strand, A.F. Isakovic, X. Lou, P.A. Crowell, B.D. Schultz, and C.J. Palmstrøm, Nuclear magnetic resonance in a ferromagnet–semiconductor heterostructure, *Appl. Phys. Lett.*, 83, 3335–3337, 2003.

34. V.G. Fleisher and I.A. Merkulov, Optical orientation of the coupled electron-nuclear spin system of a semiconductor, in *Optical Orientation*, Modern Problems in Condensed Matter Sciences, Vol. 8, F. Meier and B.P. Zakharchenya, Eds., North-Holland, Amsterdam, 1984, pp. 173–258.

35. D. Paget and V.L. Berkovits, Optical investigation of hyperfine coupling between electronic and nuclear spins, in *Optical Orientation*, Modern Problems in Condensed Matter Sciences, Vol. 8, F. Meier and B.P. Zakharchenya, Eds., North-Holland, Amsterdam, 1984, pp. 381–422.

13 Spin Photoelectronic Devices Based on Fe and the Heusler Alloy Co$_2$MnGa

M. Hickey, C.D. Damsgaard, S.N. Holmes,
A. Husmann, I. Farrer, R.F. Lee, D.A. Ritchie,
J.B. Hansen, C.S. Jacobsen, G.A.C. Jones,
and M. Pepper

CONTENTS

13.1 INTRODUCTION

Spintronics unifies a range of metallic and semiconducting devices where operation is dependent upon a spin imbalance in the structure of the semiconductor. The technology of spintronics can be applied to nanostructures for low-temperature operation of a quantum qubit in a prototype quantum computer and also for devices envisaged operating at or close to 300 K, particularly the spin-polarized light-emitting diode (spin LED). The spin LED can be used to produce photons with a predetermined polarization and such structures could form part of a quantum communication network. The spin imbalance in the semiconductor device can then be manipulated with magnetic fields, electric fields, Landé spin g-factor modulation, or microwave pulse technology.

In this chapter, we present a study of an InGaAs quantum well in the I region of a GaAs p-i-n LED device with an Fe or a Co_2MnGa Heusler alloy spin injector. These hybrid metal-semiconductor devices are grown in an ultrahigh vacuum (UHV) consisting of a double-chamber molecular beam epitaxy (MBE) system, in the case of the transition metals and two separate MBE chambers in the case of the Heusler alloys. We present optical measurements in the oblique Hanle effect geometry to measure the observed electrical spin injection. The magnetic properties of the metallic part of the structure are measured independently using the longitudinal geometry, magneto-optical Kerr effect (MOKE), and magnetization measurements using a superconducting quantum interference device (SQUID) system. A superconducting point contact probe is used to determine the Andreev reflection probability and the transport spin-polarization efficiency. Although the spin-LED device with Fe contacts has a 31% (8%) spin-injection efficiency at 5 K (300 K), the Co_2MnGa Heusler alloy spin-injection efficiency is only 14% at 5 K with the same InGaAs quantum well and AlGaAs Schottky barrier. The effects of spin injection disappear in the Heusler alloy device at temperatures > 20 K. The spin lifetime in the InGaAs quantum well is in the region of 140 ± 25 ps at 5 K and is independent of which metallic spin injector is used.

A high-efficiency semiconductor spintronic device operating at 300 K using injection of a spin-polarized current has not been satisfactorily demonstrated, although the fundamental mechanisms and limitations have now been identified. The issues include the electrical conductivity mismatch between the spin injector (metal) and the (semiconductor) device, the interface properties such as the tunneling or Schottky barrier characteristics, and the spin-polarization efficiency of the injector.[1,2] An ideal spin injector is a contact of a half-metal, which is metallic for the majority spin band and semiconducting or insulating for the minority spin band. Several metals in the class of Heusler alloys[3,4] are predicted to have this half-metallic band structure that is close to, or even 100% spin polarized. The spin-polarization efficiency (P) in the ferromagnet is defined in Equation 13.1

$$P = \frac{n^\uparrow - n^\downarrow}{n^\uparrow + n^\downarrow} \qquad (13.1)$$

where $n^\uparrow$ is the majority spin carrier density (or the density of states at the Fermi energy) and $n^\downarrow$ is the minority spin carrier density (or the density of states at the Fermi energy). Ferromagnetic Fe has a polarization of $P \approx 40\%$ and the Heusler alloy $NiMnSb$[5,6] has a polarization of 44 to 58%. This is discussed further in Section 13.5. Heusler alloys have recently attracted interest as they can be deposited epitaxially as thin layers with a close lattice match to GaAs, InP, or InAs substrates as demonstrated by the University of Minnesota group.[7,8] The Heusler alloys include the Co-based alloys Co_2MnGe,[9] Co_2MnSi,[10] Co_2CrGa,[11] the Ni-based, full Heusler alloys such as Ni_2MnGa[8] and Ni_2MnIn,[7] and the half Heusler alloy $NiMnSb$.[5,12]

Recently, advances have been made using a conventional ferromagnetic CoFe/Al_2O_3/NiFe magnetic tunnel junction (MTJ)[13] as the emitter and base of a three-terminal spin-LED device. This provides a hot electron current (with control of the injection energy) independent of the band bending (i.e., the base-collector voltage) in the LED part of the device. A polarization efficiency of 10% (at 2.5 T) is observed coming from electroluminescence in the quantum well. The polarization of the electroluminescence in an applied magnetic field follows the measured magnetic moment of the MTJ in a hard axis direction, that is, perpendicular to plane of the layers. The band bending is important at the surface as the electroluminescence is too weak to be analyzed with a base-collector voltage < 1 V. A second, recent advance has been to identify the influence of a dynamic nuclear polarization (DNP) during spin injection into an AlGaAs quantum well p-i-n structure from an Fe contact.[14] The effects of a DNP are to add an extra component (up to 0.2 T) to the applied magnetic field and to add time-dependent relaxation effects with a time scale of minutes (or longer depending on the temperature). Recent advances would not have been possible without initial research into the spin LED. A GaMnAs–GaAs structure[15] had a polarization efficiency of 1% at 6 K. Room temperature operation of this device was not possible as the Curie temperature of GaMnAs is typically < 160 K. The source of spin-polarized holes was a GaMnAs p-type layer, although recently, injected electron polarizations as high as 80% at 4.6 K have been achieved with a GaMnAs layer in a Zener diode configuration.[16] The first hybrid Fe/GaAs spin LED[17] had a polarization efficiency of 2% at 300 K and later advances[18,19] have pushed this up to 32% in an Fe/AlGaAs LED device (at 4.5 K). The latter structure derives part of its success from the AlGaAs Schottky barrier and the Al_2O_3 tunnel barrier at the hybrid interface. The IMEC group in Belgium have achieved a spin-injection efficiency of 21% (16%) at 80 K (300 K) in a CoFe/AlO$_x$/AlGaAs device[20] (see Chapter 12).

13.2 EXPERIMENTAL ISSUES FOR SPIN LEDs

Hybrid spin-LED devices involve only the minimum of lithography and processing. They can be used to determine many spin-related parameters such as the spin lifetime (T_s) and the spin scattering time (τ_s) in the semiconductor. A suitably efficient, p-i-n doping structure with an AlGaAs Schottky barrier can be used with several ferromagnetic metals and intermediate buffer layers to determine the spin-injection efficiency from the metal layer. Figure 13.1 shows the conduction and valence band profile of the InGaAs quantum well p-i-n structure used in this study. The emission from the InGaAs quantum well is below the band gap of the GaAs layers in the structure. The magnetic properties of the injecting contacts such as the anisotropy, the size of the magnetic moment, and the easy axis coercive field are measured and correlated with the optical polarization measurements. A clear understanding of the magnetic state of the ferromagnetic contact in an applied magnetic field is a prerequisite for the spin LED. Magneto-optical effects can dominate the optical polarization properties of the spin LED to the level of several percent. This has meant that more sophisticated optical measurements in addition to the perpendicular magnetic field, Faraday geometry are needed to reduce any magneto-optical effects such as magnetic circular dichroism (MCD) and Zeeman (spin) splitting of the semiconductor bands in an applied magnetic field. The present optical studies have been carried out in the oblique Hanle effect geometry, see Figure 13.2. This technique has recently been established in the field of electrical spin-injection devices, particularly by

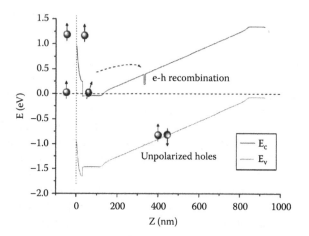

FIGURE 13.1 The conduction band and valence band profile of the MBE-grown LED structure.

the IMEC group.[20] The time development of the average electron spin (**S**) in the InGaAs quantum well can be described by Equation 13.2

$$\frac{d\mathbf{S}}{dt} = \frac{\mathbf{S}_o}{\tau} - \frac{\mathbf{S}}{T_S} + \Omega_L x \mathbf{S} \qquad (13.2)$$

where $\mathbf{S}_o$ is the average spin from a spin-injection process, τ is the electron recombination lifetime, and Ω_L is the Larmor frequency given by Equation 13.3

$$\Omega_L = g\mu_B \frac{\mathbf{B}}{\hbar} \qquad (13.3)$$

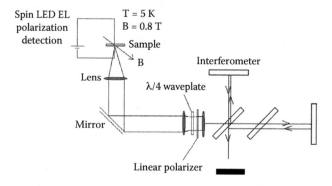

FIGURE 13.2 The experimental setup used to measure the spin-polarization efficiency of a spin-LED device in the oblique Hanle geometry. The device can be cooled to 4 K in magnetic fields of ± 0.8 T. A linear polarizer and 1/4 wave plate are used to select the sense of the circular polarization signal.

The spins precess at the Larmor frequency around the applied magnetic field (**B**), g is the Landé g factor, and μ_B is the Bohr magneton. The effective spin lifetime (T_s) depends on the spin-scattering rate and the electron recombination rate, via Equation 13.4

$$\frac{1}{T_s} = \frac{1}{\tau} + \frac{1}{\tau_s} \qquad (13.4)$$

The solutions to Equation 13.2 in the steady state (i.e., with $dS/dt = 0$) can be obtained in various geometries of spin-injection orientation and applied magnetic field direction.[20] The component of **S** out of the sample plane (S_z) is the only component that can be detected via optical recombination. This signal generally saturates in applied magnetic fields < 0.8 T. The half-width of the Hanle curve is the point where $\Omega_L T_s \sim 1$ (where the spin undergoes on average one full precessional orbit) and is much less than the magnetic field needed to saturate the magnetization. S_z is determined from the optical polarization efficiency (P_{EL}) of the electroluminescence. P_{EL} is measured experimentally and is given by Equation 13.5

$$P_{EL} = \frac{I(\sigma^+) - I(\sigma^-)}{I(\sigma^+) + I(\sigma^-)} \qquad (13.5)$$

$I(\sigma^+)$ is the intensity of positive helicity, circularly polarized radiation, and $I(\sigma^-)$ is the intensity of the negative helicity, circularly polarized radiation. The spin-injection efficiency (Π) can be determined from the optical polarization efficiency P_{EL}, the spin-scattering factor (T_s/τ), and a geometry factor to account for the oblique angle (ϕ) between the applied magnetic field and the plane of the InGaAs quantum well, see Equation 13.6

$$\Pi \frac{T_s}{\tau} = \frac{P_{EL}}{\cos(\phi)\sin(\phi)} \qquad (13.6)$$

T_s/τ, the spin-scattering factor is always < 1 from Equation 13.4. The fact that the Larmor precession frequency depends on the Landé spin g factor provides a mechanism for further spin control and manipulation in the quantum well as the Landé g factor changes both with well width and well composition. A situation could be envisaged[14] where the cross term, $\Omega_L \times$ **S** can be engineered to be ≈ 0, in the case of an AlGaAs quantum well structure where $g \approx 0$. In addition to the Hanle effect data, we first measure the energy spectrum of the electroluminescence using the Fourier transform spectrometer, shown schematically in Figure 13.2. The current-voltage characteristics and the electroluminescence intensity vs. applied voltage are measured to determine the ideal operating conditions of the spin LED. MOKE measurements are carried out on the ferromagnetic layers between 300 and 4 K in the longitudinal geometry along various in-plane crystal directions. The magnetic anisotropy symmetry can then be determined in addition to the easy axis coercive field.

13.3 THE SPIN LED BASED ON MBE-GROWN Fe

The electroluminescence emission from the InGaAs quantum well is characterized by a single peak at 1.340 eV, see Figure 13.3, which shows the electroluminescence spectrum at 4.7 K. Recombination in the InGaAs well dominates although there is a (weak) peak from the electron-to-carbon acceptor transition in a GaAs part of the structure at 1.493 eV. This information is determined to assess the LED efficiency and the current density (or applied voltage) dependence of the electroluminescence intensity. In all the measurements, we routinely replace the ferromagnetic device with a nonferromagnetic Au-contacted device to quantify any background signals unrelated to spin injection. We also determine any MCD signal component as the circularly polarized emission can be turned off at high current density. The 5 nm thick Fe contact layer is thin enough to be optically transparent but also thick enough to avoid domain pinning and excessively large in-plane switching fields at low temperatures. The easy axis (in-plane) coercivity is 50 G at 300 K and 212 G at 5 K with a small uniaxial magnetic anisotropy (UMA) characteristic. Measurement of the magnetic moment gives $2.05 \pm 0.01\mu_B$ per Fe atom at 4 K, which is close to the accepted bulk magnetization value of Fe. This also means that the probability of a nonmagnetic layer at the interface is minimal. The Fe wafer is capped with 4 nm thick Au to reduce the effects of oxidation. Recent x-ray MCD measurements[21] have demonstrated bulk like magnetic moments from 0.25 to 1 monolayer of Fe on GaAs. This is very encouraging for the spin LED and we may be able to reduce the thickness

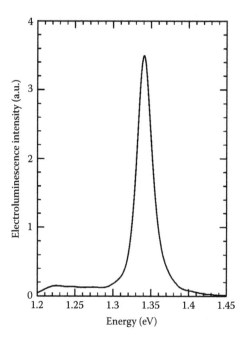

FIGURE 13.3 The electroluminescence spectrum from the Fe spin-LED at 4.7 K with an excitation current of 14 A/cm².

of the ferromagnetic component in the spin LED even further and still maintain a well-defined, bulk like magnetic moment with no nonmagnetic layer at the interface. Hanle effect data was taken with the Fe spin LED and the Au-contact spin LED and the two sets of data are compared (at 8 K) in Figure 13.4. No spin injection and hence no optical polarization is observed from the Au-contacted device. The parameter $\Pi \cdot Ts/\tau$ given by Equation 13.6 and determined by P_{EL} (measured) and ϕ (from the experimental setup), is plotted on the y axis. The spin-injection process is very dependent on current density (or applied voltage), see Figure 13.5. The applied voltage dependence of the electroluminescence polarization shows a monotonic decrease away from the maximum injected spin polarization of 16% (the value of $\Pi \cdot T_s/\tau$ for this device) at higher biases. This is due to a transition in the dominant transport regime at the interface from tunneling field emission through the Schottky barrier to thermal excitation of carriers over the barrier into the conduction band of the n-AlGaAs. Thermally excited carriers can undergo strong interfacial spin-flip scattering and further slow thermalization as they fall to the bottom of the conduction band in the semiconductor. The current densities are in the region where dynamic nuclear polarization effects can be observed.[14] This is characterized by a larger apparent magnetic field ($\mathbf{B}_{Total}$) and longer time constants due to the long spin-relaxation times of the nuclear spins. Injected spin-polarized electrons interact with the nuclear spins via hyperfine coupling and the dynamically polarized nuclear field

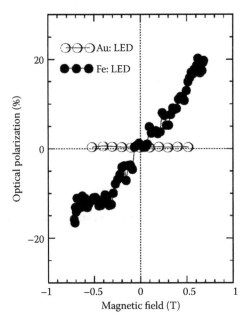

FIGURE 13.4 The oblique Hanle polarization data from the Fe spin LED shown as the difference in circular polarization intensities, scaled by a geometry factor at 8 K. The (unpolarized) signal from the LED with nonspin-injecting Au contacts is shown for reference.

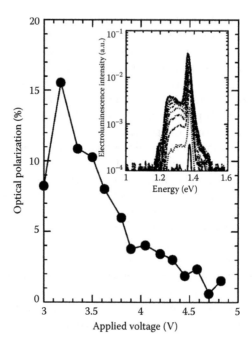

FIGURE 13.5 The voltage dependence of the spin-polarization ($\Pi \cdot T_s/\tau$) from the Fe spin LED at 4.7 K. The inset shows the electroluminescence spectra for one particular sense of circular polarization at each applied voltage level.

enhances the applied magnetic field seen by the injected electron spin in the InGaAs quantum well according to Equation 13.7

$$\mathbf{B}_{Total} = \mathbf{B}_{Applied} + C\mathbf{B}_{Applied}\left(\frac{<S> \mathbf{B}_{Applied}}{\left|\mathbf{B}_{Applied}\right|^2}\right) \qquad (13.7)$$

as suggested by previous work,[22,23] where $\mathbf{B}_{Applied}$ is the applied magnetic field, $<S>$ is the time-averaged electron spin. The parameter C depends on the overlap between the electron and the nuclear wave functions. In this work, DNP effects are present in the Hanle geometry and this is seen as a horizontal shift in the Hanle curve by up to 0.2 T. The DNP process is very temperature-dependent and disappears by 10 K in the present structures independent of whether the spin injector is an Fe or a Co$_2$MnGa layer.

In summary, the spin-injection efficiency of Fe is 8 ± 3% at 300 K, rising to 31 ± 2% at 5 K on an n-type AlGaAs Schottky barrier. The n-AlGaAs:GaAs p-i-n spin LED serves as an experimental tool for determining the spin-injection efficiency of the proposed half-metallic material Co$_2$MnGa, which is discussed in Section 13.4.

13.4 THE SPIN LED BASED ON THE HEUSLER ALLOY Co$_2$MnGa

The Co-based Heusler alloys are closely lattice matched to InGaAs-based semiconductors and a high-crystalline order can be achieved at the semiconductor interface. The exact crystalline structure of the Heusler alloy is important particularly if 100% spin polarization is required for device operation. The Co$_2$MnX Heusler alloys (where X is a nonmagnetic element such as Ga, Si, Ge, Sn, or Sb) in their half-metallic form are expected to order in the $L2_1$ structure. This $L2_1$ Heusler structure consists of a face-centered cubic (fcc) lattice with four constituent fcc sublattices. The magnetic moment can be shared between the constituents in the lattice and the site distribution of the magnetic moment is important for understanding the band-structure properties. In Co$_2$MnGa both the Co and the Mn lattice sites have a magnetic moment.[24] In the Co$_2$MnX alloys, atomic disorder is a problem and Co can occupy the Mn lattice site, even in the stoichiometric balance. This atomic disorder can be sufficient to destroy half-metal behavior. The spin-polarization efficiency is then sensitive to the local environment and control of this is a prerequisite for successful spin-LED structures. The spin polarization in the ferromagnetic layer adjacent to the GaAs substrate may determine whether spin-polarized current injection is possible even if the bulk alloy or surface is 100% spin polarized. The compositional variation, particularly at interfaces or surfaces, is the main reason why conventional tunneling magnetoresistance (TMR) and spin-valve devices in Mn-based Heusler alloys have shown a reduced magnetoresistance.

In Section 13.5, we investigate further the spin-polarization of Co$_2$MnGa using a superconducting probe system. X-ray diffraction measurements on MBE-grown Co$_2$MnGa shows the expected diffraction peaks for a cubic structure with a lattice constant of 5.770 Å and peak orders (200), (220), (400), and (422) at Bragg angles of 30.97°, 44.4°, 64.6°, and 81.7°, respectively. Although the x-ray measurements confirm a cubic lattice, the site distribution of the magnetic moment (determined by XMCD) is a more sensitive technique to determine if the structure is in the half-metallic $L2_1$ form. Magnetization data on MBE-grown Co$_2$MnGa shows that the magnetic moment is ~ 1.5 μ_B per Mn atom (at 10 K). MBE-grown Co$_2$MnGa thin films can have a weak UMA characteristic with the easy axis along the [0, −1,1] in-plane crystalline direction on the GaAs spin-LED substrate. Current-voltage characteristics are routinely taken on the spin-LED mesa, see the inset of Figure 13.6, which shows the characteristic at 300 K. There are variations in the forward bias, turn-on characteristics between different mesas, and some of the more "leaky" reverse bias diode structures show the largest spin-polarization efficiencies. Figure 13.6 shows the electroluminescence energy spectrum at 300 K with no applied magnetic field. The characteristic is dominated by electron-hole recombination in the InGaAs quantum well. The 10 nm thin films of Co$_2$MnGa were grown at 200°C with 5 nm Co–Ga buffer layers at the n-AlGaAs interface, to prevent Mn diffusion into the semiconductor. The stoichiometry in the thin film was Co$_2$MnGa to within 15%. The buffer layer should reduce the spin scattering at the semiconductor interface. Hanle effect polarization data was taken with the Co$_2$MnGa structures at 5 K, see Figure 13.7. In Figure 13.7, the difference in intensity for left and right circularly polarized light

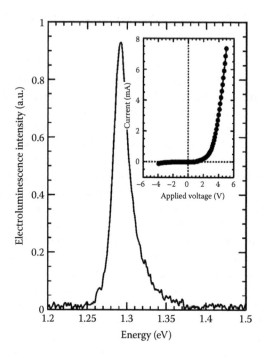

FIGURE 13.6 The electroluminescence spectrum for the Co_2MnGa spin LED at 300 K. The inset shows the current-voltage characteristic at 300 K.

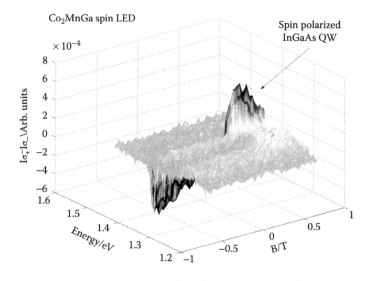

FIGURE 13.7 Oblique Hanle effect data taken at 5 K with the Co_2MnGa spin LED, showing the spectrally resolved electroluminescence intensity difference $I(\sigma^+) - I(\sigma)$, as a function of applied magnetic field.

is shown spectroscopically resolved as a function of applied magnetic field at a Hanle angle of 30°. The polarization efficiency signal (P_{EL}) has a value of 6 ± 2% at 5 K and this corresponds to a spin-injection efficiency of 14 ± 3%. The Hanle effect polarization signal from the Co_2MnGa spin LED disappears at a temperature of 20 K possibly due to excessive spin scattering at the interface between the metal and the semiconductor. The spin lifetime (T_s) is estimated to be 150 ± 50 ps and this value is close to the lifetime obtained with the Fe contact device, see Section 13.3. Although MBE-grown Co_2MnGa can be used for low-temperature spin-injection devices, the full capability of the half-metal device has not yet been demonstrated.

13.5 POINT CONTACT ANDREEV REFLECTION DETERMINATION OF SPIN POLARIZATION

Measurement of spin polarization in a ferromagnetic metal relies on the principle of Andreev reflection (AR) at the interface with a superconductor in point contact geometry. The AR probability is dependent on the minority spin population at the Fermi level in the ferromagnetic layer. This analysis is straightforward in the ballistic case with no scattering at the interface. The conductance ratio $G(0)/G_N$ is then given by Equation 13.8

$$\frac{G(0)}{G_N} = 2(1 - P_T) \qquad (13.8)$$

where P_T is the transport spin polarization in the metal. This definition of polarization can under certain circumstances be the same as the carrier density polarization given by Equation 13.1 but this is not always the case. $G(0)$ is the conductance at zero applied voltage and G_N is the (normal) conductance at high voltage ($V_{applied} >> \Delta/e$), where G is independent of bias. Δ is the superconducting band gap in the Nb tip and $V_{applied}$ is the applied voltage. This model breaks down for realistic interfaces with elastic scattering and a more realistic approach is needed to analyze $G(V_{applied})$. The modified approach is due to Blonder, Tinkham, and Klapwijk and is referred to as the "BTK" model.[25] The parameter Z is included to account for elastic interfacial scattering and can be considered in the model as a δ-function scattering center of height Z. The ballistic limit corresponds to $Z = 0$, and the tunnel junction corresponds to $Z \to \infty$. The complete experimental $G(V_{applied})$ should be understood to extract reliable P_T values. The probability of AR of electrons moving from a metal into a superconductor depends on the population of the minority spin band in the metal. In metals like Cu, which have no spin polarization in the conduction band, the normalized conductance doubles at bias voltages below the superconducting gap due to AR. In 100% spin-polarized metals, the normalized conductance at those voltages is zero as the electrons cannot form Cooper pairs with opposite spins to move into the superconductor in the ballistic limit. The spin polarization of the ferromagnetic metal in question can be extracted more accurately by fitting the normalized conductance as a function of applied voltage up to a few times (Δ/e). Proximity effects from the superconductor may induce a small superconducting layer in the metal with a smaller gap than the bulk, and an extended BTK model[26]

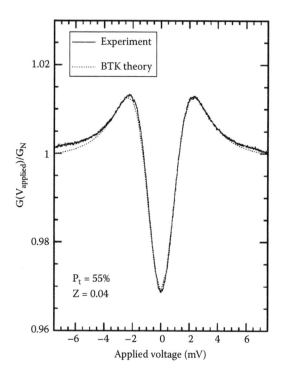

FIGURE 13.8 The normalized conductance at 4.2 K of a point contact between a supercon-ducting Nb tip and a MBE-grown Co_2MnGa wafer. The fit to the extended BTK model is shown for comparison.

incorporates this effect. However, details of the Fermi surface of the metal are not included and significant modifications to the model might be needed in complex materials such as the Heusler alloys. The highly spin-polarized state in the half-metallic ferromagnet has previously been observed in the study[26] of CrO_2 (where $P_T = 96 \pm 1\%$) and also the Heusler alloy NiMnSb (where $P_T = 44$ to 58%).[5,6] We have measured a number of MBE-grown Co_2GaMn films, see Figure 13.8, which shows the normalized conductance as a function of applied voltage for one particular wafer. The conductance is measured as a function of applied voltage using an AC-voltage excitation technique to the Nb tip and the fit is made using the BTK model. The transport spin polarization in this particular film was 55% (at $Z = 0.04$), which is higher than recent data on ferromagnetic metals such as Co and Fe.[6] However, it does not confirm the predicted half-metallic character of Co_2MnGa. This is most likely due to off-stoichiometric materials and to site disorder, which is a large problem in Mn-based Heusler alloys. High Z values will suppress the observed spin polarization due to increased spin-mixing effects, not included in the model and ideally, one is interested only in the case when $Z = 0$. Experimentally, $P_T(Z = 0)$ can be inferred from extrapolations to $Z = 0$ from measurements at finite Z. Therefore, a set of data for varying Z values gives a clearer picture of the actual spin polarization in the metal.

13.6 Co$_2$TiSn AND Co$_2$CrAl HEUSLER ALLOYS FOR SPINTRONIC DEVICES

Co$_2$TiSn and Co$_2$CrAl are Heusler alloys that do not contain Mn in the structure and they provide a more suitable material system for combining the growth of metals with GaAs-based semiconductors in the same UHV environment, where contamination of the semiconductor with unacceptable levels of Mn-dopant needs to be avoided. MTJs have already been fabricated using Co$_2$(Cr,Fe)Al[27] and a TMR value of 19% has been achieved at 300 K. Following previous work[4,28] in establishing the electronic band structure properties of the Heusler alloys, we focus our attention on Co$_2$TiSn. The Heusler alloy Co$_2$TiSn orders ferromagnetically with a Curie temperature of 370 K and a magnetic moment of 1.89 μ_B per unit cell. Previous theoretical methods used have been the augmented spherical plane wave (ASPW)[4] and full potential Korringa–Kohn–Rostoker (KKR) method[28] within the localized density approximation (LDA) in the framework of *ab initio* density functional theory (DFT). The KKR approximation method has been used to calculate the structure of Co$_2$(Cr,Fe)Al.[29] In this chapter, we use a plane wave basis code with ultrasoft pseudopotentials, treating the exchange-correlation energy in the LDA approach with the CASTEP code.[30] This code was written in the Theory of Condensed Matter Group at the Cavendish Laboratory and stands for Cambridge Serial Total Energy Package. CASTEP has been used to model the bulk band structure of Co$_2$TiSn and Figure 13.9 shows the calculated, spin-dependent, band structure in several directions of the reciprocal lattice for both up and down spins. In the center of the Brillouin zone (Γ-point) the spin-down band shows the Fermi energy (defined as $E = 0$), between two bands (i.e., half-metallic-like). The spin-up band shows the Fermi energy in a band of conduction band levels. This picture may be over simplified but it does show at least qualitatively that half-metal behavior is possible in Co$_2$TiSn. Earlier XMCD measurements[31] on the $L2_1$ crystal structure of Co$_2$TiSn also found this interpretation of the partial half-metallic character. We use an $8 \times 8 \times 8$ k-point mesh Monkhurst–Pack grid to generate a partial density of states (PDOS) for the

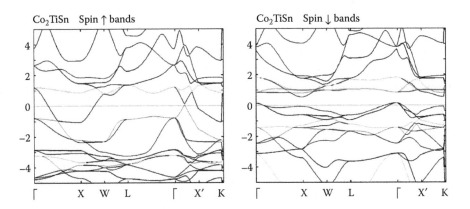

FIGURE 13.9 The spin-dependent band structure of Co$_2$TiSn calculated using CASTEP. The energy scale on the y axis is in electron volts.

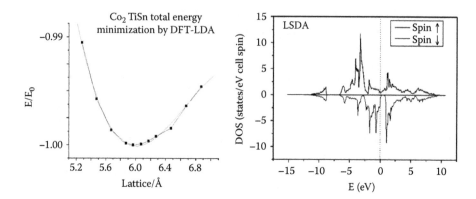

FIGURE 13.10 A total energy calculation as a function of equilibrium lattice constant in Co_2TiSn and the partial density of states that shows the band gap in the minority spin band, which is estimated to be 0.62 eV.

spin-up and spin-down electrons, see Figure 13.10. Although there is a bulk minority spin band gap of ~ 0.62 eV, the Fermi level lies at the edge of this gap while still giving a bulk spin polarization of ~ 90%. The lattice constant of Co_2TiSn, (6.077 Å) was confirmed with DFT energy minimization, see Figure 13.10. This accepted value for the equilibrium lattice constant provides the minimum total energy in the framework on the DFT-LDA approximation. The lattice constant of Co_2TiSn is close to that of the "6.1 Å" semiconductor materials, InAs, GaSb, and AlSb and coherent epitaxial growth can be expected. Voltage tunable Rashba and Dresselhaus spin-orbit coupling effects can then be utilized for spintronic devices in this group of narrow band-gap semiconductor materials.

13.7 SUMMARY AND OUTLOOK

We have presented a series of magneto-optical, transport, and magnetic characterization measurements on an Fe spin LED and a Co_2MnGa spin LED and have discussed the present issues relating to spin-injection devices. We have considered the Heusler alloys Co_2TiSn and Co_2CrAl as alternatives to Co_2MnGa for spintronic devices, although there are still many materials issues to be resolved with the Heusler alloys before they could form part of GMR or TMR device architectures. The fundamental issue for half-metal growth is achieving the correct crystal structure for the Heusler alloy and it is this obstacle that primarily needs addressing along with the properties of the interface with the semiconductor. MBE-grown Co_2MnGa ferromagnetic layers have a spin-injection efficiency of 14 ± 3% at 5 K, although the measured spin polarization on the free surface is 55%. Further studies are needed to determine why the spin-polarization signal disappears at temperatures > 20 K. The spin LED is most efficient with a doped AlGaAs Schottky barrier or Al_2O_3 tunnel barrier; however, recent measurements with Fe/MgO suggest that this combination of injector and tunnel barrier may be ideal for the spin LED. Epitaxial MTJs in Fe/MgO/Fe have already reached a TMR of 88% at 293 K where the tunneling

TABLE 13.1
Summary of the Spin-Injection and Polarization Efficiencies in Several Ferromagnetic Materials, Some of Which Have Been Used in the Fabrication of Spin LEDs

Ferromagnetic Material	Spin-Injection Efficiency (%)	Temperature (K)	Andreev Polarization (%)	Reference
Fe	2	300	40 (Ref. 6)	17
Fe	31	5	40	Present work
Fe	8	300	—	Present work
Co$_2$MnGa	14	5	55	Present work
NiMnSb	—		44–58	5, 6
CrO$_2$	—	2	96	26
GaMnAs	1	6	—	15
CoFe	21	80	—	20
CoFe	16	300	—	20

Note: Dashes indicate no data are available.

process is coherent.[32] Several candidate materials for spintronics have emerged recently and Table 13.1 summarizes the materials discussed in this chapter with respect to the spin-LED device. This field of research is expanding rapidly and other potential half-metal spin injectors and tunnel barrier materials are being grown. Time-resolved Faraday rotation spectroscopy in ferromagnetic Schottky diodes[22] has demonstrated a coupling of electron spin to nuclear spin via the hyperfine interaction. This DNP process is also evident in the spin LED presented in this work, at temperatures < 10 K. The spin-LED device could be used as a detector of dynamic nuclear spin polarization or for interrogating the spin in a few-electron quantum dot. A voltage tunable hyperfine interaction may also lead to a new range of devices whose operation involves nuclear spin processes. These processes would have applications in several schemes for quantum information processing.

REFERENCES

1. G. Schmidt, D. Ferrand, L.W. Mollenkamp, A.T. Filip, and B.J. van Wees, *Phys. Rev. B*, 62, R4790, 2000.
2. E.I. Rashba, *Phys. Rev. B*, 62, R16267, 2000.
3. F. Heusler, *Verh. Dtsch. Phys. Ges.*, 5, 219, 1903.
4. R.A. de Groot, F.M. Mueller, P.G. van Engen, and K.H.J. Buschow, *Phys. Rev. Lett.*, 50, 2024, 1983.
5. S.K. Clowes, Y. Miyoshi, Y. Bugoslavsky, W.R. Branford, C. Grigorescu, S.A. Manea, O. Monnereau, and L.F. Cohen, *Phys. Rev. B.* 69, 214425, 2004.
6. R.J. Soulen, J.M. Byers, M.S. Osofsky, B. Nadgorny, T. Ambrose, S.F. Cheng, P.R. Broussard, C.T. Tanaka, J. Nowak, J.S. Moodera, A. Berry, and J.M.D. Coey, *Science*, 282, 85, 1998.

7. J.Q. Xie, J.W. Dong, J. Lu, C.J. Palmstrøm, and S. McKernan, *Appl. Phys. Lett.,* 79, 1003, 2001.
8. J.W. Dong, L.C. Chen, C.J. Palmstrøm, R.D. James, and S. McKernan, *Appl. Phys. Lett.,* 75, 1443, 1999.
9. T. Ambrose, J.J. Krebs, and G. Prinz, *Appl. Phys. Lett.,* 76, 3280, 2000.
10. L.J. Singh, Z.H. Barber, Y. Miyoshi, Y. Bugoslavsky, W.R. Branford, and L.F. Cohen, *Appl. Phys. Lett.,* 84, 2367, 2004.
11. R.Y. Umetsu, K. Kobayashi, R. Kainuma, A. Fujita, K. Fukamichi, K. Ishida, and A. Sakuma, *Appl. Phys. Lett.*, 85, 2011, 2004.
12. W. van Roy, J. de Boeck, B. Brijs, and G. Borghs, *Appl. Phys. Lett.,* 77, 4190, 2000.
13. X. Jiang, R. Wang, S. van Dijken, R. Shelby, R. Macfarlane, G.S. Soloman, J. Harris, and S.S.P. Parkin, *Phys. Rev. Lett.,* 90, 256603, 2003.
14. J. Strand, B.D. Schultz, A.F. Isakovic, C.J. Palmstrøm, and P.A. Crowell, *Phys. Rev. Lett.,* 91, 036602, 2003.
15. Y. Ohno, D.K. Young, B. Beschoten, F. Matsukura, H. Ohno, and D.D. Awschalom, *Nature,* 402, 790, 1999.
16. P. van Dorpe, Z. Liu, W. van Roy, V.F. Motsnyi, M. Sawicki, G. Borghs, and J. de Boeck, *Appl. Phys. Lett.,* 84, 3495, 2004.
17. H.J. Zhu, M. Ramsteiner, H. Kostial, M. Wassermeier, H.-P. Schönherr, and K.H. Ploog, *Phys. Rev. Lett.,* 87, 016601, 2001.
18. A.T. Hanbicki, B.T. Jonker, G. Itskos, G. Kioseoglou, and A. Petrou, *Appl. Phys. Lett.,* 80, 1240, 2002.
19. B.T. Jonker, *Proc. IEEE,* 91, 727, 2003.
20. V.F. Motsnyi, P. van Dorpe, W. van Roy, E. Goovaerts, V.I. Safarov, G. Borghs, and J. de Boeck, *Phys. Rev. B,* 68, 245319, 2003.
21. J.S. Claydon, Y.B. Xu, M. Tselepi, J.A.C. Bland, and G. van der Laan, *Phys. Rev. Lett.,* 93, 037206, 2004.
22. E. Johnston-Halperin, D. Lofgreen, R.K. Kawakami, D.K. Young, L. Coldren, A.C. Gossard, and D.D. Awschalom, *Phys. Rev. B,* 65, 041306, 2002.
23. R.K. Kawakami, Y. Kato, M. Hanson, I. Malajovich, J.M. Stephens, E. Johnston-Halperin, G. Salis, A.C. Gossard, and D.D. Awschalom, *Science,* 294, 131, 2001.
24. P.J. Brown, K.U. Neumann, P.J. Webster, and K.R.A. Ziebeck, *J. Phys.: Condens. Matter,* 12, 1827, 2000.
25. G.E. Blonder, M. Tinkham, and T.M. Klapwijk, *Phys. Rev. B,* 25, 4515, 1982.
26. Y. Ji, G.J. Strijkers, F.Y. Yang, C.L. Chien, J.M. Byers, A. Anguelouch, Gang Xiao, and A. Gupta, *Phys. Rev. Lett.*, 86, 5585, 2001.
27. K. Inomata, N. Tezuka, S. Okamura, H. Kurebayashi, and A. Hirohata, *J. Appl. Phys.,* 95, 7234, 2004.
28. I. Galanakis, P.H. Dederichs, and N. Papanikolaou, *Phys. Rev. B,* 66, 174429, 2002.
29. Y. Miura, M. Shirai, and K. Nagao, *J. Appl. Phys.,* 95, 7225, 2004.
30. M.D. Segall, P.J.D. Lindan, M.J. Probert, C.J. Pickard, P.J. Hasnip, S.J. Clark, and M.C. Payne, *J. Phys.: Condens. Matter,* 14, 2717, 2002.
31. A. Yamasaki, S. Imada, R. Arai, H. Utsunomiya, S. Suga, T. Muro, Y. Saitoh, T. Kanomata, and S. Ishida, *Phys. Rev. B,* 65, 104410, 2002.
32. S. Yuasa, A. Fukushima, T. Nagahama, K. Ando, and Y. Suzuki, *Jpn. J. Appl. Phys.,* 43, L588, 2004.

14 Electron Spin Filtering across a Ferromagnetic Metal/Semiconductor Interface Measured by Photoexcitation

A. Hirohata, S.J. Steinmuller, and J.A.C. Bland

CONTENTS

14.1 SPIN ELECTRONICS WITH OPTICAL METHODS

14.1.1 Spin Polarization in GaAs

In GaAs, which is widely used in spin electronics, the valence band maximum and the conduction band minimum are aligned at the Γ point, the center of the Brillouin zone ($\mathbf{k} = 0$), with an energy gap $E_g = 1.43$ eV at room temperature (RT), indicating that the only transition induced by photon energy $h\nu$ occurs at Γ (direct gap semiconductor [SC]).[1,2] The valence band (p symmetry) splits into a fourfold degenerate $P_{3/2}$ state at Γ_8 and a twofold degenerate $P_{1/2}$ state, which lies at Γ_7 ($\Delta = 0.34$ eV below $P_{3/2}$), whereas the conduction band (s symmetry) is a twofold degenerate $S_{1/2}$ state at Γ_6 as schematically shown in Figure 14.1. The $P_{3/2}$ band consists of twofold degenerate bands, and the heavy hole and light hole subbands.

When $h\nu = E_g$, circularly polarized light excites electrons from $P_{3/2}$ to $S_{1/2}$. According to the selection rule ($\Delta m_j = \pm 1$), the two transitions for each photon helicity (σ^+ and σ^-) are possible; however, the relative transition probabilities for the light and heavy holes need to be taken into account to estimate the net spin polarization.[1,3] For example, if electrons are excited only from the valence band maximum (Γ_8) by circularly polarized light, three times more spins are excited from $m_j = \pm 3/2$ than from $m_j = \pm 1/2$ states as shown in Figure 14.2. Although the maximum polarization is expected to be 50% in theory, the maximum is experimentally observed to be ~40% at the threshold as shown in Figure 14.3 due to experimental limitations, such as spin depolarization in the GaAs layer and at the interfaces.[1,4]

For $E_g + \Delta < h\nu$, the polarization decreases with increasing $h\nu$ due to the mixture of the light and heavy hole states with the split-off valence band states, which have the opposite sign.[1] Such interband absorption occurs only through the spin-orbit interaction, because the electric field of exciting light only influences electron orbital motion. For $E_g + \Delta \ll h\nu$, the spin-orbit interaction becomes negligible and spin depolarization during the cascade process can be significant. Therefore, the electron optical orientation is absent.[3]

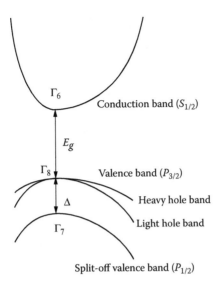

FIGURE 14.1 Schematic band structure of GaAs in the vicinity of the Γ point (center of the Brillouin zone) in **k** space. The energy gap E_g between the conduction band and the valence bands for both the heavy and light holes are shown. The spin-orbit splitting Δ also exists. (From D.T. Pierce and F. Meier, *Phys. Rev. B*, 13, 5484, 1976. Copyright 1976 American Physical Society. With permission.)

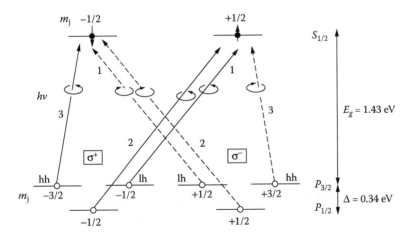

FIGURE 14.2 Schematic diagram of the allowed transitions for right (σ^+, solid lines) and left (σ^-, dashed lines) circularly polarized light in GaAs. The selection rule is $\Delta m_j = +1$ for σ^+ and $\Delta m_j = -1$ for σ^-. The numbers near the arrows represent the relative transition probabilities. The magnetic quantum numbers are also indicated at the corresponding energy levels. The heavy and light holes are abbreviated to hh and lh, respectively. (From D.T. Pierce and F. Meier, *Phys. Rev. B*, 13, 5484, 1976. Copyright 1976 American Physical Society. With permission.)

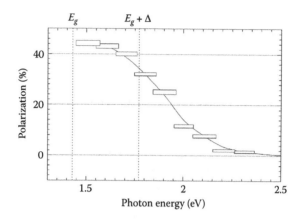

FIGURE 14.3 Photoemission spectrum of spin polarization from GaAs + CsOCs at $T \leq 10$ K. The experimental data are shown as boxes including experimental errors. (From D.T. Pierce and F. Meier, *Phys. Rev. B*, 13, 5484, 1976. Copyright 1976 American Physical Society. With permission.)

14.1.2 OPTICAL SPIN DETECTION

Circularly polarized light emission from a SC has been used as a measure of spin polarization of the SC electrons.[1] Recently spin-polarized electrons (or holes) have been injected into a SC, whose efficiency and the efficiency of this process have been estimated by using circularly polarized electroluminescence (EL) signals from a quantum well (QW) (see Table 14.1).

14.1.2.1 Spin-Polarized Light-Emitting Diodes

In dilute magnetic semiconductor (DMS)/SC heterostructures, because DMS shows a large Zeeman splitting and/or ferromagnetism,[17] the DMS can be used as a spin aligner to inject spin-polarized carriers, that is, spin-polarized electrons or holes, into the SC. The spin polarization of the injected carriers is detected optically through circularly polarized EL from the SC (e.g., GaAs), behaving as a spin-polarized light-emitting diode (spin LED). With ferromagnetic *p*-GaMnAs as a spin aligner, spin-polarized hole injection has recently been reported at low temperatures.[15] At forward bias, spin-polarized holes from the *p*-GaMnAs as well as unpolarized electrons from the *n*-GaAs layer are injected into the InGaAs QW, so that the recombination of the spin-polarized holes can create circularly polarized EL emission from the QW. However, as the spin-relaxation time for the holes is much shorter than that for the electrons,[18] the spin-polarization signal through the recombination process in the GaAs is very small (about ±1%). On the other hand, using paramagnetic *n*-BeMnZnSe with a large Zeeman splitting as a spin aligner, because the spin-diffusion length of electrons has been reported to be above 100 μm in GaAs,[19] highly efficient electron spin injection has been achieved with an applied field of ~3 T (spin polarization in EL ~90%).[4] Similar results have been obtained using CdMnTe,[5] ZnMnSe,[6,7]

TABLE 14.1
List of Recent Optical Spin-Detection Studies

Structures	Spin Polarization	References
Spin LED (spin-polarized *electron* injection)		
BeMgZnSe+BeMnZnSe/n-AlGaAs/i-GaAs QW/ ... /p-GaAs	~42% (< 5 K)	4
CdMnTe/CdTe	~30% (5 K)	5
n-ZnMnSe/AlGaAs/GaAs QW/AlGaAs	~83% (4.5 K)	6,7
Fe/GaAs/InAs QW/GaAs	~2% (25 K)	8
Fe/AlGaAs/GaAs QW/GaAs	~13% (4.5 K)	9
	~8% (240 K)	
NiFe+CoFe/AlO$_x$/AlGaAs/GaAs QW/GaAs	~9.2% (80 K)	10
FeCo/AlO$_x$/AlGaAs/GaAs QW/ ... /p-GaAs	~21% (80 K)	11
	~16% (300 K)	
(Co, Fe and NiFe)/Al$_2$O$_3$/n-AlGaAs/si-GaAs QW/ ... /GaAs	~0.8, 0.5, and 0.2% (RT)	12
CoFe/MgO/AlGaAs/GaAs QW/ ... /p-GaAs	~57% (100 K)	13
	~47% (290 K)	
Co$_{2.4}$Mn$_{1.6}$Ga/n-AlGaAs/ ... /InAs QW/GaAs	~13% (5 K)	14
Spin LED (spin-polarized *hole* injection)		
p-GaMnAs/GaAs/InAs QW	~1% (< 31 K)	15
SP-STM		
Ni STM tip/GaAs	~30% (RT)	16

ZnSe,[20] and MnGe[21] but only at low temperatures (typically T < 80 K). Because RT ferromagnetism has been theoretically predicted in several DMS compounds[22] and experimentally observed,[23] spin injection at RT with DMS may be achievable in the near future.

14.1.2.2 Schottky Diodes

A ferromagnet (FM)/SC Schottky diode, consisting of an Fe (20 nm)/GaAs/InGaAs QW LED structure, has been used to measure the circularly polarized EL by Zhu et al.[8] Spin injection from Fe to GaAs is achieved with an efficiency of about 2% at 25 K, which is independent of temperature. However, the right and left circularly polarized EL intensity does not show a clear difference. By examining the tails of Gaussian-like EL intensity distribution, a heavy hole excitation contribution is estimated. On the other hand, Hanbicki et al. have performed a similar experiment with an Fe (12.5 nm)/AlGaAs/GaAs QW LED and have observed a spin-injection efficiency of 30%.[9] They clearly observe a significant difference between the right and left circular EL intensity. The spin polarization is estimated to be 13% at 4.5 K (8% at 240 K). Taking the spin-relaxation time in the QW into account, they report a small temperature dependence in the spin-injection efficiency, which is consistent with spin-polarized electron tunneling theory.

14.1.2.3 Metal/Oxide/Semiconductor (MOS) Structures

An insulating oxide barrier has been inserted between a FM spin injector and a SC. For a CoFe/AlO$_x$/GaAs QW structure, approximately 16% of spin polarization has been reported at RT[11] (see Chapter 12). For a similar structure with a MgO barrier, about 47% of spin polarization has been observed at RT, which leads to almost 32% of spin-injection efficiency.[13] The spin polarization is expected to increase monotonically with increasing temperature; however, this does not apply for this structure. Such behavior can be induced by the spin relaxation in the QW.

14.1.2.4 Spin-Polarized Scanning Tunneling Microscopy (SP-STM) for Spin Injection

Spin-polarized electron tunneling from a Ni scanning tunneling microscopy (STM) tip into a GaAs substrate was first demonstrated by Alvarado and Renaud as shown in Figure 14.4.[16] The Ni tip is remanently magnetized by an electromagnet and is

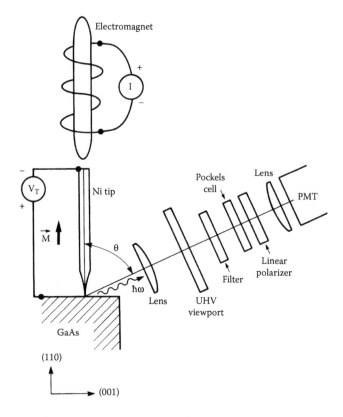

FIGURE 14.4 Schematic diagram of SP-STM by Alvarado and Renaud. Spin-polarized electrons are injected into a GaAs substrate by using a magnetized Ni STM tip, and spin polarization is measured through circularly polarized light emission from the GaAs surface. (From S.F. Alvarado and P. Renaud, *Phys. Rev. Lett.*, 68, 1387, 1992. Copyright 1992 American Physical Society. With permission.)

TABLE 14.2
List of Recent Optical Spin-Injection Studies

Structures	Spin Polarization	References
FM/SC hybrid structures		
(NiFe, Co and Fe)/n-GaAs	~2 ± 1% (RT)	28
(FeCo and Fe)/GaAs QW	~0.5% ~ MCD (10 K)	29
MOS junctions		
Co/Al$_2$O$_3$/p-GaAs	~1.2% (RT)	26
(Ni and Co)/Al$_2$O$_3$/p-GaAs	~2.5 and 1.0% (RT)	27
SP-STM		
Ni STM tip/GaAs	< 10% (RT)	30
p-GaAs STM tip/Co/mica	~10% (RT)	31
p-GaAs STM tip/NiFe/Si	~7% (RT)	32

used as a spin injector. The spin-polarized electron tunneling over the vacuum is detected as circularly polarized EL signals, in which the change is reported to be approximately −30% at RT. This value corresponds to the minority electron spin polarization of Ni(001) at the Fermi level, suggesting that the minority spin electrons dominate the tunneling current.

14.1.3 OPTICAL SPIN INJECTION

One method of injecting spin-polarized electrons is photoexcitation, which uses circularly polarized light to excite spin-polarized electrons[24,25] and detects the spin polarization as electrical signals (see Table 14.2). The possibility of passing a spin-dependent current through thin film tunnel junctions of both Co/Al$_2$O$_3$/GaAs and Co/τ-MnAl/AlAs/GaAs with photoexcited spin-polarized electrons has been first discussed by Prins et al.[26] For the former structure, a spin-dependent tunneling current is observed, but only magnetic circular dichroism (MCD) signals are seen in the latter structure. In their experiment, the sample with a 2 nm Al$_2$O$_3$ tunneling barrier shows the largest helicity asymmetry of the photoexcited current of approximately 1.2% at 1.5 eV (near the GaAs band gap). Accordingly, many studies of spin-dependent tunneling through MOS junctions have been carried out such as Reference 27, especially to realize optically pumped SP-STM as described below.

Recently, efforts have been made to observe the influence of a QW in the SC on spin-polarized electron transport across the FM/SC interface.[29] Because 20 nm thick FeCo and Fe films are used for the measurement, MCD background signals dominate the polarized photocurrent, with the result that at most 0.5% of the photocurrent is attributed to a true spin-dependent signal at 10 K.

14.1.3.1 SP-STM for Spin Detection

SP-STM was theoretically proposed using a direct-gap SC tip in 1993 by Molotkov,[33] and Laiho and Reittu.[34] After the first photoexcitation measurement by Prins et al.,[26] modulated circularly polarized light has been used to excite spin-polarized electrons in a SC (e.g., GaAs). Although optically excited electrons are scattered at the SC surface with back illumination,[35] Sueoka et al. have demonstrated the possibility of detecting spin-polarized signals by scanning a Ni STM tip over a GaAs film with circularly polarized light shone through an AlGaAs membrane.[30] Suzuki et al. have performed similar experiments but by scanning a p-GaAs STM tip over a Co film with back illumination through mica/Au/Co, as shown in Figure 14.5, and have obtained magnetic domain images.[31] GaAs tips are fabricated using photolithography and anisotropic etching to be delimited by {105} facets. A three-monolayer (ML) Co film possesses perpendicular magnetization and shows an MCD effect of less than 0.14%, which is much smaller than the observed polarization response (about 10%). Polarization modulation response images of the SP-STM show very good agreement with magnetic force microscopy (MFM) images. To avoid the MCD effect and possible light scattering through the sample structures, Kodama et al. have shone circularly polarized light onto a GaAs tip in the vicinity of the sample, which is equivalent to front illumination.[32] They detect a change of approximately 7% in I-V curves between right and left circular light irradiation with NiFe films.

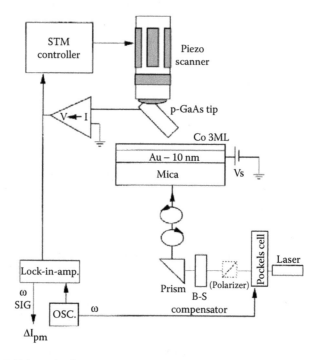

FIGURE 14.5 Schematic diagram of spin-polarized STM by Suzuki et al. Circularly polarized light shone through a Co/Au/mica structure introduces spin-polarized electrons in a GaAs STM tip. Magnetic domain images are obtained dependent upon the photon circular polarization using lock-in techniques. (From Y. Suzuki, W. Nabhan, R. Shinohara, K. Yamaguchi, and K. Mukasa, *J. Magn. Magn. Mater.*, 198–199, 540, 1999. Copyright 1999 Elsevier. With permission.)

14.2 PHOTOEXCITATION TECHNIQUES

14.2.1 POLAR PHOTOEXCITATION

A circularly polarized laser beam (with photon energies hv in the range $1.59 \leq hv \leq 2.41$ eV) was used together with an external magnetic field to investigate the spin dependence of a photoexcited electron current at RT. The light was modulated from 100% right to 100% left circular polarization using a photo-elastic modulator (PEM) at a frequency of 50 kHz. For the polarized illumination mode, the bias dependence of helicity-dependent photocurrent I through the interface was probed both (i) in the remanent state (I^0) and (ii) under the application of a magnetic field ($H = 1.8$ T) sufficient to saturate the FM mag-netization along the plane normal (I^n). A bias voltage was applied between one Al (or Au) contact on the surface of the sample and one common ohmic contact attached to the back of the GaAs substrate. The current flowing through these two pads was measured (both with and without photoexcitation), while the voltage across the sample was also measured using a separate top contact as shown in Figure 14.6.[36,37] From the I-V curves without photoexcitation, the Schottky characteristics were studied. This setup is basically the reverse exper-iment of the EL measurement at the FM/SC interfaces studied for example by Alvarado and Renaud.[16] As the polarized laser beam enters from the Au capping layer side, these structures provide a way of avoiding laser absorption at the bottom surface of the SC, as occurs when using back illumination.[35]

A key issue is the transport process (tunneling, thermionic emission, etc.) by which electrons travel from the SC to the FM. The probability for tunneling is determined by the Schottky barrier height and depletion layer width. The depletion layer width W is defined by

$$W = \{2\varepsilon_S/qN_D(V_{bi} - V)\}^{1/2} \tag{14.1}$$

where N_D, ε_S, and V_{bi} stand for the doping density, the static dielectric constant, and the built-in potential across the depletion layer, respectively.[35] For GaAs, $\varepsilon_S = 13.1 \times \varepsilon_0$ ($\varepsilon_0 = 8.85418 \times 10^{-19}$ F/m) and $V_{bi} \sim 1.3$ eV. At zero bias, W is estimated to be in the range between 3.4 nm ($N_D = 10^{25}$ m^3) and 34 nm ($N_D = 10^{23}$ m^3). When W is large, the electron tunneling process is reduced due to the reduction of the electron tunneling probability through a thick tunnel barrier; on the other hand, tunneling does not occur for very small W because of tunnel barrier breakdown. As electron tunneling through the Schottky barrier is a dominant process in the case of Si ($N_D \leq 10^{23}$ m^{-3}),[38] one can assume the tunneling process is similarly dominant in our GaAs samples. The ohmic electron transport associated with diffusive transport, which we assume to be spin independent,[39] occurs in parallel with the tunneling process and is likely to occur at local defects, indicating that the spin-polarized signals in the helicity-dependent photocurrent can be diluted by ohmic components as Schmidt et al. have already pointed out.[39]

Albrecht and Smith have recently reported that the depletion layer at the Schottky barrier is detrimental to both spin injection and detection due to strong spin

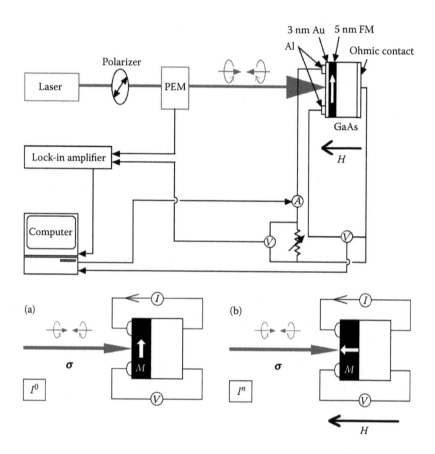

FIGURE 14.6 Schematic configuration of the polar photoexcitation experiment. The laser light (hv = 1.59, 1.96, and 2.41 eV and power 3, 5, and 30 mW, respectively) is linearly polarized in the 45° direction with reference to the modulator axis pointing along the sample plane and shone onto the sample at normal incidence. Right and left circular light is produced using a PEM. The photocurrent is measured by I-V measurement methods combined with lock-in techniques. A schematic view of the FM/GaAs hybrid structure (3 × 3 mm) is also shown in this diagram. The value of the variable resistor for the measurement is chosen to be approximately the same as that of the resistance between the FM and the GaAs substrate. The magnetization **M** in the FM and the photon-helicity σ are shown with the field H applied normal to the sample. Two configurations, (a) without (I^0) and (b) with (I^n) a magnetic field, are shown in the inset. The sheet resistance of the Au capping layer and the FM films is estimated to be 7×10^8 and 1 to 6×10^9 Ω, respectively, and therefore it is negligible.

relaxation.[40] They suggest a heavily doped region near the interface is required to form a sharp potential profile. Because the presence of the Schottky barrier is crucial in this study, the doping density needs to be controlled to observe the spin-detection signals.

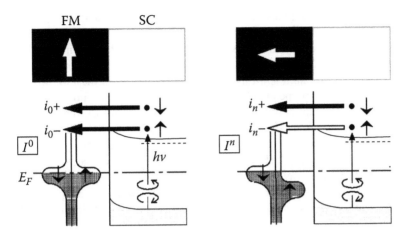

FIGURE 14.7 Schematic diagrams illustrating the spin-filtering mechanism for photoexcited electron transport in the vicinity of the Fermi level E_f. I^0 and I^n correspond to the helicity-dependent photocurrent in the remanent state and at perpendicular saturation, respectively. Up and down spin currents are indicated with + and − notations, respectively.

14.2.2 Proposed Spin-Filtering Model

The helicity-dependent photocurrent I is measured by modulating the photon helicity from right (σ^+) to left (σ^-). The two helicity values correspond to opposite spin angular momentum values of the incident photon, and the helicity gives rise to opposite spin polarizations of electrons photoexcited in the GaAs.[1] The magnetization ($\mathbf{M}$) in the FM is aligned perpendicularly ($H = 1.8$ T) or in plane ($H = 0$) by using an external field. For $\sigma//\mathbf{M}$ (or antiparallel), the electrons in the FM and the SC share the same spin quantization axis, while for $\sigma\perp\mathbf{M}$, the two possible spin states created by the circularly polarized light are equivalent when projected along the magnetization direction in the FM (see Figure 14.7). Consequently, in the remanent state ($\sigma\perp\mathbf{M}$), when the magnetization $\mathbf{M}$ is orthogonal to the photoexcited spin polarization, both up and down spin-polarized electrons in the SC can flow into the FM. At perpendicular saturation ($\sigma//\mathbf{M}$), on the other hand, the up spin electron current from the SC is filtered due to the spin-split density of states (DOS) in the FM.[41–43] Spin filtering is therefore turned on and off by controlling the relative axes of σ and $\mathbf{M}$, and is detected as the helicity-dependent photocurrent I. With $\sigma\perp\mathbf{M}$, there is no spin filtering, but spin filtering is turned on by rotating to $\sigma//\mathbf{M}$. The helicity-dependent photocurrents I^0 and I^n correspond to the magnetization configurations $\sigma\perp\mathbf{M}$ (see Figure 14.6a) and $\sigma//\mathbf{M}$ (see Figure 14.6b), respectively.

The helicity-dependent photocurrents, I^0 and I^n, measured are proportional to the difference between the current components for right (σ^+) and left (σ^-) circularly polarized light for each magnetization configuration: $I^0 = p^0|i_0^+ - i_0^-|$ and $I^n = p^n|i_n^+ - i_n^-|$, where p^0 and p^n are phase factors for $\sigma\perp\mathbf{M}$ and $\sigma//\mathbf{M}$, respectively. We shall discuss the measurement of the phase shift as a function of applied field later (see Section 14.3.2.2). Because the autophase mode of the lock-in amplifier is used for the measurements of I^n and I^0, the phase factor p is adjusted to be 1 in each case. As shown in Figure 14.7, $i_0^+ = i_0^-$ is expected for the case of the remanent state because there is no particular spin splitting in the averaged DOS in the FM; whereas, $i_n^+ \neq i_n^-$ for the case of perpendicular saturation due to the spin polarization of the DOS in the FM. In principle, the helicity-dependent photocurrent I^0 should be zero and I^n should reflect the spin polarization both in the SC and the FM.

In addition, Andresen et al. have recently investigated the Schottky barrier characteristics in the SC.[44] They consider a simple model of the band bending at different bias conditions as schematically shown in Figure 14.8. Pinning of the Fermi level in the FM within the band gap of the SC gives rise to depletion and accumulation of carriers.[35] Thus, the band bending takes place according to the Poisson equation

$$-\nabla^2 \varphi = e(n_{\text{dope}} - n_e + n_h) \qquad (14.2)$$

where φ is the electric potential, e is the electron charge, n_{dope} is the doping density, and n_e and n_h are the electron and hole densities, respectively. At the zero bias condition, electron depletion occurs with the resulting band bending as illustrated in Figure 14.8c. This implies a negative photocurrent originating from electrons diffusing into the SC and holes tunneling into the FM. As illustrated in Figure 14.8b, a certain value of the forward bias voltage V_{flat} creates a flat band configuration with neither electron depletion nor accumulation. The disappearance of a vanishing internal electric field gives rise to a vanishing photocurrent. Figure 14.8a shows that electron accumulation near the barrier takes over at forward bias exceeding V_{flat}. Electrons tunneling into the FM and holes diffusing into the SC yield a positive photocurrent at forward bias moderately exceeding V_{flat}. As the forward bias is further increased, the electron accumulation region is narrowed, and the photocurrent decreases rapidly. In contrast, hole accumulation occurs over an extensive reverse bias range as shown in Figure 14.8d, causing less pronounced variations in the photocurrent as the bias is changed.

14.2.3 IN-PLANE PHOTOEXCITATION

An in-plane photoexcitation setup was also employed to measure a spin-dependent photocurrent across spin-valve structures as shown in Figure 14.9. Circularly polarized He–Ne laser light ($h\nu = 1.96$ eV) incident at an angle θ from the sample plane normal with front illumination was used to excite electrons with a spin polarization in the SC. The bias dependence of the current across the Schottky barrier ($-2.5 < V < +1.5$ V) was measured both with and without photoexcitation. An optical magnetocurrent I, dependent upon the relative magnetization **M** configuration in plane between the two FM layers (both parallel [I^p] and antiparallel [I^a] configurations), was measured for the incidence angle range of $0 \leq \theta \leq 45°$.

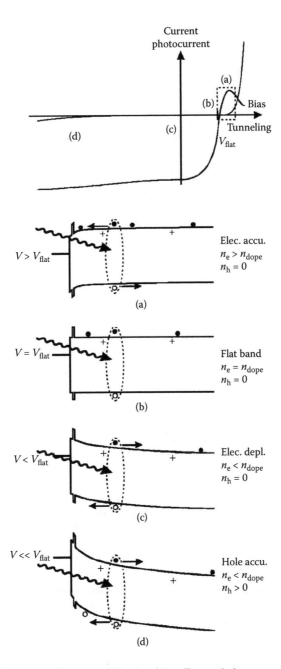

FIGURE 14.8 Schematic diagrams of the band bending and the corresponding current-voltage (*I-V*) and photocurrent-voltage characteristics in the case of (a) forward bias electron accumulation, (b) forward bias flat band configuration, (c) zero bias electron depletion, and (d) reverse bias hole accumulation. (From S.E. Andresen, S.J. Steinmuller, A. Ionescu, G. Wastlbauer, C.M. Gürtler, and J.A.C. Bland, *Phys. Rev. B,* 68, 073303, 2003. Copyright 2003 American Physical Society. With permission.)

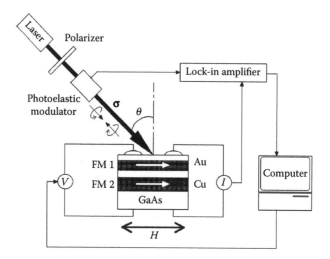

FIGURE 14.9 Schematic configuration of the in-plane photoexcitation experiment. The laser is polarized in the 45° direction with reference to the modulator axis pointing along the sample plane and shone onto the sample at normal incidence. Right and left circular light is produced using a PEM. A computer controlled bias voltage is applied between one Al (or Au) contact and the bottom common ohmic contact, and the voltage drop across another Al (or Au) contact and the bottom contact is measured using lock-in techniques.

For the case of $\theta = 0$, the photon helicity σ is orthogonal to the FM magnetization ($\sigma \perp M$) and no spin dependence is expected. For $\theta > 0°$, on the other hand, σ possesses an in-plane component, which excites SC electrons with a component of in-plane spin polarization. This suggests that the spin-dependent electron transmission should increase with increasing θ.[45]

14.3 SPIN FILTERING IN FERROMAGNET/ SEMICONDUCTOR SCHOTTKY DIODES

Three key experiments to verify spin-polarized electron transport across the FM/GaAs Schottky barrier interfaces have been carried out: the effects of (i) various FM metals, (ii) doping density of the GaAs substrates on the Schottky characteristics and spin filtering, and (iii) FM thickness dependence of spin filtering. The former two experiments confirm that the Schottky barrier can behave as a tunnel barrier for spin-polarized electron transport in FM/SC hybrid structures, which can be controlled by the Schottky characteristics such as barrier height and depletion layer width. This suggests that the Schottky diodes are a mixture between a FM/SC hybrid structure and a magnetic tunnel junction (MTJ) with the advantage of easy controllability of the barrier characteristics. In the latter measurement (iii), the spin-transport effects show a similar tendency with the spin polarization in the GaAs, indicating that spin-polarized electrons are excited in the SC and filtered by the FM only when the magnetization is aligned along the photon helicity.

14.3.1 Sample Preparation

Ultrahigh vacuum (UHV) deposition techniques were used to fabricate 5 nm thick FM layers ($Ni_{80}Fe_{20}$, Co and Fe as the FM, and antiferromagnetic [AF] Cr as reference) directly onto GaAs substrates, capped with 3 nm thick Au layers. The ohmic contacts on the back of the n- and p-type substrates were prepared by evaporating 100 nm thick GeAuNi and AuBe, respectively, and then annealed at 770 K for 2 min. The GaAs substrates were cleaned for 2 min using an oxygen plasma together with chemical cleaning with acetone and isopropanol and then loaded into the UHV chamber.[46] The FM films were grown at a rate of approximately one ML per minute by electron-beam (e-beam) evaporation. The substrate temperature was held at 300 K and the pressure was approximately 7×10^{-10} mbar during the growth. The deposition rate was monitored by a quartz microbalance that was calibrated using atomic force microscopy (AFM). Two Al (or Au) electrical contacts (0.5 mm $\times$ 0.5 mm $\times$ 550 nm) were then evaporated onto the Au capping layer for the transport measurements.

14.3.2 NiFe/GaAs Schottky Diodes

14.3.2.1 Semiconductor Doping Density Dependence

To test the role of the Schottky barrier, we have investigated NiFe samples with several doping densities of the GaAs. Figure 14.10 shows the I-V curves of the $Ni_{80}Fe_{20}$/GaAs samples without photoexcitation measured at RT. A degree of linearity in the I-V curves around zero bias suggests that these samples also contain weak ohmic components. This is associated with diffusive transport, which we assume to be spin-independent[39] and to occur in parallel with spin-polarized electron transport across the FM/SC interface. As the doping density increases, the Schottky barrier

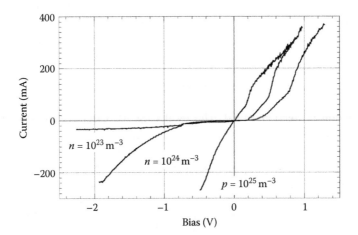

FIGURE 14.10 Bias dependence of currents through the $Ni_{80}Fe_{20}$/GaAs(100) ($n = 3.0 \times 10^{23}$, 1.5×10^{24}, and $p = 1.8 \times 10^{25}$ m^{-3}) interface obtained without photoexcitation at RT.

height ϕ_b is observed to decrease from approximately 0.8 ($n = 10^{23}$ m^{-3}) to 0.2 eV ($p = 10^{25}$ m^{-3}) as expected.[36] The difference in the helicity-dependent photocurrents $\Delta I = I^n - I^0$ is found to decrease with increasing SC doping density, clearly indicating that the spin filtering is only observed when the Schottky barrier is present.[36]

14.3.2.2 Ferromagnetic Layer Thickness Dependence

The bias dependence of the helicity-dependent photocurrent for the permalloy samples (thickness: $2.5 \leq t \leq 7.5$ nm) is presented in Figure 14.11. The helicity-dependent photocurrent is almost two orders of magnitude smaller than the unpolarized photocurrent as anticipated from our model, but the bias dependence follows that of the unpolarized photocurrent. In our experiment the light has to pass through the FM layer before reaching the GaAs substrate, so there will be a magneto-optical background due to MCD. The asymmetry in σ^+ and σ^- light intensity induced by MCD leads to an asymmetry of the photocurrent magnitude, therefore obscuring or even mimicking any real spin-filtering effect and complicating the data analysis in earlier studies.[28,36,37] In Section 14.3.2.3, we will demonstrate a method to separate MCD effects from our measurements. For the NiFe/GaAs structures, spin filtering is observed at forward bias; whereas at reverse bias only magneto-optical effects are seen.[47] When the applied bias is increased above the Schottky barrier height ϕ_b (i.e., when the Schottky diode is switched on), the spin-filtering effect disappears. It is therefore essential for achieving spin-dependent electron transport to operate the FM/SC devices in the bias range below ϕ_b, which was misleadingly referred to as the reverse bias range in earlier publications.[28,47,48] The helicity-dependent, spin-filtered photocurrent strongly depends on the applied bias and the details of the Schottky barrier (which are determined by the GaAs doping density), suggesting that electron tunneling is the relevant spin-dependent transport mechanism. Further support to this interpretation is added by temperature-dependent measurements of NiFe/AlGaAs barrier/GaAs structures[44] (see Section 14.3.2.3) that allow for a precise control of the bias and temperature range under which tunneling occurs.

To investigate the FM layer thickness dependence further and to explicitly exclude a possible MCD effect in the SC, we measure the applied magnetic field dependence of the spin-filtering effect (see Figure 14.12). First, the phase shift of the lock-in amplifier is measured, showing almost a 180° change upon reversing the saturation magnetization direction in the NiFe samples as presented in Figure 14.12a. This indicates unambiguously that the magnetization alignment of the FM layer with respect to the photon-helicity σ controls the spin-filtering effect as expected from the discussion in Section 14.2.2. The results for AF Cr/GaAs are also shown in Figure 14.12a for reference, but they show no magnetic field dependence, which also confirms that the signals we observe originate from the FM layer.

Similarly, the applied field dependence of ΔI without bias is shown in Figure 14.12b for the corresponding samples. Most importantly, the field dependence of ΔI matches that of the polar magneto-optical Kerr effect (MOKE) signals (see Figure 14.12c), which suggests that there are no significant background effects due to the Zeeman splitting in the GaAs,[8,29] for instance. Although the Cr samples show a small offset, they do not possess any magnetic field dependence as seen in Figure 14.12b, which also confirms that the Zeeman splitting effect in the SC is negligible in our measurement.

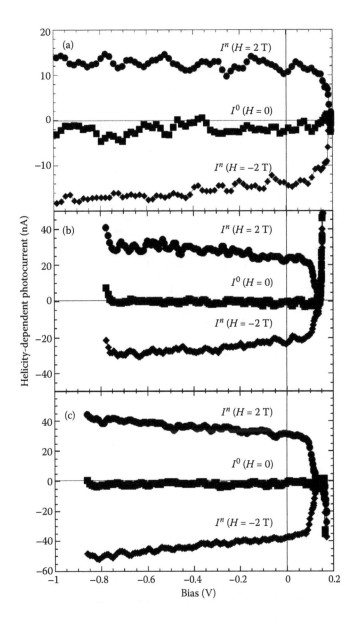

FIGURE 14.11 Bias dependence of the helicity-dependent photocurrent without (closed squares, I^0) and with the applied magnetic field of 2 T (closed circles and closed rhombuses, I^n) in the case of NiFe/GaAs(100) ($n = 10^{23}$ m^{-3}) with $t =$ (a) 2.5, (b) 5.0, and (c) 7.5 nm.

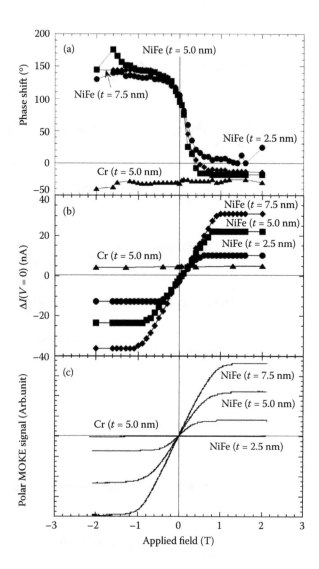

FIGURE 14.12 Applied magnetic field dependence of (a) phase shifts, (b) ΔI, and (c) MOKE signals for NiFe/GaAs(100) ($n = 10^{23}$ m^{-3}; $t = 2.5$, 5.0, and 7.5 nm) and 5 nm thick Cr/GaAs(100) ($n = 10^{23}$ m^{-3}) samples.

14.3.2.3 Spin Filtering at Ferromagnet/Semiconductor Interfaces

The dependence of the helicity-dependent photocurrent on the magnetic field follows the hysteresis loop of the magnetic film with a small constant offset as observed in the Section 14.3.2.2. The offset is independent of the magnetic field magnitude, which proves that it arises purely due to optical effects. To remove the contribution from the nonmagnetic offset to the helicity-dependent photocurrent, Andresen et al.

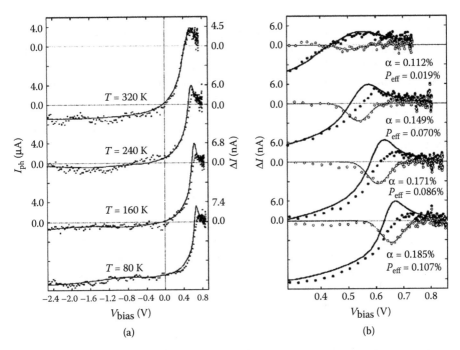

FIGURE 14.13 (a) Comparative plots of the photocurrent I_{ph} (solid lines) and the magnitude of the helicity-dependent photocurrent ΔI (points) as a function of a bias voltage V_{bias} measured at different temperatures. (b) Separation of the total helicity-dependent photocurrent ΔI (solid circles) into the magneto-optical contribution ΔI_{MCD} (solid lines) and the contribution due to spin filtering ΔI_{SF} (open circles). The dashed line is a guide to the eye, and the sequence of temperatures is the same as that shown in (a) Values for the MCD coefficient α and the effective P_{eff} are also given for the investigated temperature range. (From S.E. Andresen, S.J. Steinmuller, A. Ionescu, G. Wastlbauer, C.M. Guertler, and J.A.C. Bland, *Phys. Rev. B*, 68, 073303, 2003. Copyright 2003 American Physical Society. With permission.)

have applied a magnetic field of 26 mT (sufficient to saturate the FM films in plane) and subtracted the helicity-dependent photocurrent for opposite field directions by using the in-plane photoexcitation configuration.[44] Now, the magnitude of the helicity-dependent photocurrent (ΔI) is given by the relation (using a slightly different notation here)

$$\Delta I = (I^+ - I^-)/2 \tag{14.3}$$

with the helicity-dependent photocurrent I^+ and I^- for positive and negative saturation, respectively.

Figure 14.13a shows the bias dependence of both the photocurrent and the magnitude of the helicity-dependent photocurrent for a Au (3 nm)/NiFe (5 nm)/*n*-GaAs (2 nm)/AlGaAs (2 nm)/*n*-GaAs(100) sample measured at different temperatures.[44] The bias dependence of the two currents at reverse bias is almost the same for all temperatures (see Figure 14.13a); however that at forward bias is significantly different, coinciding with the photocurrent peak at 0.4 to 0.8 V, which corresponds to the bias regime where electron tunneling occurs (see Section 14.2.2).

Different transport mechanisms (hole diffusion into the FM, thermionic emission of electrons over the AlGaAs barrier, electron tunneling across the AlGaAs barrier) will contribute to the unpolarized photocurrent, depending upon the applied bias. Significant spin-filtering effects are expected to occur at reverse bias for the case of spin-dependent hole transport and at forward bias for the case of spin-dependent electron transport, respectively. The difference in magnitude the helicity-dependent photocurrent ΔI is a superposition of magneto-optical (ΔI_{MCD}) and spin-filtering (ΔI_{SF}) effects

$$\Delta I = \Delta I_{SF} + \Delta I_{MCD} \qquad (14.4)$$

with ΔI_{MCD} being proportional to the unpolarized photocurrent ($\Delta I_{MCD} = \alpha I_{ph}$).

The well-defined structure of their sample allows a clear separation of all these contributions. As seen in Figure 14.13, a significant difference between the bias dependences of the unpolarized photocurrent and the helicity-dependent photocurrent is only observed at forward bias (0.4 to 0.8 V), where electron tunneling occurs, whereas the bias dependences of both currents match each other closely at reverse bias. The latter finding shows that spin-dependent hole transport does not play an important role and that the helicity-dependent photocurrent at reverse bias arises mainly from MCD. The spin-filtering efficiency can then be quantified in terms of an effective polarization P_{eff}, defined as

$$P_{eff} = (\Delta I - \alpha I_{ph})/2I_{ph} \qquad (14.5)$$

For their case, at about 300 K, where thermionic emission significantly contributes to the transport process, P_{eff} approaches zero. As shown in Figure 14.13b, practically no spin filtering is observed at this temperature.

14.3.3 FE/GAAS SCHOTTKY DIODES

Figure 14.14a shows the I-V curve of a 5 nm thick Fe/GaAs diode. The curve with $n = 10^{24}$ m^{-3} possesses a weaker Schottky barrier, but surprisingly, a lower impedance as compared with the measured NiFe samples. The helicity-dependent photocurrent with the corresponding Fe/GaAs diode displays a clear difference between I^n and I^0 related to the presence of the significant Schottky barrier at the Fe/GaAs interface (see Figure 14.14b). The bias dependence of the helicity-dependent photocurrent for the Fe sample is similar to the Permalloy samples.

14.3.4 CO/GAAS SCHOTTKY DIODES

Figure 14.15a shows the I-V curves of the 5 nm thick Co samples without photoexcitation. These curves indicate that the Schottky barrier height ϕ_b again falls with increasing doping density. The helicity-dependent photocurrent is shown in Figure 14.15b and Figure 14.15c with (I^n) and without (I^0) perpendicular saturation for the values of doping (b) $n = 10^{24}$ and (c) $p = 10^{25}$ m^{-3}. For the case of $n = 10^{24}$ m^{-3}, I^0

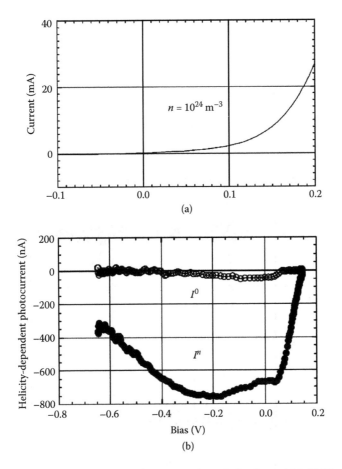

FIGURE 14.14 (a) Bias dependence of a current through the 5 nm thick Fe/GaAs(100) (n = 10^{24} m^{-3}) interface obtained without photoexcitation (I-V curve). (b) Bias dependence of the helicity-dependent photocurrent without (open circles, I^0) and with perpendicular saturation (closed circles, I^n) with the corresponding Fe/GaAs(100) sample.

is almost constant (~ -0.16 µA), while I^n is -0.20 ± 5.6 µA for $-0.2 < V < 0.6$ V, in other words, fluctuations occur. The difference $\Delta I = I^n - I^0$ in the helicity-dependent photocurrents is difficult to analyze but is different from that previously reported with the Co/Al$_2$O$_3$/GaAs system, in which $\Delta I/I$ is of the order of a few percent at reverse bias and diverges gradually at zero bias without showing any peak at forward bias.[26,27] The observed behavior of ΔI suggests that almost no spin filtering occurs in the investigated Co/GaAs structure. As MCD in Co has been reported to be $\sim 0.15\%$,[27] the MCD effects may dominate in the Co samples. Two peaks also appear at $V = 0.6$ and -0.2 V, respectively, as previously reported;[48] however the origin of the peaks is not clear at this stage. With $p = 10^{25}$ m^{-3} as shown in Figure 14.15c, ΔI is approximately -35 nA, which is much smaller than the fluctuation in ΔI with $n = 10^{24}$ m^{-3}, and corresponds to a decrease of spin-polarized current with increasing doping density.

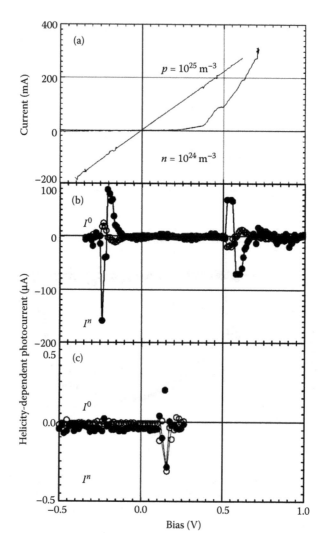

FIGURE 14.15 (a) Bias dependence of a current through the 5 nm thick Co/GaAs(100) ($n = 10^{24}$ and $p = 10^{25}$ m^{-3}) interface obtained without photoexcitation. Bias dependence of the helicity-dependent photocurrent without (I^0) and with perpendicular saturation (I^n) for the corresponding Co/GaAs(100) samples are shown in b and c.

14.3.5 ANTIFERROMAGNETIC CR/GAAS SCHOTTKY DIODES

Figure 14.16a shows the *I-V* curve of a Cr sample without photoexcitation, which indicates that the sample is a very good Schottky diode (the Schottky barrier height ϕ_b is very small, < 0.02 eV, and the ideality factor $n = 1.53$) with a small offset at reverse bias. The origin of the small offset in Cr/GaAs and NiFe/GaAs is discussed elsewhere.[49] Because all the samples, including NiFe/GaAs, Fe/GaAs, Co/GaAs, and Cr/GaAs, are prepared using the same procedure, the good Schottky characteristics

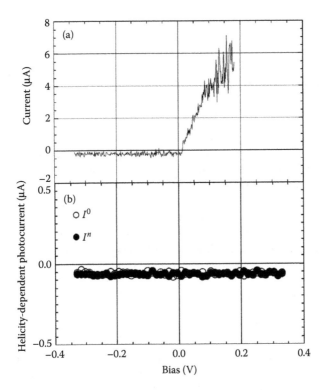

FIGURE 14.16 (a) Bias dependence of a current through the 5 nm thick Cr/GaAs(100) ($n = 10^{23}$ m^{-3}) interface obtained without photoexcitation. (b) Corresponding bias dependence of the helicity-dependent photocurrent without (open circles, I^0) and with perpendicular saturation (closed circles, I^n).

obtained for the Cr/GaAs suggest that the sample preparation procedures are appropriate for epitaxial growth, but may not be ideal for good Schottky characteristics with some metals.

The helicity-dependent photocurrent is shown in Figure 14.16b with (I^n) and without (I^0) perpendicular saturation. There is no difference between I^n and I^0, indicating that there is no spin-polarized electron current flowing across the AF Cr/GaAs interface. This is one of the crucial tests for the validity of this photoexcitation study as discussed in Section 14.3.2.2.

14.4 BALLISTIC SPIN TRANSPORT IN SPIN-VALVE STRUCTURES

Evidence for spin detection has been discussed in Section 14.3: spin filtering of electrons transmitted across the single layer FM/SC Schottky interface has been observed using photoexcitation techniques at RT.[28,37,47,48] To test the validity of the spin-filtering mechanism, a spin-valve structure is employed here, in which the FM layers can be switched independently, as demonstrated in spin-valve transistors.[51,52]

14.4.1 SAMPLE PREPARATION AND CHARACTERIZATION

14.4.1.1 Sample Preparation

Samples with a spin-valve structure, such as Au (2 nm)/Co (2 nm)/Cu (3 nm)/NiFe (4 nm)/n-GaAs(100), were fabricated for the in-plane photoexcitation measurements in an UHV system. The preparation of the ohmic contacts on the bottom of the GaAs substrates and the substrate cleaning were performed in the same manner as described in Section 14.3.1. The FM layers were grown at a rate of approximately 0.1 nm/min by e-beam evaporation, and electrical contacts were fabricated in using the same procedures as described in Section 14.3.1. The samples were characterized by MOKE, magnetoresistance (MR), and I-V measurements.

14.4.1.2 MR and MOKE

Figure 14.17a shows a current-in-plane (CIP) MR curve for the case of the Au/Co/Cu/NiFe/n-GaAs spin-valve sample measured by using a four-terminal method. The result clearly indicates a large resistance for the antiparallel configuration as is typical for spin-valve structures and a minimum resistance for the parallel configuration.[53] The parallel to antiparallel configuration switching field is estimated to be 3.8 mT, but the field required to align the magnetizations in both the FM1 (Co) and the FM2 (NiFe) layers is approximately 7.0 mT as is also confirmed by the hysteresis loop obtained by MOKE (see Figure 14.17b).

14.4.1.3 Schottky Characteristics

Corresponding I-V curves for both the parallel and antiparallel configurations are shown in Figure 14.18a for 1.5 and 300 K. The I-V curves for both configurations are almost the same at each temperature and possess large ohmic components below the Schottky barrier height ϕ_b. Because this sample shows leaky I-V characteristics, pinholes may play an important role in current distributions within each spin-valve layer, causing the current flow in plane to concentrate at the pinholes to travel into the neighboring layer. Using a surface electrode and a back ohmic contact (across the spin valve structure, see Figure 14.9) current-perpendicular-to-the-plane (CPP) MR is also measured as shown in Figure. 14.18b. The CPP-MR ratio shows a slight increase at negative bias and decreases rapidly above $V \sim \phi_b$. The CPP-MR behavior indicates that the Schottky diode is switched "on" with $\phi_b < V$, causing the MR to vanish due to current shunting. With $V < \phi_b$, the Schottky diode is "off" and conventional spin-valve effects are seen. This behavior suggests that the MR is predominantly due to the CIP-MR as the CPP-MR is almost zero because of a large effective sample area.

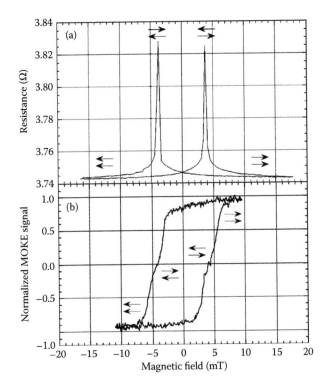

FIGURE 14.17 (a) CIP-MR curve and (b) MOKE loop of the Au (2 nm)/Co (2 nm)/Cu (3 nm)/ NiFe (4 nm)/GaAs(100) ($n = 1.5 \times 10^{24}$ m^{-3}) spin-valve sample showing both parallel and antiparallel configurations measured at 300 K.

14.4.2 OPTICAL MAGNETOCURRENT WITH SPIN-VALVE STRUCTURES

14.4.2.1 Angular Dependence

The optical magnetocurrent in the CPP configuration is measured using the in-plane photoexcitation technique. The magnetization in the two FM layers is first aligned parallel in plane by the application of an external field ($H > 20$ mT), and the optical magnetocurrent in the parallel configuration (I^p) is measured. The external field is then reversed to -3.8 mT to reverse the FM2 (NiFe) magnetization to align for the antiparallel configuration (I^a) measurement. These measurements are repeated sequentially several times. The spin dependence of the average difference in the optical magnetocurrent between the two configurations ($\Delta I = I^p - I^a$) is then determined. Here the optical magnetocurrents I^p and I^a are proportional to the difference between the current components for right (σ^+) and left (σ^-) circularly polarized light for each magnetization configuration: $I^p = p^p |i_p^+ - i_p^-|$ and $I^a = p^a |i_a^+ - i_a^-|$, where p^p and p^a are phase factors for the parallel and antiparallel configurations, respectively. The phase factor is adjusted to be 1 for the parallel configuration, and I^p is measured first. Without further adjustment of the phase factor, I^a is then measured.

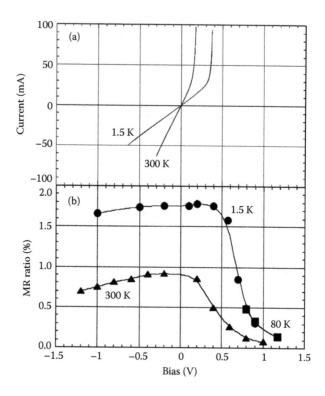

FIGURE 14.18 (a) *I-V* curves in both parallel and antiparallel configurations (overlapping) and (b) CPP-MR curves of the Au (2 nm)/Co (2 nm)/Cu (3 nm)/NiFe (4 nm)/GaAs(100) (n = 1.5×10^{24} m^{-3}) spin-valve sample measured at 1.5 (▲), 80 (■), and 300 (●) K.

Figure 14.19 shows the angular dependence of the difference in the optical magnetocurrent $\Delta I = I^p - I^a$. ΔI is found to increase significantly with increasing θ as expected (see Section 14.2.3). For the case of $\theta = 0$ ($\sigma \perp M$), no change in the optical magnetocurrent is detected as shown in Figure 14.19a, and therefore $\Delta I = 0$. Because the circularly polarized light excites electrons in the SC with spin polarization along the plane normal, these spins are not filtered at the FM2/SC interface due to the 90° rotation of the spin-quantization axis. With increasing θ, the in-plane component of σ becomes larger, with the result that σ excites electrons that share the spin-quantization axis with those in the FM layers. This causes an increase in the magnitude of both I^a and I^p. Finally, at $\theta = 45°$, the magnitude of the optical magnetocurrent approaches almost 200 nA (see Figure 14.19b) .[53] This gives a very large difference in the optical magnetocurrent ($\Delta I \sim 350$ nA) observed at zero bias, corresponding to the optical magnetocurrent passing from the SC to the spin valve. This large difference in the optical magnetocurrent indicates that a spin-valve structure has a high capability of detecting a small spin imbalance in a current flow across the structure.

In our picture, spin-polarized electrons, photoexcited in the GaAs, travel across the Schottky barrier at the NiFe/GaAs interface and ballistically propagate through

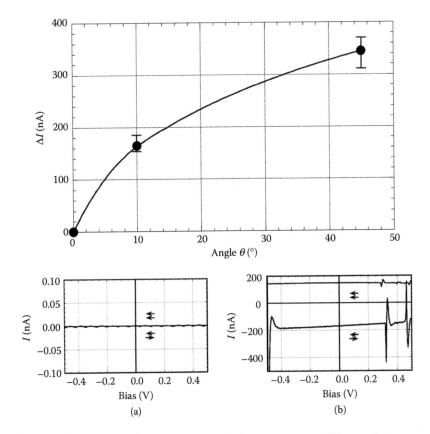

FIGURE 14.19 Angular dependence of the optical magnetocurrent difference between parallel and antiparallel configurations at zero bias $\Delta I = I^p - I^a$ for the case of the Au (2 nm)/Co (2 nm)/Cu (3 nm)/NiFe (4 nm)/GaAs(100) ($n = 1.5 \times 10^{24}$ m^{-3}) spin-valve sample. The optical magnetocurrents at $\theta =$ (a) 0 and (b) 45° are also shown.

the spin valve. We attribute the observed spin-filtering effect to electronic band-structure differences at the FM/SC and FM/nonmagnetic (NM) metal interfaces. Due to the requirement of momentum conservation, electrons have to maintain their transverse **k** when propagating across an interface. If there is no state with a matching **k** available in the band structure of the neighboring material, they will be reflected.[55] Because the reflection and transmission probabilities depend on the details of the Fermi surfaces in the different materials, this transport process is spin-dependent. As a result of this ballistic spin filtering, the optical magnetocurrent through the spin-valve trilayer is controlled by the relative orientation of the two FM layers.

This effect resembles the reverse effect reported by Rippard and Buhrman,[56] in which spin filtering of ballistic electrons transmitted through ultrathin Co films is identified from *I-V* measurements on Co/Cu/Co structures. For very thin (approximately 0.3 nm) Co layers, minority spin-polarized electron transport across the

Co/Cu interfaces has been reported below ~1.2 eV, and majority spin transport has been observed for both very thick Co layers with energy above ~1.4 eV and thicker Co layers (> 0.3 nm). These spin-polarized electron transport effects are due to both the spin split DOS in Co and the electron spin attenuation length.

It should be emphasized that electrons entering the FM must be spin-polarized because both I^p and I^a fall to zero with $\theta = 0$, indicating that the polarization of the magnetocurrent, so differing from the conventional giant MR (GMR) mechanism. Moreover, the optical magnetocurrent change depends on the relative alignment of the magnetization in the FM layers. In combination with the observation that I^p and I^a only show a very weak dependence on the bias (see Figure 14.19b), this finding also suggests a ballistic transport mechanism: spin-polarized electrons from the GaAs enter the FM2 and travel above the E_F across the spin-valve structure. Such ballistic electrons encounter effective potentials as described above according to their spin orientation and the alignment of the layers within the spin valve. Consequently, a strong optical magnetocurrent effect is seen according to the alignment of the spin valve.

14.4.2.2 Photon-Energy Dependence

Our picture of ballistic electron spin filtering is further supported by the photon energy dependence of the optical helicity-dependent photocurrent I studied by Steinmuller et al.[56] Figure 14.20 shows the variation of I with the applied magnetic field for three different photon energies of (a) 1.96, (b) 1.85, and (C) 1.58 eV. In the first two cases, the energy of the photoexcited electrons lies above the Schottky barrier height; whereas in the latter case, it lies below, as depicted in Figure 14.20 (top). For the first case, they observe up to 2400% increase in the optical helicity-dependent photocurrent by switching the spin valve from parallel to antiparallel configurations. The relative height of the optical helicity-dependent photocurrent peaks at the antiparallel alignment decreases with decreasing photon energy (see Figure 14.20a and Figure 14.20b), although the spin polarization of the electrons excited in the GaAs is increased to about 20%.[1] For $h\nu = 1.58$ eV (see Figure 14.20c), the peaks disappear, which suggests that either very few spin-polarized electrons travel across the SC/FM interface or that at this energy, the electron transport process is only weakly sensitive to the relative alignment of the initial spin polarization in the GaAs and the magnetization of the Co layer. This clearly indicates that the ballistic electron spin filtering is dominant in the spin-valve/SC structures. In this case, spin-polarized electrons are excited in the GaAs, enter the spin valve above the Schottky barrier, and ballistically propagate through the metal layers. Some of these electrons are reflected at the FM/NM interfaces in the spin valve, due to band-structure mismatches and the requirement of transverse momentum conservation. The reflection and transmission probabilities depend on the details of the Fermi surfaces in the different materials and consequently are spin-dependent.[55] The strong variation of the optical helicity-dependent photocurrent with photon energy is therefore likely to be related to the energy dependence of the electronic band structure in the different metal layers.

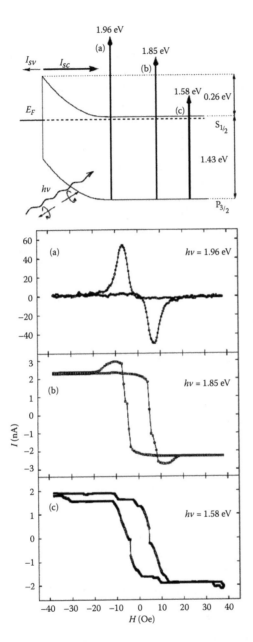

FIGURE 14.20 Schematic diagram of photoexcited electrons in the GaAs close to the Schottky barrier for three different photon energies (top). Here $S_{1/2}$ and $P_{3/2}$ denote the conduction and light and heavy hole valence bands, respectively. Optical helicity-dependent photocurrent as a function of an applied magnetic field are also shown for the photon energies of (a) 1.96, (b) 1.85, and (C) 1.58 eV for the spin-valve sample, consisting of Au (2 nm)/Co (2 nm)/Cu (5 nm)/NiFe (3 nm)/n-GaAs(100). (From S.J. Steinmuller, T. Trypiniotis, W.S. Cho, A. Hirohata, W.S. Lew, C.A.F. Vaz, and J.A.C. Bland, *Phys. Rev. B*, 69, 153309, 2004. Copyright 2004 American Physical Society. With permission.)

14.5 SUMMARY

We have discussed in detail evidence for RT spin filtering of spin-polarized electrons at the FM/SC interface, in the context of optical studies of spin detection. Hybrid FM/GaAs Schottky barrier structures with different FM layer materials, thicknesses, and GaAs doping densities were investigated, as well as an AF Cr sample for reference. The magnetic field dependence of the helicity-dependent photocurrent follows that of the polar MOKE signal in all cases, showing that ΔI is determined by the magnetic properties of the FM layer. Any magnetic background from the SC substrate can therefore be ruled out. This is confirmed by the fact that no magnetic field-dependent signal was observed for the Cr sample. A careful analysis of the data enabled us to separate MCD from the measured helicity-dependent photocurrent and to isolate the true spin-filtering signal. We found that at reverse bias, when most of the photoexcited electrons travel into the bulk of the SC, ΔI arises purely due to MCD. At forward bias, however, we observed clear evidence for the spin filtering of spin-polarized electrons propagating across the FM/SC interface. The bias and GaAs doping density dependence of ΔI suggest that electron tunneling is the spin-dependent transport mechanism. Further proof to this picture was added by temperature-dependent measurements of band-gap engineered NiFe/AlGaAs barrier/SC structures: spin-dependent effects were only observed in the bias and temperature range where electron tunneling occurs. This finding provides clear evidence that significant spin-filtering effects can only be expected for tunneling electrons.

In addition, strong optical magnetocurrent effects at RT have been observed in spin-valve/SC structures. The difference in the optical magnetocurrent obtained between parallel and antiparallel spin-valve configurations was extremely large (up to 2400%) and showed a clear variation with the angle between the photon helicity σ and the magnetization $\mathbf{M}$ in the FM layers, reaching zero for $\sigma \perp \mathbf{M}$. This indicated that the spin-dependent electron transport across the spin-valve structures was determined by the relative spin alignment of the FM layers and the initial spin polarization of the photoexcited electrons. The photon energy dependence of the optical helicity-dependent photocurrent further suggested that the photoexcited electrons propagate ballistically through the spin valve.

These results unambiguously indicate that spin-polarized electrons are transmitted from the SC to the FM with high efficiency. We conclude that the spin transmission is associated with spin filtering of electrons entering the FM layers. This spin-filtering effect could be used for future spintronic devices, such as an optically assisted magnetic sensor.[58]

ACKNOWLEDGMENTS

The authors would like to thank Dr. Y.B. Xu, Dr. C.M. Gürtler, Dr. S.N. Holmes, K. Cooper, Dr. W.S. Lew, and Dr. G. Wastlbauer for their help with the sample preparation. Thanks are also due to Dr. W.S. Cho and Professor G.X. Chen for their assistance in experiments. It is a pleasure to thank Dr. M. Johnson, Dr. B.T. Jonker, Professor G. Güntherodt, Dr. G. Schmidt, and Dr. W.F. Egelhoff, Jr. for enlightening and fruitful discussions. This work was supported by Engineering and Physical Sciences Research Council (United Kingdom), the European Union (ESPRIT program), and Toshiba Research Europe.

REFERENCES AND NOTES

1. D.T. Pierce and F. Meier, *Phys. Rev. B*, 13, 5484, 1976.
2. S. Adachi, *GaAs and Related Materials*, World Scientific, Singapore, 1994.
3. F. Meier and B.P. Zakharchenya, Eds., *Optical Orientation*, North-Holland Physics Publishing, Amsterdam, 1984.
4. R. Fiederling, M. Keim, G. Reuscher, W. Ossau, G. Schmidt, A. Waag, and L.W. Molenkamp, *Nature*, 402, 787, 1999.
5. M. Oestreich, J. Hübner, D. Hägele, P.J. Klar, W. Heimbrodt, W.W. Rühle, D.E. Ashenford, and B. Lunn, *Appl. Phys. Lett.*, 74, 1251, 1999.
6. B.T. Jonker, Y.D. Park, B.R. Bennett, H.D. Cheong, G. Kioseoglou, and A. Petrou, *Phys. Rev. B*, 62, 8180, 2000.
7. B.T. Jonker, A.T. Hanbicki, Y.D. Park, G. Itskos, M. Furis, G. Kioseoglou, A. Petrou, and X. Wei, *Appl. Phys. Lett.*, 79, 3098, 2001.
8. H.J. Zhu, M. Ramsteiner, H. Kostial, M. Wassermeier, H.-P. Schönher, and K.H. Ploog, *Phys. Rev. Lett.*, 87, 016601, 2001.
9. A.T. Hanbicki, B.T. Jonker, G. Itskos, G. Kioseoglou, and A. Petrou, *Appl. Phys. Lett.*, 80, 1240, 2002.
10. V.F. Motsnyi, J. De Boeck, J. Das, W. Van Roy, G. Borghs, E. Goovaerts, and V.I. Safarov, *Appl. Phys. Lett.*, 81, 265, 2002.
11. V.F. Motsnyi, P. Van Dorpe, W. Van Roy, E. Goovaerts, V.I. Safarov, G. Borghs, and J. De Boeck, *Phys. Rev. B*, 68, 245319, 2003.
12. T. Manago and H. Akinaga, *Appl. Phys. Lett.*, 81, 694, 2002.
13. X. Jing, R. Wang, R.M. Shelby, R.M. Macfarlane, S.R. Bank, J.S. Harris, and S.S.P. Parkin, *Phys. Rev. Lett.*, 94, 056601, 2005.
14. M.C. Hickey, C.D. Damsgaard, I. Farrer, S.N. Holmes, A. Husmann, J.B. Hansen, C.S. Jacobsen, D.A. Ritchie, R.F. Lee, and G.A.C. Jones, *Appl. Phys. Lett.*, 86, 252106, 2005.
15. Y. Ohno, D.K. Young, B. Beschoten, F. Matsukura, H. Ohno, and D.D. Awschalom, *Nature*, 402, 790, 1999.
16. S.F. Alvarado and P. Renaud, *Phys. Rev. Lett.*, 68, 1387, 1992.
17. H. Ohno, *Science*, 281, 951, 1998.
18. M.E. Flatté and J.M. Byers, *Phys. Rev. Lett.*, 84, 4220, 2000.
19. J.M. Kikkawa and D.D. Awschalom, *Nature*, 397, 139, 1999.
20. I. Malajovich, J.M. Kikkawa, D.D. Awschalom, J.J. Berry, and N. Samarth, *Phys. Rev. Lett.*, 84, 1015, 2000.
21. Y.D. Park, A.T. Hanbicki, S.C. Erwin, C.S. Hellberg, J.M. Sullivan, J.E. Mattson, T.F. Ambrose, A. Wilson, G. Spanos, and B.T. Jonker, *Science*, 295, 651, 2002.
22. T. Dietl, H. Ohno, F. Matsukura, J. Cibert, and D. Ferrand, *Science*, 287, 1019, 2000.
23. G.A. Medvedkin, T. Ishibashi, T. Nishi, K. Hayata, Y. Hasegawa, and K. Sato, *Jpn. J. Appl. Phys.*, 39, L949, 2000.
24. If the exciting energy is large (typically more than 4 eV), the magnetization precession in a FM cannot be neglected (see Reference 25).
25. W. Weber, S. Riesen, and H.C. Siegmann, *Science*, 291, 1015, 2001.
26. M.W.J. Prins, H. van Kempen, H. van Leuken, R.A. de Groot, W. van Roy, and J. de Boeck, *J. Phys.: Condens. Matter*, 7, 9447, 1995.
27. K. Nakajima, S.N. Okuno, and K. Inomata, *Jpn. J. Appl. Phys.*, 37, L919, 1998.
28. A. Hirohata, S.J. Steinmueller, W.S. Cho, Y.B. Xu, C.M. Gürtler, G. Wastlbauer, J.A.C. Bland, and S.N. Holmes, *Phys. Rev. B*, 66, 035330, 2002.
29. A.F. Isakovic, D.M. Carr, J. Strand, B.D. Schultz, C.J. Palmstrøm, and P.A. Crowell, *Phys. Rev. B*, 64, R161304, 2001.

30. K. Sueoka, K. Mukasa, and K. Hayakawa, *Jpn. J. Appl. Phys.*, 32, 2989, 1993.
31. Y. Suzuki, W. Nabhan, R. Shinohara, K. Yamaguchi, and K. Mukasa, *J. Magn. Magn. Mater.*, 198–199, 540, 1999.
32. H. Kodama, T. Uzumaki, M. Oshiki, K. Sueoka, and K. Mukasa, *J. Appl. Phys.*, 83, 6831, 1999.
33. S.N. Molotkov, *Surf. Sci.*, 287/288, 1098, 1993.
34. R. Laiho and H.J. Reittu, *Surf. Sci.*, 289, 363, 1993.
35. S.M. Sze, *Physics of Semiconductor Devices*, 2nd ed., John Wiley and Sons, New York, 1981.
36. A. Hirohata, Y.B. Xu, C.M. Gürtler, and J.A.C. Bland, *J. Appl. Phys.*, 87, 4670, 2000.
37. J.A.C. Bland, A. Hirohata, Y.B. Xu, C.M. Gürtler, and S.N. Holmes, *IEEE Trans. Magn.*, 36, 2827, 2000.
38. L.J. Brillson, Ed., *Contacts to Semiconductors Fundamentals and Technology*, Noyes Publications, New York, 1993.
39. G. Schmidt, D. Ferrand, L.W. Molenkamp, A.T. Filip, and B.J. van Wees, *Phys. Rev. B*, 62, R4790, 2000.
40. J.D. Albrecht and D.L. Smith, *Phys. Rev. B*, 68, 035340, 2003.
41. I.I. Mazin, *Phys. Rev. Lett.*, 83, 1427, 1999.
42. B. Nadgorny, R.J. Soulen, Jr., M.S. Osofsky, I.I. Mazin, G. Laprade, R.J.M. van de Veerdonk, A.A. Smits, S.F. Cheng, E.F. Skelton, and S.B. Qadri, *Phys. Rev. B*, 61, R3788, 2000.
43. C. Li, A.J. Freeman, and C.L. Fu, *J. Magn. Magn. Mater.*, 75, 53, 1988.
44. S.E. Andresen, S.J. Steinmuller, A. Ionescu, G. Wastlbauer, C.M. Gürtler, and J.A.C. Bland, *Phys. Rev. B*, 68, 073303, 2003.
45. For $\theta \neq 0$, the Fresnel constants r_{sp} and r_{pp} can be different, which may provide small changes in the optical magnetocurrent.
46. Y.B. Xu, E.T.M. Kermohan, D.J. Freeland, A. Ercole, M. Tselepi, and J.A.C. Bland, *Phys. Rev. B*, 58, 890, 1998.
47. S.J. Steinmuller, C.M. Gürtler, G. Westlbauer, and J.A.C. Bland, *Phys. Rev. B.*, 72, 045301, 2005.
48. A. Hirohata, Y.B. Xu, C.M. Gürtler, J.A.C. Bland, and S.N. Holmes, *Phys. Rev. B*, 63, 104425, 2001.
49. A. Hirohata, Y.B. Xu, C.M. Gürtler, and J.A.C. Bland, *IEEE Trans. Magn.*, 35, 2910, 1999.
50. A. Hirohata, C.M. Gürtler, W.S. Lew, Y.B. Xu, J.A.C. Bland, and S.N. Holmes, *J. Appl. Phys.*, 91, 7481, 2002.
51. D.J. Monsma, J.C. Lodder, T.J.A. Popma, and B. Dieny, *Phys. Rev. Lett.*, 74, 5260, 1995.
52. R. Jansen, P.S. Anil Kumar, O.M.J. van't Erve, R. Vlutters, P. de Haan, and J.C. Lodder, *Phys. Rev. Lett.*, 85, 3277, 2000.
53. U. Hartmann, Ed., *Magnetic Multilayers and Giant Magnetoresistance*, Springer, Berlin, 2000.
54. Small features seen at the bias of 0.25 and 0.43 V in both I^p and I^a might be related to the barrier height at the NiFe/GaAs and Co/Cu interfaces, because the magnitude of these features becomes larger in antiparallel configuration (I^a), which is similar to the enhancement of the helicity-dependent photocurrent at peak A in NiFe/GaAs samples (see Reference 48).
55. M.D. Stiles, *J. Appl. Phys.*, 79, 4805, 1996.
56. W.H. Rippard and R.A. Buhrman, *Phys. Rev. Lett.*, 84, 971, 2000.
57. S.J. Steinmuller, T. Trypiniotis, W.S. Cho, A. Hirohata, W.S. Lew, C.A.F. Vaz, and J.A.C. Bland, *Phys. Rev. B*, 69, 153309, 2004.
58. J.A.C. Bland and A. Hirohata, Basic British Patent BP 0,022,328.9.

15 Ferromagnetic Metal/III-V Semiconductor Hybrid Spintronic Devices

Ping Kwan J. Wong and Yongbing Xu

CONTENTS

15.1 INTRODUCTION

The success of semiconductor (SC) electronics has been built on the charge degree of freedom of electrons in SCs. The biggest impact on the solid state electronic industry at that time stemmed from the first discovery of transistor action in Bell Labs in 1947. It was unexpected that the discovery took just 5 years to be commercialized and now becomes an indispensable component in every computer. Since then, many other SC devices such as a field-effect transistor, light-emitting diode (LED), integrated circuit (IC), and SC laser have become available on the market. This revolution brought about the beginning of miniaturization in electronics. In the current overwhelmingly digital world, demands for higher processing speed, lower energy dissipation, and denser storage medium have been unavoidable and the successes of the SC electronics in the past do not offer clear solutions for the future development of the industry.

In 1965, Gordon Moore, a cofounder of Intel® predicted that from the invention of ICs the density of components in a circuit would increase exponentially or more exactly, double every year.[1] Technological advancement in the SC industry in which the density doubling approximately every 18 months has proved the validity of his prediction. But still, continuing miniaturization has to rely on different approaches such as finding new fabrication techniques, creating new materials, and improving design geometries so as to obtain performance of desired functions in smaller physical spaces. As one may realize, such miniaturization in ICs will eventually fail to continue when a fundamental limit in dimension is reached.

When the physical dimensions scale down to say a nanometer, quantum-mechanical phenomena can be observed and start to predominate the behavior. It is these effects that give rise to a new paradigm for the next generation of electronic devices. In conventional electronics, only carrier charge has been actively involved in device functionality and carrier spin, a quantum-mechanical object, has only drawn much attention recently. When the fundamental limit is reached, spin-dependent effects such as exchange interaction among spins can no longer be neglected and researchers will have to work with these carrier spins. By active manipulation of carrier spin and its transport, it is envisioned that more advanced devices potentially giving increased data processing speed, lower power consumption, higher integration densities, and, ideally, nonvolatility storage capabilities could not be only a dream.[2-5] Such active control of spin, which takes advantage of novel quantum-mechanical effects enabling functionalities and technologies that are not feasible in the current standard microelectronics, has triggered development of a new discipline of research area, namely spin electronics or magnetoelectronics, which is a fast developing area integrating spin-dependent effects with the well-developed SC functionality.

Spintronics is based on SC technologies with the additional use of magnetic materials, which is what distinguishes it from standard microelectronics. The use of magnetic materials in the electronics industry is indeed not new but has been explored since the discovery of the well-known giant magnetoresistance (GMR) effect in alternating layers of magnetic and nonmagnetic thin films in 1988.[6] GMR has been widely applied in magnetic information storage due to its large change in resistance

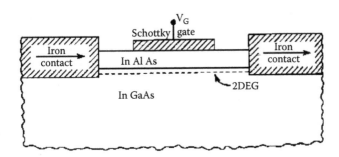

FIGURE 15.1 Datta and Das spin FET. (From S. Datta and B. Das, *Appl. Phys. Lett.,* 56, 665, 1990. With permission.)

in the presence of a relatively small magnetic field. This unique characteristic has been used in read heads for magnetic hard disk drives.[1] Both GMR-based read heads and sensors are of fundamental relevance to spintronics as they make use of spin-dependent scattering in layered structures. In other words, these early applications gave birth to magnetoelectronics. Note that, however, spintronic devices aim to use carrier spin actively to provide their potential functionalities, whereas in the case of either GMR read heads or sensors, the role of spin has been passive in determining the device characteristics.[1,5,7]

Much literature in this field was indeed inspired by the earliest proposed spin-polarized field-effect transistor (spin FET) by Datta and Das in 1990, Figure 15.1.[8] In this prototype ferromagnet (FM)/SC hybrid device, the source and drain for injecting and detecting spin-polarized electrons, respectively, are made of FM materials. Transport has been restricted in a high mobility channel, two-dimensional electron gas (2DEG). Basically, a spin FET has two means of controlling the flow of spin-polarized carriers. Without bias from the gate, the relative magnetization directions in the source and drain dominate the conductivity in the device. While applying a gate voltage across the channel, the spin-polarized electrons "see" an effective magnetic field due to the Rashba spin-orbit interaction leading to precession of spin of the electrons[9] and thus the change in polarization of the injected current. Such a three-terminal device is intriguing because at first glance, its structure is similar to a conventional transistor but has an almost completely different working mechanism. On the other hand, fabrication of the spin FET is built on the basis of standard SC technology, which has been a familiar issue with scientists and engineers to date.

A main difference between metal- and SC-based devices is that the former does not amplify signals; whereas the latter can offer amplification and serves as a multifunctional device. Among the reported spintronic applications to date, most of them are based on the GMR effect.[6] One of the spin-based devices applying such an effect is a spin-valve transistor (SVT), which was first demonstrated by Monsma et al. in 1998.[10] In a related aspect, the discovery of room temperature (RT) tunneling magnetoresistance (TMR) by Moodera et al.[11] and Miyazaki and Tezuka,[12] respectively, in 1995 triggered intensive study in magnetic tunnel junctions (MTJs), which in turn facilitated the development of high-performance magnetic random access memory (MRAM) and also a new kind of transistor, magnetic tunneling transisor (MTT).[13] Both of these transistors rely on spin-dependent hot-electron filtering[14,15] and have the potential to dominate in the future. Putting forward these ideas into reality would undoubtedly bring about subtle problems. However, clear evidence reported, for example, an extraordinary large RT MR in a hybrid metal/SC structure reaching 750,000% at a relatively high magnetic field of 4 T,[16] and a similarly huge value obtained in hybrid metal/SC granular films[17] has proved that hybrid structures play a tremendous role toward an entirely new generation of technology with new capabilities and opportunities. Until now, magnetic materials and SCs have been developed quite separately, with magnetic materials mainly used for data storage such as computer hard disks and SC devices for data processing such as the FET and metal-oxide semiconductor FET (MOSFET) in the information technology industry. Currently, there are two classes of material systems exploited for use in second-generation spintronic devices integrating magnetic materials with SC devices: (i) dilute magnetic semiconductors (DMSs) such as $Ga_{1-x}Mn_xAs$, $Cd_{1-x}Co_xSe$, and so on, and (ii) hybrid magnetic-semiconductor (HMS) materials, such as Fe/GaAs, Co/Si, Fe_3O_4/GaAs, CoMnGa/GaAs, and so forth. Even though DMS structures might offer opportunities of easy integration with conventional SC devices, it is however a great challenge to improve the quality of DMSs and, in particular, to increase their Curie temperatures. HMS structures have the advantages of high Curie temperatures, well-controlled magnetic properties, and easy integration with current magnetic technologies. The magnetic-SC hybrid materials and devices will be the theme of this chapter. This review chapter is organized as follows: Section 15.1 is an introduction to spintronics. Section 15.2 discusses the growth and properties of HMS materials with a particular focus on III-V SCs. Section 15.3 covers spin injection and detection in hybrid devices with an introduction to the fabrication of a vertical spintronic hybrid device and its magnetotransport properties. Section 15.4 provides a brief summary and outlook of this exciting research area.

15.2 GROWTH AND PROPERTIES OF HYBRID MATERIALS

Triggered partially by the prototype hybrid spintronic device of Datta and Das,[8] intensive experimental effort has been put into the growth of high-quality FMs on SCs, recently opening up opportunities in both new research areas and scientific advancement. It has been proved that spin injection will not be possible unless a modification of the interface between the FM and the SC of either a Schottky barrier or a tunnel junction is present, due to the restriction imposed by the mismatch of conductivities between the FM and SC.[18,19] Forming a uniform, defect-free FM/SC interface is thus a substantially essential core of spintronics. In particular, we will focus our discussion on the growth of transition FM on III-V SC in this section and at the end of this part, a summary of the FM/SC hybrid systems discussed will be given in Table 15.1.

TABLE 15.1
Summary of Different FM/SC Hybrid Systems

System	Crystal Structure	Reconstruction	Film Thickness/nm	Spin Magnetic Moments/(μ_B)	Orbital Magnetic Moments/(μ_B)	Growth Temperature (K)	Growth Rate (A min⁻¹)	Growth Technique	Reference
Fe/GaAs(001)	bcc	(4 × 6)	1.1	2.03 ± 0.14	0.26 ± 0.03	RT	1.43	MBE	23
			4.7	2.07 ± 0.14	0.12 ± 0.02	RT	1.43	MBE	23
		(2 × 4)						MBE	24
		c(4 × 4)						MBE	24
Fe/GaAs(100)		(4 × 6)	0.04	1.96 ± 0.5	1.23 ± 0.1	RT	1.43	MBE	25
			0.07	1.84 ± 0.21	0.25 ± 0.05	RT	1.43	MBE	25
			0.14	1.84 ± 0.11	0.23 ± 0.04	RT	1.43	MBE	25
Fe/InAs(100)	bcc	(4 × 2)	1.1	1.22 ± 0.12	0.22 ± 0.03	448	1.43	MBE	30
			3.6	1.90 ± 0.15	0.16 ± 0.01	448	1.43	MBE	30
		c(8 × 2)/(4 × 2)	3.9	M_s: 1.2 × 10³ emu/cm³		RT	1.3 –1.5	MBE	31
			40			RT	3	MBE	122
			40			448	3	MBE	122
Co/GaAs(110)	bcc		35.7	μ: 1.53		448–498	3.3	MBE	32
			20.2	μ: 1.3–1.4		448–498	3.3	MBE	42

(continued)

TABLE 15.1 (CONTINUED)
Summary of Different FM/SC Hybrid Systems

System	Crystal Structure	Reconstruction	Film Thickness/nm	Spin Magnetic Moments/(μ_B)	Orbital Magnetic Moments/(μ_B)	Growth Temperature (K)	Growth Rate (Å min^{-1})	Growth Technique	Reference
Co/GaAs(100)		(2×2)	21.6		μ: 1.3–1.4	448–498	3.3	MBE	42
		c(8×2)	0.2–15			RT	1	MBE	33
Co/O/GaAs(100)	hcp		1.0–2.0	M_s: 1407 emu/cm³		RT	10	RF Sputtering	69
Co/GaAs(001)	hcp	(4×6)	5,15			413	1	MBE	34
	bcc	(4×2)	0.8–2			423	2	MBE	36
	bcc/hcp	(4×2)	2.0–6.0			423	2	MBE	36
	bcc	(2×4)	3	μ: 1.2		423	0.16	MBE	34
			8	μ: 1.7		423	0.16	MBE	34
	bcc		3.0–15			448–498		MBE	37
	bcc/hcp		15–50			448–498		MBE	37
	bcc	c(4×4)	5.6			263–498	0.71	MBE	38
	bcc	(4×6)	1.1	M_s: 717 emu/cm³				MBE	39
	bcc/hcp	(4×6)	1.6	M_s: 1194 emu/cm³				MBE	39
		(4×6)	5	M_s: 1314 emu/cm³				MBE	39
		(4×6)	7	M_s: 1377 emu/cm³				MBE	39

Material	Structure	Reconstruction		Comment	Temp	Moment	Method	Ref
Co/S/GaAs(001)	bcc/hcp		50,560		RT		MBE	41
	hcp	(2 × 1)	2.0–5.0		RT		MBE	50
Co/Sb/GaAs(110)	bcc	(1 × 1)	0.3–2.4	2.3 times higher than on GaAs	443		MBE	44
Co/InAs(111)		(111)A–(2 × 2)	0.07–0.28		RT	0.71	MBE	51
		(111)B–(1 × 1)	0.07–0.28		RT	0.71	MBE	51
Ni/GaAs(001)	bcc		0.2–2.5		RT	2	MBE	52
	bcc/fcc		2.5–6		RT	2	MBE	52
	fcc		5	M_s: 387 emu/cm^3	RT		Electrode position	55
			> 10	M_s: 484 emu/cm^3	RT		Electrode position	55
	bcc	(4 × 6)	< 3.5	μ: 0.52+/−0.08	170		MBE	53
Fe$_{34}$Co$_{66}$/GaAs(001)	bcc	(4 × 6)	1.43–14.3	M_s: 1710 emu/cm^3	RT	0.5 for FeCo	MBE	58
Fe$_{1-x}$Co$_x$/GaAs(001)*		(4 × 2)	0.43–11.4		RT	0.6—0.7 for Fe	MBE	59
Fe$_{1-x}$Co$_x$/GaAs(001)		(2 × 4)	20		368		MBE	60
Fe$_{48}$Co$_{52}$/GaAs(100)		(2 × 4)/c(2 × 8)β2	20		368		MBE	61
Fe/GaSb(001)	bcc	(1 × 3)	1.7		RT	0.05	MBE	62
		(1 × 3)	80		RT	0.1	MBE	62
Fe/InP(001)	bcc	(2 × 4)	0.37		150	2.15	MBE	63
		(2 × 4)	0.51		300	2.15	MBE	63

(continued)

TABLE 15.1 (CONTINUED)
Summary of Different FM/SC Hybrid Systems

System	Crystal Structure	Reconstruction	Film Thickness/nm	Spin Magnetic Moments/(μB)	Orbital Magnetic Moments/(μB)	Growth Temperature (/K)	Growth Rate/(A min^{-1})	Growth Technique	Reference
Fe/GaN(001)	bcc	(1 × 1)	65			RT	10	MBE	66
Fe/GaN(0001)	bcc	(7 × 7)	5			RT	2	MBE	67
	bcc	(1 × 1)	5–50			RT	11	MBE	67
	bcc	(1 × 1)	5–70			RT	0.002	MBE	68
	bcc	(1 × 1)	5–70			523	0.002	MBE	68

*ᵃ x = 0.22, 0.33.

RT: Room temperature.

M_s: Saturation magnetization.

μ: Average magnetic moment.

15.2.1 Fe/GaAs

Fe/GaAs has been a model system for studying FM/SC hybrid structure due to two fundamental facts: (i) Fe has a Curie temperature well above RT, and (ii) body-centered cubic (bcc) Fe can be epitaxially grown on GaAs because the lattice constant of bcc Fe (2.866 Å) is almost exactly half that of GaAs (5.654 Å) as shown in Figure 15.2. In the aspect of spin-based devices, the interface formed between the chosen FM and SC should be magnetic and of high quality. In the study of interface magnetism in this system, there is a long-lasting debate over the presence of a magnetic dead layer between the Fe/GaAs interface, which causes a strong reduction of magnetization of the Fe film.[20] Possible reasons for this detrimental effect have been attributed to antiferromagnetic Fe_2As formation due to As outdiffusion.[21] It was later confirmed by superconducting quantum interference device (SQUID) and alternating gradient force magnetometer (AGFM) that the lowering of magnetization was due to the presence of half-magnetized $Fe_3Ga_{2-x}As_x$ at the interface.[22] Surprisingly, Xu et al. studied the magnetic properties and structure of Fe grown on GaAs(001)-(4 × 6) at RT and demonstrated that no dead layer was at the interface and that the Fe thin film shows a bulklike moment by using *in situ* magneto-optical Kerr effect (MOKE) and low-energy electron diffraction (LEED) and *ex situ* AGFM.[23]

It is believed that the first 4 monolayers (MLs) of Fe on GaAs grown at RT follows a three-dimensional (3D) mode (Volmer–Weber) growth. This is supported by the lack of Fe LEED patterns within these MLs. A significant MOKE signal was not detected until a thickness of 3.5 ML. With further Fe deposition, the MOKE-loop curves become *s* shaped at 4.3 ML. The lack of hysteresis indicates that the ferromagnetic phase has not yet developed. The magnetization curves indicate the presence of either paramagnetism or superparamagnetism. The onset of the

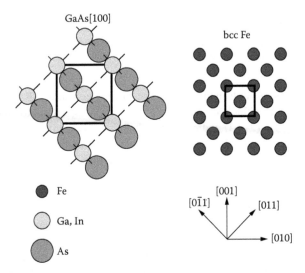

FIGURE 15.2 Crystal structures of bcc Fe and bcc Fe on GaAs[100] and InAs[100].

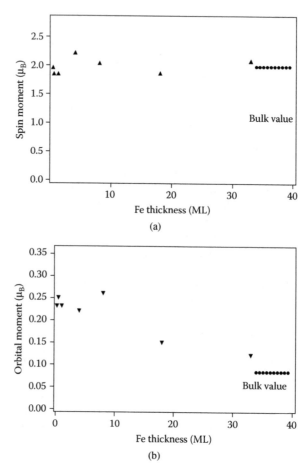

FIGURE 15.3 A: The evolution of spin moment as a function of the Fe thickness. B: The evolution of orbital moment as a function of the Fe thickness. (From J.S. Claydon, Y.B. Xu, M. Tselepi, J.A.C. Bland, and G. van der Laan, *Phys. Rev. Lett.,* 93, 37206, 2004. With permission.)

ferromagnetic phase is confirmed by MOKE measurements after 4.8 ML of Fe, which is less than the reported thicknesses for Fe grown on GaAs(001)-(4x6) at 448 K,[24] and for Fe on GaAs(001)-(2 × 4) and c(4 × 4) substrates.[20] Recently, the spin and orbital magnetic moments of monolayer and submonolayer Fe atoms at the interface with a Co capping laser have been studied by Claydon et al. using the x-ray magnetic circular dichroism (XMCD) technique (Figure 15.3).[25] The spin moments are close to that of the bulk values and the orbital moments are enhanced. Most importantly, the bulklike spin moment detected throughout the range of thicknesses studied indicated that there is no magnetic dead layer or half-magnetic layer as previously reported. These results further suggest that Fe/GaAs is a suitable hybrid system for spintronic applications.

Another equally fundamental issue is the origin of the in-plane uniaxial magnetic anisotropy (UMA) observed in Fe/GaAs. It has been proposed by Kneedler et al. that the unidirectional nature of Fe-As or Fe-Ga bonds is responsible for the UMA.[20]

On the other hand, uniaxial magnetoelastic coupling due to anisotropic lattice relaxation may be another possible source.[26,27] The magnetoelastic effect is present because compression of the Fe film occurs in order to fit the GaAs lattice. So far, from our knowledge, there is no clear evidence ambiguously distinguishing the two aforementioned effects upon the study of the UMA in Fe/GaAs. More effort in this area is required to fully explain the experimental findings.

15.2.2 Fe/InAs

Metals on narrow gap SCs, such as InAs, which has a direct band gap as small as 0.36 eV at 300 K, form ohmic contacts.[28] InAs also has a higher low-field mobility than GaAs and InP, which makes it an excellent choice for high-speed electronics. Xu et al. demonstrated that bcc Fe could be grown epitaxially on InAs(100)-(4 × 2) at 448 K and room temperature although the lattice mismatch between bcc Fe and InAs (6.058 Å) is 5.4%.[29] The epitaxial relationship was found to be Fe(100)<001> ‖InAs(100)<001>, the same as that in the Fe/GaAs(100). The cubic anisotropy dominates after coverage of 16 MLs, which is lower than the 50 MLs in the case of Fe/GaAs. Apart from the cubic anisotropy, UMA is also present as in the case of Fe/GaAs. However, the two systems show very different behavior in which the easy axis of the UMA of Fe/InAs(100)-(4 × 2) was found to be along [011].[26] That is, the easy axis has been rotated 90° with respect to that in Fe/GaAs (100)-(4 × 6).[23] Reflected high-energy electron diffraction (RHEED) of 50 MLs of Fe on InAs is shown in Figure 15.4. The presence of UMA has been attributed to the anisotropic lattice relaxation in this system where relaxation along [0$\bar{1}$1] direction is significantly faster than that along the [0$\bar{1}$1] direction.[27] The lattice constant along the [0$\bar{1}$1] direction changes rather sharply with increasing thickness and approaches the bulk value around 10 MLs, while the lattice constant along the [011] direction changes much more slowly and levels off around 25 MLs. In a scanning tunneling microscopy (STM) study, it has been revealed that the In rows are along [011], which will present an additional energy barrier to the motion of

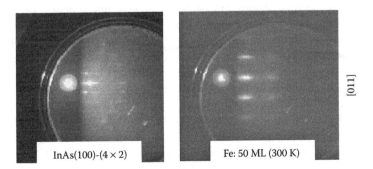

FIGURE 15.4 RHEED patterns of InAs(100)-(4 × 2) and 50 MLs of Fe on InAs grown at 300 K.

the interfacial dislocations along [011] direction and therefore the different thickness
dependences of the lattice relaxation along the two <011> directions.

As has been found in Fe/GaAs, the Fe films on InAs show bulklike spin
moments of $1.90 \pm 0.15\mu_B$ with the thickness above about 20 MLs and a large
enhancement of around 260% of the ratio of orbital vs. spin moment in the
ultrathin region.[30]

15.2.3 Co/GaAs

Co is another interesting FM to form a hybrid system with SC due in part to its
relatively high spin imbalance at the Fermi level and thus a high spin polarization.
Co has its stable and metastable phases as hexagonal-close-packed (hcp) and face-
centered-cubic (fcc) structures, respectively. Prinz, in 1985, first demonstrated that
bcc Co could actually exist in thin film form by growing it upon GaAs(110) via
molecular beam epitaxy (MBE).[32] Since then, there have been many reported
experiments on Co/GaAs but unfortunately these results have shown inconsistency,
making this area controversial.

Co thin films have been epitaxially grown on GaAs(001) [equivalent to
GaAs(100)[32–40] and GaAs(110)[32,41–44]]. It has been found that the bcc phase develops
at the initial growth while hcp Co starts building up at higher coverage in general.
The difficulty in unambiguously confirming the phase evolution during growth is
due to the fact that the Co lattice parameter is nearly half that of GaAs and experimental data from techniques such as RHEED and LEED may easily be interpreted
incorrectly. By using x-ray absorption fine structure (EXAFS), Idzerda et al. were
able to confirmed the bcc structure of 357 Å Co on GaAs(110).[43] In the literature,
there are debates over the lattice structure and magnetic properties of Co thin films
grew on GaAs. The theory predicts that the bcc phase is not a metastable phase but
a forced structure originating from imperfections.[45] Mangan et al. demonstrated the
coexistence of bcc and hcp phases in Co on GaAs(001) with the help of transmission
electron microscopy (TEM) observations.[37] This is in agreement with the results of
Wu et al.[36] Generally speaking, the onset of the hcp phase is different for most of
the reported results and it is believed that the hcp phase establishment depends on
the substrate preparation. Another inconclusive issue of this system is the magnetic
anisotropy of the Co film. The epitaxial Co films grown on a GaAs(001) substrate
in general have a cubic anisotropy below the thickness of 40 MLs[34–36] and show an
uniaxial anisotropy above 40 MLs. However, this boundary is not an absolute value
but depends on different samples. For these samples, the hard and easy axes of the
uniaxial anisotropy are along the [110] and the [$\bar{1}$10] directions, whereas the hard
and easy axes of the cubic anisotropy are along the <110> and <100> directions.
The reported magnetization was observed to be less than the bulk Co hcp value.[46]
Experimentally, the magnetic moments of bcc Co have been observed to be 1.3 to
1.5 μ_B/atom,[32,42,47] which is less than the theoretical calculations for bulk bcc Co,
which give a value of about $1.7\mu_B$/atom.[48,49] The reduction of this magnetic moment
has been argued to be related to the nonmagnetic layer that appears at the interface.[47]
To overcome this detrimental interdiffusion between the film and the substrate,

passivating layers such as S and Sb have been used.[44,50] Both approaches effectively reduce the chemical interaction at the Co/GaAs interface, and in particular, the latter case limits the lattice distortion to only 4% and gives a factor of 2.3 enhancement of the magnetic moment compared to the film deposited on bare GaAs(110) substrate has been acquired.[44]

15.2.4 Co/InAs

Only a few studies have been reported in this combination. Nevertheless, a study on Co/InAs(111) by Palmgren et al.[51] shows that InAs is a good candidate for high-speed applications, and uniquely, it has a large Rashba coefficient and forms ohmic contacts with most metals. High-resolution x-ray photoelectron spectroscopy (XPS) and STM were used in the study. Deposition of Co on the InAs(111) was done at RT resulting in a In-Co layer. Confirmed by the In $4d$ and As $3d$ core level spectra, the interfacial reaction between In and Co is relatively strong even below monolayer coverage showing that Co/GaAs may not be as good as Fe/GaAs and Fe/InAs for hybrid spintronic applications.

15.2.5 Ni/GaAs

Epitaxial growth of Ni on GaAs(001) is one of the most interesting hybrid FM/SC systems for heterointerface studies because bcc Ni, which does not exist in nature, could be formed on GaAs(001) by MBE growth. This was first demonstrated by Tang et al.[52] By growing the Ni film at RT, the presence of a bcc phase up to 2.5 nm has been confirmed by *in situ* RHEED and *ex situ* MOKE showing a fourfold in-plane magnetic anisotropy with the easy axes along the <100> directions. Still, it is expected that the evolution of the bcc phase is strongly dependent on the sample preparation recipe as in the case of bcc Co on GaAs discussed. Beyond 2.5 nm, the fcc phase starts developing and coexists with the bcc Ni. Moreover, uniaxial anisotropy has been observed to contribute to the total magnetic anisotropy of the film. Recently, a systematic study has been carried out by the same group.[53] Bcc Ni has been epitaxially grown on GaAs(001)-(4 × 6) at 170 K instead of 300 K, and a thicker bcc Ni film obtained in this condition. The bcc Ni lattice constant was found to be 0.282 nm, which is in good agreement with the reported results.[54] A series of MOKE measurements for Ni film thicknesses ranging from 0.8 to 3.0 nm confirmed the absence of a magnetic dead layer in the Ni/GaAs interface. The magnetization of the bcc Ni was determined by *ex situ* SQUID at 5 K yielding a value of $0.52 \pm 0.08 \, \mu_B$/atom, which is similar to the theoretical predicted value of $0.54 \, \mu_B$/atom.[54] The magnetic anisotropy in the system studied has been attributed to both cubic and uniaxial anisotropies with $K_1 = 4.0 \times 10^5 \, ergs \, / \, cm^3$. It has been proposed that the uniaxial term may be a result of the in-plane shear strain induced in the system.[53]

Another approach for growing high-quality epitaxial Ni films on GaAs(001) has been given by Scheck et al., who use electrodeposition with the intention to overcome the intermixing at the interface.[55,56] Interestingly, the Ni film in this case is in the fcc phase instead of the bcc phase previously reported. The deposition of Ni was

carried out at RT and the epitaxial relation achieved by this technique is given as
Ni(001)[100]∥GaAs(001)[110]. Characterized by XPS, the absence of an As peak
for a 14 nm film suggests that both As segregation and As diffusion do not take
place. On the other hand, the presence of the As $3d$ peak in the 6 nm Ni film reveals
the occurrence of As diffusion into the Ni layer destroying the magnetic properties
of the fcc Ni film and leading to a 20% reduction of the magnetization compared
to the bulk value. In the study of the evolution of interface properties of the electrodeposited Ni film upon annealing, interdiffusion between Ni and GaAs has been
found not to be pronounced for annealing up to 523 K. Significant increase of As
outdiffusion has been observed for annealing temperatures up to 623 K accompa-
nying a rise in Schottky barrier height, which has been attributed to the Ni-Ga-As
compound formation.[56]

15.2.6 FeCo/GaAs

FeCo is a ferromagnetic alloy showing a relatively high spin polarization. It is
known that Fe_xCo_{1-x} with $x > 0.2$ has a stable bcc phase with GaAs(001).[57] The
main advantage of a magnetic alloy is that its magnetic properties can be engineered by doping. Single crystalline $Fe_{34}Co_{66}$ has been successfully grown on
GaAs(001)-(4 × 6) by means of MBE at ambient temperature.[58] A 3D growth
mode has been suggested by the disappearance of the RHEED pattern at 1 ML.
Above 4 MLs, bcc FeCo with a lattice constant identical to that of bulk Fe (2.87 Å)
was established. For films ranging 10 to 100 MLs, both unaxial and fourfold terms
contributed to the in-plane magnetic anisotropy with the latter term dominating
thick samples. The unaxial and fourfold anisotropies coexist with the respective
easy axis along [110]. Compared to bcc and bulk Fe, the K_1 of the FeCo is negative.
An interesting phenomenon appears when the cubic anisotropy changes sign from
positive to negative at a film thickness from 6 to 7 MLs. This has been attributed
to the fact that the surface and volume anisotropy constants are of opposite sign
for all alloy compositions.[59] Chen et al. also attempted to grow epitaxial bcc Fe_xCo_{1-x}
on GaAs(100) aiming at controlling the magnetic anisotropy of the heterostructure
by modifying the surface symmetries.[60] The uniaxial anisotropy dominates with
the easy axis along [011], which is in agreement with the results in Reference 60.
However, Park et al. determined, by polarized neutron reflectometry, the presence
of a magnetic dead layer of 6 Å for most film thicknesses, which might be a
consequence of the formation of interfacial alloys such as $(FeCo)_3GaAs$ on the
GaAs side.[61]

15.2.7 Fe/GaSb

To date there has only been one reported result on this heterostructure by Lépine
et al., who grew epitaxial bcc Fe on Sb-rich GaSb(001)-(1 × 3) at RT.[62] The epitaxial
relationship is Fe(001)<100>∥GaSb(001)<100>. They have not detected any contribution of unaxial anisotropy down to 1.7 nm. They proposed that this is due to the
large lattice mismatch ($\eta = -5.95$) in which the film was not strained even at low
coverage.

15.2.8 Fe/AlAs

The lattice mismatch in Fe/AlAs(001) is $\eta = 1.25\%$, which is very close to Fe/GaAs(001). This system has rarely been studied, Lépine et al. reported that in 1.7 nm Fe/AlAs(001), uniaxial in-plane anisotropy dominates with its axis along [110].[62] The same argument discussed above may explain such behavior.

15.2.9 Fe/InP

This system is believed to have potential in applications in high-speed optoelectronics and spintronics. Zavaliche et al. are perhaps the first to systematically study the magnetic properties of this heterostructure by growing Fe on P-rich InP(001)-(2×4) at 150 and 300 K.[63] From their results, it has been shown that the initial coverage of Fe exhibits a 3D morphology with islands that look irregular and varying heights by STM, with no preferred direction. This is confirmed by the disappearance of the LEED pattern from 1 to 25 MLs. Considering the interface, only In segregating to the Fe surface has been observed for 300 K growth samples and no outdiffusion was detected for 150 K ones. The most surprising finding in Reference 63 is the magnetic properties of the film thicknesses of 2.6 and 3.6 MLs studied. Both films demonstrated dominant uniaxial anisotropy with an easy axis along [$\bar{1}$10] persisting up to 13 to 15 MLs regardless of the growth temperature.[63]

15.2.10 Fe/GaN

GaN is a SC being studied intensively nowadays for solid state spintronics because theoretical predictions reveal that the electron spin can survive longer in pure GaN than in GaAs at all temperatures by three orders of magnitude.[64] It appears that spin transport and dynamics can then be comprehensively studied according to such long spin coherence length and indeed Beschoten et al. have proved that the lifetimes are around 20 ns at 5 K.[65] Nevertheless, literature on FM/GaN is still rare. Lallaizon et al. demonstrated epitaxial growth of bcc Fe(001) at RT on cubic GaN(001) between which the lattice mismatch is in principle very large (26%).[66] However, from what they found in the XPS analysis, the major axis of Fe has been rotated by 45° with respect to that of GaN giving the epitaxial relationship Fe(001)<100>||GaN(001)<110>. This can be realized by the fact that the mismatch can now become −10.3% after the rotation. Magnetically, a 65 nm Fe sample still keeps its cubic anisotropy with the easy axis along <100> as in the case of bulk Fe. From the materials science point of view, GaN in wurtzite form is more thermodynamically stable than in the cubic structure. Calarco et al. experimentally demonstrated for the first time that fcc Fe(110) could indeed be grown on wurtzite GaN(0001) by both MBE and metal-organic vapor-phase epitaxy (MOVPE).[67] STM studies suggest that the Fe growth is in island form with peak-to-peak roughness of about 5 to 10 nm. As confirmed by SQUID, the films exhibit ferromagnetism at RT and the coercivities of the films grown by both techniques decrease with increasing temperature. To determine the epitaxial relationship, Meijers et al. compared the simulated LEED patterns with the two possible Fe layer growth orientations, Nishiyama–Wassermann and Kurdjumov–Sachs orientations, respectively.[68] Further confirmed by pole figure, the epitaxial relationship is proposed as Fe(110)[001]‖ GaN(0001)[11$\overline{2}$0]. Again the films exhibit ferromagnetic order up to RT; however in this case, an effective hexagonal in-plane magnetic anisotropy with easy axes 60° apart has been observed as a direct consequence of the three crystalline domains that are rotated by 120° relative to each other.

15.3 SPIN INJECTION AND DETECTION IN HYBRID DEVICES

15.3.1 HISTORICAL PERSPECTIVE

Early experimental evidence of creation of nonequilibrium spin by means of transport, optical, or resonance methods has opened up the possibility of incorporating these techniques on SC spintronics.[70–72] Since the proposal of Aronov[73] that spin-polarized currents could be driven across a FM/SC interface, with polarized carriers diffusing into the SC for a finite distance, there has been considerable effort in demonstrating electrical spin injection of spin-polarized carriers across the interfaces of FM/normal metal (NM) and FM/SC over the past few decades. Following successful observation of spin accumulation in metals,[74–77] a novel spin FET with a FM source and drain has been proposed by Datta and Das.[8] The source-drain conductance is proportional to the relative orientation of the magnetization of the drain and the direction of the carrier spin. A gate that provides an interfacial electric field in the direction perpendicular to the direction of the traveling carriers and thus a transverse effective magnetic field[9] has been included to control the spin precession of the carriers in the 2DEG. To put spintronics closer to the realm of practical applications, one needs to overcome three major issues: (i) a mechanism for electrical injection of spin-polarized carriers into the SC, (ii) a means of spin manipulation and transport, and (iii) a detection scheme for the resulting spin polarization. Attempts to modulate such phenomenon by an applied voltage have been studied and experimentally confirmed by Nitta et al.[78] Recently, the excitation of spin-polarized carriers into the conduction band of GaAs and their subsequent detection has been demonstrated by synchronous optical pump-probe techniques. Pulses of circularly polarized light were used to generate a population of spin-polarized carriers and polarization analysis of reflected probe pulses gave a measurement of the carrier spin lifetime.[79,80] Spin-diffusion lengths of the order of several micrometers were also measured and it was shown that an electric field could push the spin-polarized carriers over distances of the order of 100 μm, which is larger than typical source-drain separations. Photon-excitation technique has also been used to create spin-polarized electrons in the study of Schottky barrier height dependence of photoexcited electron transport at FM/SC interface[81] (see Chapter 14).

In the various aspects of incorporating spintronics in the mainstream SC technologies, the only missing element is the efficient electrical spin injection in FM/SC, which has been proved to be an elusive task hampering the progress in spintronics.[82–87] Schmidt et al. have attributed such inefficient electrical spin injection in FM/SC to the conductivity mismatch between the FM and SC[18] based on the model developed earlier by van Son et al.[88] As a separate, but not unrelated work, Rashba extended the study and found that the presence of a suitable energy barrier takes control of the current polarization and dominates the effect of resistances on both sides of the FM/SC interface, and thus leads to a highly efficient spin injection by quantum mechanical tunneling process.[19] However, it should be noted that the theoretical estimation on the effect of spin accumulation at the FM/SC interface has assumed an ideal interface that does not include any spin-flip scattering. One has to keep in mind that the realization of a clean interface or good tunneling barriers on SCs is rather challenging from a technological point of view.

Optical pumping, which is sometimes referred to as optical orientation, is a technique for generating highly polarized carriers by irradiation with circularly polarized light of either right- or left-circular polarization.[89] In the equilibrium state, there are no excess spins in the medium such as a SC. While constantly irradiating the SC with circularly polarized light, a population of spin will exist defined by competition between the creation of nonequilibrium spin and factors such as carrier recombination and spin relaxation in the system. The recombination in this case is the recombination of the photoexcited spin-polarized electrons and the unpolarized holes. Note that holes are generally assumed to be unpolarized as their spin-relaxation time is relatively short.[90] Due to the polarization of electrons, the recombination process between the electrons and holes gives partially polarized luminescence. By applying the well-known optical selection rule, the interband transition probabilities for the polarized electrons can be calculated and thus useful quantities such as spin-relaxation time, recombination time, and spin orientation can be obtained and have recently been used as detection schemes for spin polarization.[91–95]

Electrical spin injection was first demonstrated experimentally by Clark and Feher in 1963.[72] In their experiment, a DC current was driven through a sample of InSb in the presence of a constant external magnetic field. This work revealed the coupling between the electron and nuclear spins contributing dynamic nuclear polarization. Aronov and Pikus studied electrical spin injection from FMs into nonmagnetic metals, SCs, and superconductors for the first time.[73] In their pioneering work, spin-polarized carriers from a FM were predicted to contribute to a net magnetization current entering a nonmagnetic region and to build up a nonequilibrium magnetization in the nonmagnetic region occupying a distance depending on the spin-diffusion length. The nonequilibrium magnetization generated is equivalent to a nonequilibrium spin accumulation in the nonmagnetic region where a competition between spin-relaxation process and the spins due to the presence of the magnetization current occurs. The prediction was not unambiguously proved until 1985 when Johnson and Silsbee demonstrated measurement of spin accumulation in the form of either a voltage or resistance change due to electrical injection of spin into NMs.[74–77] The spin injection and transport across FM/NM interface have subsequently been studied in detail by other groups.[88,96,97]

15.3.2 Theoretical Predictions

Successful demonstrations of spin injection across hybrid FM/SC junctions recently have been encouraging and drawn a lot of attention in the area of spintronics. However, these experimental results remain to be understood. Obviously, the degree of current polarization injected into a nonmagnetic region from a FM will be complicated by the presence of disorder, surface roughness, and different scattering mechanisms.[98–101] Nevertheless, theoretical calculations provide guidelines on which FM/SC hybrid systems are going to be useful and have high efficiencies for spin injection and detection. So far, from our knowledge, the existing theories solely treat the FM contact and the SC simply as uniform conductive media and do not address critical issues of the real structures used in experiments that typically include band bending in a depletion region for the case of a Schottky barrier or a *p-n* junction.[102,103] With respect to these, much effort has been devoted to identifying the dependence of spin injection on different aspects such as the barrier height, the SC doping, and the applied current density.[100,102–105] However, these studies are somehow contradicting each other and a clearly defined consensus among the theories is still elusive. In general, most of the calculations agree with the idea that in the presence of a suitably high-energy barrier at the FM/SC interface, efficient spin injection will happen. This is true for both diffusive and ballistic systems. When a spin-dependent interfacial resistance such as a tunneling barrier is present, a discontinuity in the electrochemical potential difference is introduced at the interface and thus generates a much higher difference in the potentials in the SC than in the FM, which in turn enhances spin injection into the SC.

Spin injection from Fe into GaAs(001) in a ballistic scheme has been realized by Wunnicke et al. via an *ab initio* calculation in which sophisticated factors such as band structures and interface issues have been taken into account.[106,107] Extremely high polarization of current of nearly 100% has been obtained in their Fe/GaAs(001) model and such a high value originates from the mismatch of the bands of the Fe minority spin and the GaAs, which has the highest symmetry in the (001) orientation. Inclusion of the effect from the surface termination has also been calculated by Wunnicke et al. showing the influence of surface termination of GaAs(001) on the spin polarization with the change in Schottky barrier thickness, see Figure 15.5. The reduction observed in Fe/As term-GaAs(001) is due to the presence of resonant interface states near the Fermi level. Spin injection across the Fe/InAs(001) interface has been predicted by Zwierzycki et al. driven by the fact that the contact between Fe and InAs is ohmic without any energy barrier formation at the interface.[101] For both terminations, a large spin asymmetry has been predicted and the specular interfaces formed between the Fe and InAs(001) act as efficient spin filters giving rise to relatively high-current polarizations, 98 and 89%, respectively, in In-terminated and As-terminated surfaces.

15.3.3 ELECTRICAL SPIN INJECTION AND DETECTION IN VERTICAL SPINTRONIC HYBRID DEVICES

Hybrid FM/SC spintronics devices where FM materials are used with SC materials emerging as a significant area of research. The hybrid structures offer easy integration of FM materials with digital SC processing and are inspired by the long spin lifetimes and long spin coherence length in SCs. Although it has been found that spin-injection efficiency from FM to SC is very low due to the conductivity mismatch between these materials,[18] high spin injection is expected to be happened when a suitable energy barriers such as the Schottky barrier are present at the interface.[19,108–110] To study spin

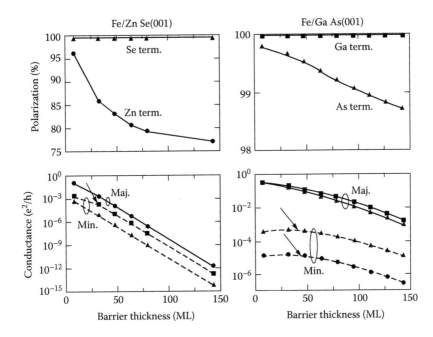

FIGURE 15.5 Theoretical prediction of spin polarization due to spin injection from Fe into different surface terminations of GaAs(001) and ZnSe(001). (From O. Wunnicke, Ph. Mavropoulos, R. Zeller, and P.H. Dederichs, *J. Phys.: Condens. Matter.*, 16, 4643, 2004. With permission.)

injection from FM to SC, a novel vertical spin-injection device, which has a thin GaAs membrane with thickness of about 0.1 to 0.5 μm, has been developed by York Spintronics Laboratory.[111] Such a membrane has also been used for the Lorentz microscopy of the domain structure of Fe dots and studies of the microscopic magnetic reversal behavior of epitaxial Fe/GaAs(001) thin films.[112] Two layers of FM, sandwiching a GaAs(100) substrate, are grown by MBE and thermal evaporation, respectively. A window with size of around 200×200 μm is defined on the GaAs by optical lithography and then the GaAs is selectively etched by chemicals. By designing the device in this way, one of the FMs serves as an emitter and the other as a collector for electron tunneling.

15.3.3.1 Fabrication of Vertical Spintronic Hybrid Devices

A 15 ML of a Co layer was deposited on a 10×10 mm As-adsorbed GaAs(100) substrate. The GaAs(100) substrate contains epilayers of $Al_{0.3}Ga_{0.7}As$ and GaAs with the following structure, As-capping/GaAs(50nm, *n*-type, $10^{18}/cm^{-3}$)/$Al_{0.3}Ga_{0.7}As$(200 nm, *n*-type, $10^{18}/cm^{-3}$)/GaAs(100). The As was desorbed at a temperature of 573 K. The Co was then grown epitaxially in an ultrahigh vacuum chamber with a base pressure of 1×10^{-10} mbar. Then the Co layer was capped with 20 MLs of Cr. The sample was then covered by 350 nm of Al deposited by thermal evaporation. This layer is used for bonding at later stage. The sample was then mounted upside down on a piece of glass with a low-melting-point wax and thinned down to around 100 μm using H_2SO_4:H_2O_2:H_2O (1:8:8) mixture, a popular etchant usually used for etching GaAs. The processed sample surface was smooth and flat under the optical microscope. Optical lithography was performed on the thinned sample from the backside to open a 200×200 μm square window. Selective chemical etching was done through the backside window with a $NH_4OH:H_2O_2$ (1:40) solution, which selectively etches GaAs and stops on AlGaAs. The surface of the AlGaAs appears to be very flat when observed under optical microscope. When the AlGaAs layer is reached, the etch pit starts broadening the GaAs sidewise. Figure 15.6 shows schematically the steps involved in producing the membrane-based device. Once the AlGaAs layer is reached, the sample was put into a thermal evaporator. A 30 nm of $Ni_{0.8}Fe_{0.2}$ layer followed by 10 nm of Cr was evaporated and liftoff was performed. To electrically connect the NiFe in the etched pit to a chip holder for magnetotransport measurement, a 350 nm of Al was deposited by thermal evaporation on the backside of the sample. The sample was then attached upside down on the chip

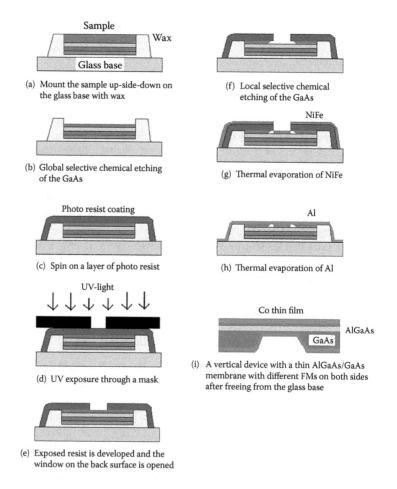

(a) Mount the sample up-side-down on the glass base with wax

(b) Global selective chemical etching of the GaAs

(c) Spin on a layer of photo resist

(d) UV exposure through a mask

(e) Exposed resist is developed and the window on the back surface is opened

(f) Local selective chemical etching of the GaAs

(g) Thermal evaporation of NiFe

(h) Thermal evaporation of Al

(i) A vertical device with a thin AlGaAs/GaAs membrane with different FMs on both sides after freeing from the glass base

FIGURE 15.6 The schematic processing steps of the membrane fabrication in the vertical spintronic device.

holder and electrical connections on the back surface were made by Al wire bonding using an ultrasonic bonding machine. The top surface, which was attached to the chip holder, is the other conducting path.

15.3.3.2 Magnetotransport Measurement

The *I-V* characteristics of the device are highly rectifying and depend on the applied field. Although the device conducts at about 0.5 V, the actual field dependency becomes apparent above 5 V. The current is observed to be decreasing with increasing field values. The effect is identical for both polarities of applied field. Breakdown occurs at a reversed bias voltage of about 30 V and the leakage current begins to build up at a voltage of 20 V. In this device, the current is established between the NiFe and Co layers such that electrons are injected into the GaAs/AlGaAs layer through the Co side by applying a certain bias voltage between the NiFe (positive) and Co (negative). Ideally, it is expected that Schottky barriers would be formed at the NiFe/AlGaAs and GaAs/Co interfaces. In a simplistic equivalence, two head-on Schottky diodes, we could then imagine that the electrons entering the SCs could be spin polarized as they pass through the magnetic material, here the NiFe layer. It is also reasonable to consider that the Schottky barriers at two interfaces are not identical. When applying a bias voltage from NiFe to Co layer, the Schottky diode at the NiFe end becomes forward biased and the Co end reverse biased. In the absence of any magnetic materials, we would expect to see that with increasing bias voltage the current flowing through the NiFe side would overcome the reverse-biased barrier of the Schottky diode at the Co end and would start conducting beyond the barrier voltage, which is typically less than a volt. This analogue is not far from the actual case, as we could see that the conduction in the forward bias region starts around 0.5 V. However, no field dependence is observed at this range. The current starts building up once the bias voltage reaches 5 V and a rapid increase is observed beyond 10 V. In addition, the influence of the applied magnetic field becomes apparent. The external field is applied normal to the plane of current and along the easy direction of the Co film. The rapid increase at and beyond 10 V could be an indication that one of the interfaces is putting up an unusually high resistance to the current than a Schottky barrier could. It is apparent from the fabrication procedure that the barrier at the NiFe side is the one producing the high resistance to the current because the AlGaAs side of the sample was exposed to ambient atmosphere before it is transferred to the thermal evaporator and the NiFe was deposited by thermal evaporation. Therefore, the NiFe/AlGaAs interface is not perfectly clean and thus there is the probability of incorporating an insulating-oxide layer giving rise to the unusually high barrier. This high barrier could possibly be responsible for hot-electron injection affecting the *I-V* characteristics at high voltages. The MR is measured with the field applied along the easy axis of the Co layer. Here a certain current is passed through the device and the voltage is measured at different fields. Here we define MR as $\Delta R/R$, where $\Delta R = R(H) - R_{min}$, and $R = R_{min}$. A reversed sense of MR curve with the minimum at 200 Oe is observed. The reason for the decrease of the current at high fields becomes apparent as the MR is observed to be increasing with high fields.

The magnetic behavior of the device could not be measured as the membrane is fragile and incompatible with standard magnetic measurement techniques. On the other hand, the magnetic property of the NiFe could only be obtained by using advanced techniques such as nano-MOKE in which the laser spot could be limited to 1 μm diameter. It is possible that the magnetic properties of the individual ferromagnetic layers of the sandwich structure could be influenced by the fact that it is of the form of a membrane. The membrane usually forms a convex surface that could influence the magnetic properties of the sandwich structure by straining the individual FM layers. The occurrence of the minimum of the MR curve at the field value of 200 Oe could be related to this possible change in the magnetic property of the sandwich structure.

The effect of biasing current on the MR has also been observed in the device. In general, the MR increases with increasing field at a biasing current higher than 2 μA. A maximum of > 12% increase is observed at the highest applied field at RT for a current of 15 μA. The increase in the MR could be related to a Lorentz force that has been exerted on the electron during the traverse in a direction normal to the applied field, or to a spin polarized current (if present) that would be expected to increase with high biasing current. A very unique behavior is depicted in Figure 15.7, where the current dependency of the MR is presented. At low fields, the MR did not show significant variation; whereas at high fields, the MR increases sharply with increasing biasing current and becomes somewhat stable or saturated beyond the current of 10 μA. A maximum change of greater than 10% has been observed for biasing higher than a critical current at the highest applied field. This is a large change compared to the ordinary anisotropic MR effect measured at RT. The change in the MR becomes stable beyond the critical current at high fields. It provides an indication that the MR cannot be of the Lorentz–MR type as that would increase with increasing current in a fixed field. However, the Lorentz MR could contribute in the region up to 10 μA where we observed an increase of MR for a fixed field. But beyond this current, the MR became stable or saturated indicating that there could be a

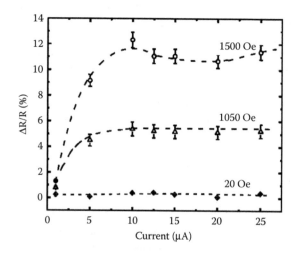

FIGURE 15.7 Field-dependent MR of the device. (From E. Ahmad, A. Valavanis, J.S. Claydon, Y.X. Lu, and Y.B. Xu, *IEEE Trans. Magn.*, 41, 2592, 2005. With permission.)

phenomenon other than the Lorentz force responsible for this. As we have a sandwich structure with a SC layer between two FM layers, we could not rule out the possibility that a contribution to the MR could come from spin injection and detection through the FM layers. However, the overall behavior is still an open question and might be dependent on a more detailed scenario involving the band structures of individual materials.

The most interesting behavior of this vertical device is the resemblance with FET characteristics. This similarity evokes the possibility that a two-terminal vertical device such as ours could serve the purpose of a three-terminal device such as a FET. A transistor could be turned to "on" and "off" states by controlling the gate voltage. Likewise, our two-terminal vertical device could be turned "on" (high MR) or "off" (low MR) just by varying the applied magnetic field for a critical current of greater than 5 µA. Thus, a device like this could be of immense interest in spintronic logics or field sensors.

15.3.4 Comparison of Spin Injection and Detection Techniques

15.3.4.1 Optical and Electrical Injection

Far from the theoretical search for methods for efficient spin injection, experimental evidence for RT injection of polarized spins into GaAs from polycrystalline and epitaxial Ni, respectively, in the STM experiment was demonstrated a few years ago.[113,114] An important implication of the STM experiments is that a family of transition FM metals such as Fe, Co, and Ni, which exhibit a relatively high spin polarization above RT, could be utilized as a spin-injection contact although the experiment itself is unlikely to be implemented in spin transfer devices. Zhu et al. demonstrated, for the first time, that the transition metal Fe could indeed inject spin-polarized electrons into a GaAs/(In,Ga)As quantum well (QW) structure grown by MBE via a Schottky barrier formed at the Fe/GaAs interface.[94] The polarization was detected by the circular polarization of the electroluminescence (EL) in the active region of the QW in Faraday geometry, resulting in constant spin-injection efficiency of 5% at temperatures from 25 to 300 K, which is an encouraging result because Fe/GaAs is one of the most well-understood hybrid structures to date.[115] Reported by the same research group, high-quality FM metal[116] MnAs and a Heusler alloy[117] Fe_3Si were also grown by MBE and utilized as a spin injector into the similar LED structure as in Reference 94. The detection gave rise to similar spin-injection efficiencies of which, however, the detected polarizations are dependent on spin-relaxation processes as functions of temperature in the active region of the spin LED. Hanbicki et al. reported spin-injection efficiency via a Fe/AlGaAs Schottky barrier as high as 30% at RT detected in a spin LED in reversed bias (Figure 15.8)[118] and confirmed it by assessing the Rowell criteria that the transport was dominated by the tunneling process.[119] An extended study of the structure has shown that both Fe/Al_2O_3 and Fe/AlGaAs produced electron spin polarization of similar values but in the case of the tunnel barrier, the EL intensity was significantly lower than that of the Schottky one. This could be explained by the higher density of scatterers present at the Al_2O_3/AlGaAs interface in the LED.[120] Among the numerous articles for spin

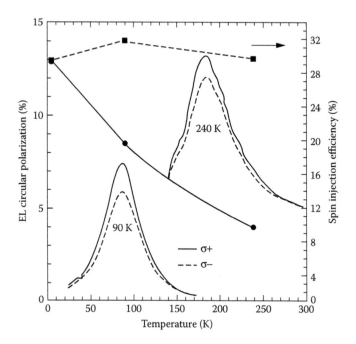

FIGURE 15.8 Temperature dependence of the circular polarization and the corresponding spin-injection efficiency of the Fe/AlGaAs LED. (From A.T. Hanbicki, B.T. Jonker, G. Itskos, G. Kioseoglou, and A. Petrou, *Appl. Phys. Lett.*, 80, 1240, 2002. With permission.)

injection from FMs into GaAs, there are only a few regarding InAs, which is also an interesting SC for studying spin-based devices. Nevertheless, Yoh et al. detected spin polarization by means of circularly polarized EL from their patterned Fe/InAs samples without any tunneling barrier.[121] A maximum polarization of about 12% was obtained at 6.5 K. Later the authors reported that samples with single-crystal Fe thin films grew on InAs at 300 K, which showed an increased degree of spin polarization of 36 to 40% at 6.5 K.[122] This is attributed to the fact that at low temperature, interface reaction and outdiffusion of the SC are not as pronounced as at high temperature.

15.3.4.2 Optical and Electrical Detection

Hirohata et al.[81] at Cambridge have clearly observed a clear difference in the helicity-dependent photocurrent through the NiFe/GaAs interface according to the magnetization orientation either parallel or perpendicular to the samples with different doping densities, see Figure 15.9. The presence of Schottky-type barriers has been identified by *I-V* measurement and the polarization of the photocurrent detected at different biasing voltage ranges from 16 to −23% for doping densities from 10^{17} to 10^{19} cm^{-3}. So far, due to the constraint imposed by the Faraday geometry, the spin momentum has to be aligned with the light-emitted direction (parallel to the growth direction) by using a relatively high magnetic field, which is typically more than 2 T. Such a high field might actually induce side effects such as magnetic circular

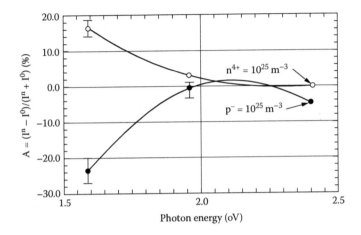

FIGURE 15.9 Photon energy dependence of the asymmetry at zero biasing voltage with different substrate doping densities. (From A. Hirohata, Y.B. Xu, C.M. Gürtler, J.A.C. Bland, and S.N. Holmes, Phys. Rev. B, 63, 104425, 2001. With permission.)

dichroism (MCD) and Zeeman splitting in SC giving spurious contributions to the measurement. To address these problems, Motsnyi et al. applied an oblique Hanle effect in which a moderate magnetic field was applied at some angle of maximum 45° against the FM film plane normal to detect the presence of spin injection. They have successfully measured spin polarization of around 16% at RT in their designed metal-insulator-semiconductor (MIS) tunnel junction using Co and AlO_x as the spin injector and the insulating layer, respectively (see Chapter 12).[123,124]

Electrical detection of injected polarized carriers is essential for studying spin injection. To eliminate spurious contributions to measurements, a careful design of a detection scheme is a prerequisite. Probing spin injection in the form of TMR between FM contacts that sandwich a SC in a MTJ has been routinely used.[125–127] In particular, Mattana et al. demonstrated the first clear evidence of electrical spin detection of spin injected from GaMnAs into a GaAs layer through an AlAs barrier.[125] Not only has this manifested a tunnel barrier as a necessity for efficient spin injection, but also suggests a feasible method for detection. Apart from the well-studied AlO_x barrier, Jiang et al. have obtained a highly efficient spin polarization of 52% at 100 K by a CoFe/MgO (100) tunnel injector.[128] The percentage could be further boosted up to 55% after annealing at 613 K.[129] Such promising results have revealed the fact that tunnel junctions using MgO are far outperforming conventional junctions with AlO_x barriers and with better thermal stability. Nevertheless, Vaz et al. have recently studied the magnetic properties of Fe/AlO_x/GaAs hybrid structure by means of polarized neutron reflection.[130] It has been found that the magnetic moment of Fe $(2.164 \pm 0.026)\mu_B$ is very close to the bulk value $2.185\mu_B$ manifesting the absence of a magnetically dead layer formed between Fe and AlO_x. This also suggests that AlO_x is still a good candidate as a tunnel barrier for efficient spin injection.

Crooker et al.[134] have unambiguously studied, by scanning Kerr microscopy, spin-injection, transport, accumulation, and detection in a lateral device that has an epitaxially grown Fe source and drain at the two ends of a 300 × 100 μm GaAs channel. The experiment was carried out at 4 K. At the first glance, the authors demonstrated clearly that both Fe/GaAs Schottky barriers at the source and drain could be used electrically for injection and detection of spins supporting the relevance of electrical injection and detection in spin-based devices such as the vertical spin-injection device introduced in Section 15.3.3.[111] From both scientific and industrial points of view, spin polarization of electrons in SC by reflection at the interface with a FM as shown experimentally in Reference 134 (Figure 15.10) has opened up

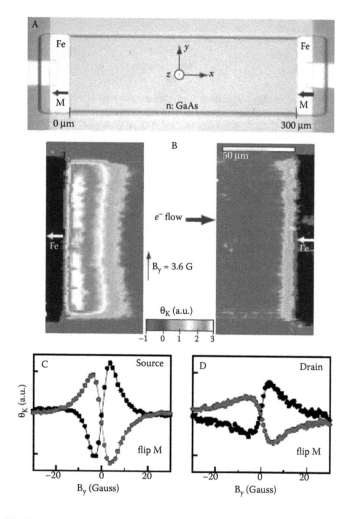

FIGURE 15.10 Micrograph of the lateral Fe/GaAs/Fe structure studied by scanning Kerr microscopy. The magnetization of Fe on both ends is along [011]. The images of Kerr rotation angle near the source and drain biased at 0.4 V indicates that the spin polarizations near the source and drain are antiparallel to M. (From S.A. Crooker, M. Furis, X. Lou, C. Adelmann, D.L. Smith, C.J. Palmstrøm, and P.A. Crowell, *Science,* 309, 2192, 2005. With permission.)

the opportunity for an alternative way to produce spin accumulation at the FM/SC interface,[135,136] and after all, all-electrical creation and detection of spin-polarized carriers in SCs are foreseeable to be incorporated in real spin-electronic devices.

Table 15.2 summarizes the findings discussed in Section 15.3.4.1 and Section 15.3.4.2.

15.4 SUMMARY AND OUTLOOK

In this chapter, we have reviewed some of the important FM/III-V SC hybrid systems in terms of their epitaxial growth, structure, and magnetic properties. At the moment, it is still not clear which hybrid structure will be an obvious choice for the next generation device applications due to the fact that there are still questions to be understood. As a whole, the most essential criterion for one to be incorporated in spintronic devices is to form a sharp interface between the FM and SC. As we have briefly discussed, most of the FM/SC systems exhibit interfacial compounds during growth and thus reduce the magnetic moment of the FM at the interface. In this sense, one should realize that the sharpness of the interface is actually limited by the amount of antiferromagentic or nonmagnetic compounds formed at the interface. Approaches for tackling these are to have low substrate temperatures during growth and to grow a surfactant layer to reduce segregation of atoms from the substrate. As quite a confusing issue, the large amount of literature reporting different onset of ferromagnetism of FM thin films in terms of film coverage and evolution of different types of anisotropy reveal the fact that the magnetic properties and crystal structure of the film are strongly dependent on the sample preparation recipes.

The step toward spin-based devices has been fueled by many of the encouraging theoretical and experimental demonstrations that we have described in this chapter. Spin injection and detection are still the most challenging problems physicists and engineers have to overcome before one can put the spin degree of freedom closer to the realm of application. The seemingly insurmountable problem of efficient spin injection from a FM into a SC in the diffusive regime has been relieved by many of the recent theoretical and experimental efforts. It has been proved that an energy barrier would be helpful for maintaining the difference of the chemical potential between the two spin species or filtering low-energy electrons. On the other hand, one should keep in mind that if a tunnel barrier is responsible for electron tunneling, the presence of any additional barrier formed between the metal/SC interface may not be actually beneficial. Thus the control over the interface characteristic between the FM/SC is again extremely important.

From the industrial point of view, there are always two different approaches, either (i) perfecting the existing technology by searching for new materials or improving device functioning efficiency or (ii) concentrating on creating novel effects. Obviously both approaches are not easy to achieve, but one should be clear that it will always be easier to integrate SC-based devices with traditional SC technology. On the other hand, with the advances in SC science and technology, the control and manipulation of spin in SC becomes increasingly possible. All-metal GMR devices such as magnetic field sensors or reading heads and TMR memory cells are already on the market or close to industrial realization; however, devices

TABLE 15.2
Theoretical and Experimental Spin Injection in Different Hybrid Systems

			Theoretical		
System	Injection	Detection	Polarization[*a]	Temperature	Reference
Fe/GaAs(001)			~100%		106,107
Fe/InAs(001)			98% (In-term)		101
			89% (As-term)		101
			Experimental		
Poly Ni/GaAs(110)	II	I	−31%	RT	113
Single Ni/GaAs(110)	II	I	25%	100 K	114
Fe/GaAs-LED	II	I	5%	RT	118
Fe/AlGaAs/GaAs(001)-LED	II	I	4%	240 K	119
	II	I	32%	4.5 K	119
Fe/Al$_2$O$_3$/GaAs(001)-LED	II	I	30%	5 K	120
MnAs/GaAs(001)-LED	II	I	1.50%	80 K	116
CoFe/AlO$_x$/GaAs- LED	II	I	9%	80 K	123
	II	I	21%	80 K	124
Co/Al$_2$O$_3$/GaAs-LED	II	I	~1%	RT	131
CoFe/MgO/GaAs-LED	II	I	32%	290 K	128
	II	I	55%	100 K	129
NiFe/GaAs	I	II	16–23%	RT	81
Fe/InAs	II	I	12%	6.5 K	121
	II	I	36–40%	6.5 K	122
			MR		
NiFe/InAs 2DEG	II	II	~1%	RT	83
	II	II	0.20%	RT	82
NiFe/AlGaAs/GaAs/Co	II	II	12%	RT	111
GaMnAs/AlAs/GaAs/ AlAs/GaMnAs	II	II	38%	4 K	125

[*a] In order to avoid confusion, polarization of detection signal has been cited instead of spin-injection efficiency.

I: Optical.

II: Electrical.

based on spin injection and detection in SC are still far from commercial usage. As we have reviewed the extensive literature for spin injection and detection, the optical scheme has been adopted more frequently for detecting the presence of successful spin injection due to the ease of ruling out spurious effects such as MCD and the Zeeman effect. Nevertheless, such an approach is indeed not as suitable as the electrical one due to the ease of integration with practical devices. So, one can anticipate that the use of electrical detection is going to be dominating although we will have to face problems such as spurious effects masking the output we are looking for.

From the progress of uncovering the true potential of spintronics, we can see that this area is bright and intriguing. Solid state electronics with less power dissipation, faster switching, and denser and virtually everlasting memory are no longer a dream. At this stage, we are not sure how robust and powerful the next-generation devices will be in reality; however, we are anticipating with confidence that the electron spin is going to impact electronics just as the GMR effect did 18 years ago.

REFERENCES

1. G. Moore, *Electronics*, 38, 8, 1965.
2. G. Bourianoff, "The future of nanocomputing," *Computer*, August 2003, p. 44.
3. G. Prinz, *Science*, 282, 1660, 1995.
4. S.A. Wolf and D. Treger, *IEEE Trans. Magn.*, 36, 2748, 2000.
5. S.A. Wolf, D.D. Awschalom, R.A. Buhrman, J.M. Daughton, S. von Molnár, M.L. Roukes, A.Y. Chtchelkanova, and D.M. Treger, *Science*, 294, 1488, 2000.
6. M.N. Baibich, J.M. Broto, A. Fert, F. Nguyen Van Dau, F. Petroff, P. Eitenne, G. Creuzet, A. Friederich, and J. Chazelas, *Phys. Rev. Lett.*, 61, 2472, 1988.
7. B.T. Jonker, S.C. Erwin, A. Petrou, and A.G. Petukhov, *MRS Bulletin*, October 2003, p. 740.
8. S. Datta and B. Das, *Appl. Phys. Lett.*, 56, 665, 1990.
9. Y.A. Bychkov and E.I. Rashba, *JETP Lett.*, 39, 78, 1984.
10. D.J. Monsma, R. Vlutters, and J.C. Lodder, *Science*, 281, 407, 1998.
11. J.S. Moodera, L.R. Kinder, T.M. Wong, and R. Meservey, *Phys. Rev. Lett.*, 74, 3273, 1995.
12. T. Miyazaki and N. Tezuka, *J. Magn. Magn. Mater.*, 139, L231, 1995.
13. S.S.P. Parkin, K.P. Roche, M.G. Samant, P.M. Rice, R.B. Beyers, R.E. Scheuerlein, E.J. O'Sullivan, S.L. Brown, J. Bucchigano, D.W. Abraham, Yu Lu, M. Rooks, P.L. Trouilloud, R.A. Wanner, and W.J. Gallagher, *J. Appl. Phys.*, 85, 5828, 1999.
14. R. Jansen, *J. Phys. D: Appl. Phys.*, 36, R289, 2003.
15. S. van Dijken, X. Jiang, and S.S.P. Parkin, *Phys. Rev. Lett.*, 90, 197203, 2003.
16. S.A. Solin, T. Thio, D.R. Hines, and J.J. Heremans, *Science*, 289, 1530, 2000.
17. H. Akinaga, *Semicond. Sci. Technol.*, 17, 322, 2002.
18. G. Schmidt, D. Ferrand, L.W. Molenkamp, A.T. Filip, and B.J. van Wees, *Phys. Rev. B*, 62, R4790, 2000.
19. E.I. Rashba, *Phys. Rev. B*, 62, R162367, 2000.
20. E.M. Kneedler, B.T. Jonker, P.M. Thibado, R.J. Wagner, B.V. Shanabrook, and L.J. Whitman, *Phys. Rev. B*, 56, 8163, 1997.
21. G.A. Prinz, G.T. Rado, and J.J. Krebs, *J. Appl. Phys.*, 53, 2087, 1982.

22. A. Filipe, A. Schuhl, and P. Galtier, *Appl. Phys. Lett.,* 70, 129, 1997.
23. Y.B. Xu, E.T.M. Kernohan, D.J. Freeland, A. Ercole, M. Tselepi, and J.A.C. Bland, *Phys. Rev. B,* 58, 890, 1998.
24. M. Gester, C. Daboo, R.J. Hicken, S.J. Gray, A. Ercole, and J.A.C. Bland, *J. Appl. Phys.,* 80, 347, 1996.
25. J.S. Claydon, Y.B. Xu, M. Tselepi, J.A.C. Bland, and G. van der Laan, *Phys. Rev. Lett.,* 93, 37206, 2004.
26. Y.B. Xu, D.J. Freeland, M. Tselepi, and J.A.C. Bland, *J. Appl. Phys.,* 87, 6110, 2000.
27. Y.B. Xu, D.J. Freeland, M. Tselepi, and J.A.C. Bland, *Phys. Rev. B,* 62, 1167, 2000.
28. E.H. Rhoderick and R.H. Williams, *Metal/Semiconductor Contact,* 2nd ed., Oxford University Press, Oxford, 1988.
29. Y.B. Xu, E.T.M. Kernohan, M. Tselepi, J.A.C. Bland, and S. Holmes, *Appl. Phys. Lett.,* 73, 399, 1998.
30. Y.B. Xu, M. Tselepi, J. Wu, S. Wang, J.A.C. Bland, Y. Huttel, and G. van der Laan, *IEEE Trans. Magn.,* 38, 2652, 2002.
31. L. Ruppel, G. Witte, Ch. Wöll, T. Last, S.F. Fischer, and U. Kunze, *Phys. Rev. B,* 66, 245307, 2002.
32. G.A. Prinz, *Phys. Rev. Lett.,* 54, 1051, 1985.
33. F. Xu, J.J. Joyce, M.W. Ruckman, H.-W. Chen, F. Boscherini, D.M. Hill, S.A. Chambers, and J.H. Weaver, *Phys. Rev. B,* 35, 2375, 1987.
34. S.J. Blundell, M. Gester, J.A.C. Bland, C. Daboo, E. Gu, M.J. Baird, and A.J.R. Ives, *J. Appl. Phys.,* 73, 5948, 1993.
35. E. Gu, M. Gester, R.J. Hicken, C. Daboo, M. Tselepi, S.J. Gray, J.A.C. Bland, L.M. Brown, T. Thomson, and P.C. Riedi, *Phys. Rev. B,* 52, 14704, 1995.
36. Y.Z. Wu, H.F. Ding, C. Jing, D. Wu, G.L. Liu, V. Gordon, G.S. Dong, X.F. Jin, S. Zhu, and K. Sun, *Phys. Rev. B,* 57, 11935, 1996.
37. M.A. Mangan, G. Spanos, T. Ambrose, and G.A. Prinz, *Appl. Phys. Lett.*, 75, 346, 1999.
38. K. Lüdge, B.D. Schultz, P. Vogt, M.M.R. Evans, W. Braun, C.J. Palmstrøm, W. Richter, and N. Esser, *J. Vac. Sci. Technol. B,* 20, 1591, 2002.
39. M. Madami, S. Tacchi, G. Gubbiotti, G. Carlotti, and G. Socino, *Surf. Sci.,* 566–568, 246, 2004.
40. K. Lüdge, P. Vogt, W. Richter, B.-O. Fimland, W. Braun, and N. Esser, *J. Vac. Sci. Technol. B,* 22, 2008, 2004.
41. Z. Ding, P.M. Thibado, C. Awo-Affouda, and V.P. LaBella, *J. Vac. Sci. Technol. B,* 22, 2068, 2004.
42. S. Subramanian, X. Liu, R.L. Stamps, R. Sooryakumar, and G. A. Prinz, *Phys. Rev. B,* 52, 10194, 1995.
43. Y.U. Idzerda, W.T. Elam, B.T. Jonker, and G.A. Prinz, *Phys. Rev. Lett.,* 62, 2480, 1989.
44. M. Izquierdo, M.E. Dávila, J. Avila, H. Ascolani, C.M. Teodorescu, M.G. Martin, N. Franco, J. Chrost, A. Arranz, and M.C. Asensio, *Phys. Rev. Lett.,* 94, 187601, 2005.
45. A.G. Aronov and G.E. Pikus, *Sov. Phys. Semicond.,* 10, 698, 1976.
46. M. Johnson and R.H. Silsbee, *Phys. Rev. Lett.,* 55, 1790, 1985.
47. M. Johnson and R.H. Silsbee, *J. Appl. Phys.,* 63, 3934, 1988.
48. A.Y. Liu and D.J. Singh, *J. Appl. Phys.,* 73, 6189, 1993.
49. M. Bozorth, *Ferromagnetism,* Van Nostrand, Reinhold, New York, 1951, 264.
50. J.A.C. Bland, R.D. Bateson, P.C. Riedi, R.G. Graham, H.J. Lauter, J. Penfold, and C. Shackleton, *J. Appl. Phys.,* 69, 4989, 1991.
51. D. Bagayoko, A. Ziegler, and J. Callaway, *Phys. Rev. B,* 27, 7046, 1983.
52. V.L. Moruzzi, P.M. Marcus, K. Schwarz, and P. Mohu, *J. Magn. Mater.,* 54, 955, 1986.
53. K.G. Nath, F. Maeda, S. Suzuki, and Y. Watanabe, *J. Appl. Phys.,* 90, 1222, 2001.

54. P. Palmgren, K. Szamota-Leandersson, J. Weissenrieder, T. Claesson, O. Tjernberg, U.O. Karlsson, and M. Göthelid, *Surf. Sci.*, 574, 181, 2005.
55. W.X. Tang, D. Qian, D. Wu, Y.Z. Wu, G.S. Dong, X.F. Jin, S.M. Chen, X.M. Jiang, X.X. Zhang, and Z. Zhang, *J. Magn. Magn. Mater.*, 240, 404, 2002.
56. C.S. Tian, D. Qian, D. Wu, R.H. He, Y.Z. Wu, W.X. Tang, L.F. Yin, Y.S. Shi, G.S. Dong, X.F. Jin, X.M. Jiang, F.Q. Liu, H.J. Qian, K. Sun, L.M. Wang, G. Rossi, Z.Q. Qiu, and J. Shi, *Phys. Rev. Lett.*, 94, 137210, 2005.
57. G.Y. Guo and H.H. Wang, *Chin. J. Phys.*, 38, 949, 2000.
58. C. Scheck, P. Evans, G. Zangari, and R. Schad, *Appl. Phys. Lett.*, 82, 2853, 2003.
59. C. Scheck, Y.-K. Liu, P. Evans, R. Schad, and G. Zangari, *J. Appl. Phys.*, 95, 6549, 2004.
60. C.J. Gutierrez, G.A. Prinz, J.J. Krebs, M.E. Filipkowski, V.G. Harris, and W.T. Elam, *J. Magn. Magn. Mater.*, 126, 232, 1993.
61. M. Dumm, M. Zölfl, R. Moosbühler, M. Brockmann, T. Schmidt, and G. Bayreuther, *J. Appl. Phys.*, 87, 5457, 2000.
62. M. Dumm, B. Uhl, M. Zölfl, W. Kipferl, and G. Bayreuther, *J. Appl. Phys.*, 91, 8763, 2002.
63. L.C. Chen, J.W. Dong, B.D. Schultz, C.J. Palmstrøm, J. Berezovsky, A. Isakovic, P. A. Crowell, and N. Tabat, *J. Vac. Sci. Technol. B*, 18, 2057, 2000.
64. S. Park and M.R. Fitzsimmons, X.Y. Dong, B.D. Schultz, and C.J. Palmstrøm, *Phys. Rev. B*, 70, 104406, 2004.
65. B. Lépine, C. Lallaizon, S. Ababou, A. Guivarc'h, S. Députier, A. Filipe, F. Nguyen Van Dau, A. Schuhl, F. Abel, and C. Cohen, *J. Cryst. Growth*, 201/202, 702, 1999.
66. F. Zavaliche, W. Wulfhekel, and J. Kirschner, *Phys. Rev. B*, 65, 245317, 2002.
67. S. Krishnamurthy, M. van Schilfgaarde, and N. Newman, *Appl. Phys. Lett.*, 83, 1761, 2003.
68. B. Beschoten, E. Johnston-Halperin, D.K. Young, M. Poggio, J.E. Grimaldi, S. Keller, S.P. DenBaars, U.K. Mishra, E.L. Hu, and D.D. Awschalom, *Phys. Rev. B*. 63, 121202R, 2001.
69. C. Lallaizon, P. Schieffer, B. Lépine, A. Guivarc'h, F. Abel, C. Cohen, G. Feuillet, B. Daudin, and F. Nguyen Van Dau, J. Cryst. Growth, 240, 236, 2002.
70. R. Calarco, R. Meijers, N. Kaluza, V.A. Guzenko, N. Thillosen, Th. Schäpers, H. Lüth, M. Fonin, S. Krzyk, R. Ghadimi, B. Beschoten, and G. Güntherodt, *Phys. Stat. Sol. A*, 202, 754, 2005.
71. R. Meijers, R. Calarco, N. Kaluza, H. Hardtdegen, M.v.d. Ahe, H.L. Bay, H. Lüth, M. Buchmeier, and D.E. Bürgler, *J. Cryst. Growth*, 283, 500, 2005.
72. V. Delmouly, A. Bournel, G. Tremblay, and P. Hesto, *Thin Solid Film*, 384, 282, 2001.
73. G. Lampel, *Phys. Rev. Lett.*, 20, 491, 1968.
74. R.R. Parsons, *Phys. Rev. Lett.*, 23, 1152, 1969.
75. W.G. Clark and G. Feher, *Phys. Rev. Lett.*, 10, 134, 1963.
76. M. Johnson and R.H. Silsbee, *Phys. Rev. Lett.*, 60, 377, 1988.
77. M. Johnson and R.H. Silsbee, *Phys. Rev. B*, 37, 5326, 1988.
78. J. Nitta, T. Akazaki, H. Takayanagi, and T. Enoki, *Phys. Rev. Lett.*, 78, 1335, 1997.
79. J.K. Kikkawa and D.D. Awschalom, *Phys. Rev. Lett.*, 80, 4313, 1998.
80. J.K. Kikkawa and D.D. Awschalom, *Nature*, 397, 139, 1999.
81. A. Hirohata, Y.B. Xu, C.M. Guertler, J.A.C. Bland, and S.N. Holmes, *Phys. Rev. B*, 63, 104425, 2001.
82. S. Gardelis, C.G. Smith, C.H.W. Barnes, E.H. Linfield, and D.A. Ritchie, *Phys. Rev. B*, 60, 7764, 1999.

83. P.R. Hammar, B.R. Bennett, M.J. Yang, and M. Johnson, *Phys. Rev. Lett.,* 83, 203, 1999.
84. F.G. Monzon and M.L. Roukes, *J. Magn. Magn. Mater.,* 199, 632, 1999.
85. A.T. Filip, B.H. Hoving, F.J. Jedema, B.J. van Wees, B. Dutta, and S. Borghs, *Phys. Rev. B,* 62, 9996, 2000.
86. F.J. Jedema, A.T. Filip, and B.J. van Wees, *Nature,* 410, 345, 2001.
87. F.G. Monzon, H.X. Tang, and M.L. Roukes, *Phys. Rev. Lett.,* 84, 5022, 2000.
88. P.C. van Son, H. van Kempen, and P. Wyder, *Phys. Rev. Lett.,* 58, 2271, 1987.
89. F. Meier and B.P. Zakharchenya, Eds., *Optical Orientation,* Worth-Holland, New York, 1984.
90. I. Žutić, J. Fabian, and S. Das Sarma, *Phys. Rev. B,* 64, 121201, 2001.
91. R. Fiederling, M. Keim, G. Reuscher, W. Ossau, G. Schmidt, A. Waag, and L. W. Molenkamp, *Nature,* 402, 787, 1999.
92. Y. Ohno, D.K. Young, B. Beschoten, F. Matsukura, H. Ohno, and D.D. Awschalom, *Nature,* 402, 790, 1999.
93. B.T. Jonker, Y.D. Park, B.R. Bennett, H.D. Cheong, G. Kioseoglou, and A. Petrou, *Phys. Rev. B,* 62, 8180, 2000.
94. H.J. Zhu, M. Ramsteiner, H. Kostial, M. Wassermeier, H.-P. Schönherr, and K.H. Ploog, *Phys. Rev. Lett.,* 87, 016601, 2001.
95. M. Oestreich, J. Hübner, D. Hägele, P.J. Klar, W. Heimbrodt, W.W. Rühle, D.E. Ashenford, and B. Lunn, *Appl. Phys. Lett.,* 74, 1251, 1999.
96. T. Valet and A. Fert, *Phys. Rev. B,* 48, 7099, 1993.
97. S. Hershfield and H.L. Zhao, *Phys. Rev. B,* 56, 3296, 1997.
98. K.M. Schep, J.B.A.N. van Hoof, P.J. Kelly, G.E.W. Bauer, and J.E. Inglesfield, *Phys. Rev. B,* 56, 10805, 1997.
99. M.D. Stiles and D.R. Penn, *Phys. Rev. B,* 61, 3200, 2000.
100. S. Kreuzer, J. Moser, W. Wegscheider, D. Weiss, M. Bichler, and D. Schuh, *Appl. Phys. Lett.,* 80, 4582, 2002.
101. M. Zwierzycki, K. Xia, P.J. Kelly, G.E.W. Bauer, and I. Turek, *Phys. Rev. B,* 67, 092401, 2003.
102. J.D. Albrecht and D.L. Smith, *Phys. Rev. B,* 66, 113303, 2002.
103. J.D. Albrecht and D.L. Smith, *Phys. Rev. B,* 68, 035340, 2003.
104. V.V. Osipov and A.M. Bratkovsky, *Phys. Rev. B,* 70, 205312, 2004.
105. V.V. Osipov and A.M. Bratkovsky, *Phys. Rev. B,* 72, 115322, 2004.
106. O. Wunnicke, Ph. Mavropoulos, R. Zeller, P.H. Dederichs, and D. Grundler, *Phys. Rev. B,* 65, 241306, 2002.
107. Ph. Mavropoulos, O. Wunnicke, and P.H. Dederichs, *Phys. Rev. B,* 66, 024416, 2002.
108. D.L. Smith and R.N. Silver, *Phys. Rev. B,* 64, 045323, 2001.
109. A. Fert and H. Jaffies, *Phys. Rev. B,* 64, 184420, 2001.
110. S. Takahashi and S. Maekawa, *Phys. Rev. B,* 67, 052409, 2003.
111. E. Ahmad. A. Valavanis, J.S. Claydon, Y.X. Lu, and Y.B. Xu, *IEEE Trans. Magn.,* 41, 2592, 2005.
112. E. Gu, J.A.C. Bland, C. Daboo, M. Gester, L.M. Brown, R. Ploessl, and J.N. Chapman, *Phys. Rev. B,* 51, 3596, 1995.
113. S.F. Alvarado and P. Renaud, Phys. Rev. Lett., 68, 1387, 1992.
114. S.F. Alvarado, *Phys. Rev. Lett.,* 75, 513, 1995.
115. G. Wastlbauer and J.A.C. Bland, *Adv. Phys.,* 54, 137, 2005.
116. M. Ramsteiner, H.Y. Hao, A. Kawaharazuka, H.J. Zhu, M. Kästner, R. Hey, L. Däweritz, H.T. Grahn, and K.H. Ploog, *Phys. Rev. B,* 66, 081304, 2002.

117. A. Kawaharazuka, M. Ramsteiner, J. Herfort, H.-P. Schönherr, H. Kostial, and K.H. Ploog, *Appl. Phys. Lett.,* 85, 3492, 2004.
118. A.T. Hanbicki, B.T. Jonker, G. Itskos, G. Kioseoglou, and A. Petrou, *Appl. Phys. Lett.,* 80, 1240, 2002.
119. A.T. Hanbicki, O.M.J. van 't Erve, R. Magno, G. Kioseoglou, C.H. Li, B.T. Jonker, G. Itskos, R. Mallory, M. Yasar, and A. Petrou, *Appl. Phys. Lett.,* 82, 4092, 2003.
120. O.M.J. van 't Erve, G. Kioseoglou, A.T. Hanbicki, C.H. Li, B.T. Jonker, R. Mallory, M. Yasar, and A. Petrou, *Appl. Phys. Lett.,* 84, 4334, 2004.
121. K. Yoh, H. Ohno, Y. Katano, K. Mukasa, and M. Ramsteiner, *J. Cryst. Growth,* 251, 337, 2003.
122. K. Yoh, H. Ohno, K. Sueoka, and M. Ramsteiner, *J. Vac. Sci. Technol. B,* 22, 1432, 2004.
123. V.F. Motsnyi, J. de Boeck, J. Das, W. van Roy, G. Borghs, E. Goovaerts, and V.I. Safarov, *Appl. Phys. Lett.,* 81, 265, 2002.
124. V.F. Motsnyi, P. van Dorpe, W. van Roy, E. Goovaerts, V.I. Safarov, G. Borghs, and J. de Boeck, *Phys. Rev. B,* 68, 245319, 2003.
125. R. Mattana, J.-M. George, H. Jaffrès, F. Nguyen Van Dau, A. Fert, B. Lépine, A. Guivarc'h, and G. Jézéquel, *Phys. Rev. Lett.,* 90, 166601, 2003.
126. M. Tanaka and Y. Higo, *Phys. Rev. Lett.,* 87, 026602, 2001.
127. W.H. Butler, X.-G. Zhang, T.C. Schulthess, and J.M. MacLaren, *Phys. Rev. B,* 63, 054416, 2001.
128. X. Jiang, R. Wang, R.M. Shelby, R.M. Macfarlane, S.R. Bank, J.S. Harris, and S.S. P. Parkin, *Phys. Rev. Lett.,* 94, 056601, 2005.
129. R. Wang, X. Jiang, R.M. Shelby, R.M. Macfarlane, and S.S.P. Parkin, S.R. Bank, and J.S. Harris, *Appl. Phys. Lett.,* 86, 052901, 2005.
130. C.A.F. Vaz, A. Ionescu, T. Trypiniotis, and J.A.C. Bland, R.M. Dalgliesh, and S. Langridge, *J. Appl. Phys.,* 97, 10J119, 2005.
131. T. Manago and H. Akinaga, *Appl. Phys. Lett.,* 81, 694, 2002.
132. V.P. LaBella, D.W. Bullock, Z. Ding, C. Emery, A. Venkatesan, W.F. Oliver, G.J. Salamo, P.M. Thibado, and M. Mortazavi, *Science,* 292, 1518, 2001.
133. O. Wunnicke, Ph. Mavropoulos, R. Zeller, and P.H. Dederichs, *J. Phys.: Condens. Matter.,* 16, 4643, 2004.
134. S.A. Crooker, M. Furis, X. Lou, C. Adelmann, D.L. Smith, C.J. Palmstrøm, and P.A. Crowell, *Science,* 309, 2192, 2005.
135. C. Ciuti, J.P. McGuire, and L.J. Sham, *Phys. Rev. Lett.,* 89, 156601, 2002.
136. J. Stephens, J. Berezovsky, J.P. McGuire, L.J. Sham, A.C. Gossard, and D.D. Awschalom, *Phys. Rev. Lett.,* 93, 097602, 2004.

16 The Spin-Valve Transistor*

R. Jansen

CONTENTS

16.1 INTRODUCTION

Two exciting scientific breakthroughs within the last two decades, and their exceptionally rapid implementation into products, demonstrate the impact of using the electron spin degree of freedom. In 1988, the giant magnetoresistance (GMR) effect[2] was discovered in multilayer structures that contain layers of ferromagnetic metals separated by a thin spacer of normal metal. The resistance of such structures was found to depend greatly on the relative magnetic orientation of neighboring magnetic layers, making it attractive for application in highly sensitive magnetic field sensors.[3–5] As early as 1997, GMR was incorporated into the read heads of magnetic hard disk recording systems, where it has been one of the main factors enabling the tremendous

---

* *Note:* Combining ferromagnetic and semiconductor materials is a challenging route to create new options for electronic devices in which the spin of the electron is employed. The spin-valve transistor (SVT) is the first of such hybrid devices shown to work successfully. This chapter describes the basic science and technology of the SVT and derived devices, such as the magnetic tunnel transistor (MTT). It is an updated version of a previously published topical review.[1]

increase in storage density over the past decade. The second important breakthrough[6] was the demonstration in 1995 of large tunnel magnetoresistance (TMR) in tunnel junctions in which two ferromagnetic electrodes are separated by an ultrathin insulator. Key features[7] are the large spin dependence of the tunneling process, even at room temperature, and the reproducible fabrication of reliable tunnel barriers typically using 1 to 2 nanometers of Al_2O_3 or MgO. These features facilitated the development of a magnetic random access memory (MRAM), which is nonvolatile.[8-10] With low power consumption and fast switching speed it has the potential of becoming a universal memory. The first introduction of low-density tunneling-based MRAM into the market has begun, and TMR has also been introduced in read-head sensors of magnetic hard disks.

Although GMR and TMR demonstrate the opportunities of using electron spin, it is still an open question how pervasive spintronics technology will be and what classes of material it will involve. In GMR and TMR structures, ferromagnets are combined with thin layers of normal metals and insulators, respectively. Hence, in these structures the most essential materials in contemporary electronics, namely, semiconductors, are still absent. The core element of electronics is the transistor with its amplification, which relies on extremely pure semiconductor materials in which electrical conduction can be controlled by manipulating the carrier density using electric fields supplied by a gate. Semiconductors allow precise tuning of carrier concentrations, band-gap engineering, and, interestingly, also exhibit extremely long electron spin lifetimes.[11,12] Spin electronics (or spintronics) based on semiconductors is therefore an active and promising research field,[4] with several approaches being explored in parallel. These are briefly reviewed below.

Spin transport in (quantum) structures built exclusively from nonmagnetic or paramagnetic semiconductors is explored actively, but a concern is that the creation of a spin polarization at present involves optical techniques or large magnetic fields to induce a sizable Zeeman splitting of the spin states. It would be advantageous to use conventional magnetic materials such as Fe, Ni, Co, and their alloys, in which ferromagnetism is robust and Curie temperatures are high. This supports a stable, virtually everlasting memory function, while switching between different magnetization states is extremely rapid, typically at nanosecond or even picosecond timescales. It is thus attractive to examine different ways (Figure 16.1) to combine ferromagnets and semiconductors.

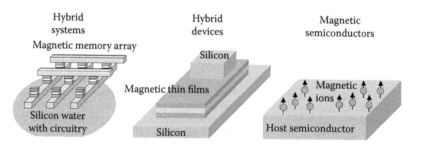

FIGURE 16.1 Schematic representation of the different ways in which semiconductors and ferromagnetic materials can be combined in novel electronics.

The most straightforward approach is the one employed in MRAM, where an array of magnetic memory elements is placed on top of a semiconductor wafer containing transistors and other circuitry required to drive the memory. This requires a modest, yet not trivial integration at the system level, but does not take advantage of the unique properties of semiconductors in manipulating spin. The most intimate form of integration is to put magnetic properties into semiconductor materials, thus creating ferromagnetic semiconductors.[13] Such materials can be obtained by doping with a certain amount of magnetic atoms, as in the case of the notable example of GaMnAs. Interplay between ferromagnetism and carrier densities via electrical gating has been achieved in some of the materials.[14,15] The hunt is now on for compounds that exhibit both semiconducting and ferromagnetic properties at temperatures well above room temperature.

An intermediate option is to create hybrid device structures in which ferromagnets and semiconductors are combined. One could think of many different geometries of such hybrid structures. Two main categories will be distinguished, based on whether the control and manipulation of the spins occurs in the semiconductor or in the ferromagnetic material. In the first category, electron spins that originate from a ferromagnetic source material are injected into a semiconductor, in which they are transported and manipulated, followed by some means of spin detection at the "other end" of the device.[16–18] Much progress has recently been made on the first important step of spin injection into the semiconductor,[19–24] but implementation into working devices that operate at room temperature remains to be demonstrated. The opposite is true for the second class of hybrid devices, where a device concept has been successfully demonstrated by Monsma et al.[25] with the introduction of the SVT in 1995, and the subsequent observation of huge magnetic response at room temperature[1, 26,27] a few years later. Although the spin dependence of the transport is in the ferromagnetic materials, the semiconductors are used to create energy barriers in the electron's potential landscape that are essential to the operation of the device.

This chapter will cover this second category of semiconductor/ferromagnet hybrid structures, including the SVT and derived hot-electron devices such as the MTT. On the one hand, this chapter aims to establish the physical basis that underlies the device operation and to examine the fundamental scattering processes involved in the spin-dependent transport of nonequilibrium electrons in ferromagnets. On the other hand, the device characteristics are discussed providing insight into the strengths and weaknesses. Such discussion is relevant for spin electronics in general, because other spintronic devices, such as those based on spin injection into semiconductors, in the end also have to address issues such as signal levels, noise, room temperature operation, device scaling, and so forth. It is important to realize the magnitude of the challenge this poses, as contemporary semiconductor electronics is a highly mature technology with exceptional performance. Spin electronics will be successful only if it results in greatly enhanced performance, or, in a completely novel electronic functionality. Hence, merely creating a transistor based on a different material (ferromagnet, carbon nanotube, organic crystal, or molecule) is not sufficient, because silicon transistors with power amplification and on-off ratios of more than 10^8 are readily available.

The chapter is organized as follows. In Section 16.2, the state-of-the-art of the SVT is presented. It starts with a description of the device structure, the principle of operation, as well as essential fabrication technology. It then continues to cover two main topics in more detail: (i) the fundamental physics of the spin-dependent electronic transport and (ii) important device characteristics and performance. In Section 16.3, the MTT is discussed. Because the principle of operation and the underlying physics is, to a large extent, the same as for the SVT, the focus is on those aspects of the MTT that are different from the SVT. Section 16.4 provides a brief summary and identifies some promising avenues for further research.

16.2 SVT

16.2.1 Basic Device Structure and Properties

The SVT was introduced in 1995 and is the first working hybrid device in which ferromagnets and semiconductors have been closely integrated, and both materials are essential in controlling the electrical transport through the device. The three-terminal device has the typical emitter/base/collector structure of a (bipolar) transistor, but is different in that the base region is metallic and contains at least two magnetic layers separated by a normal metal spacer (see Figure 16.2). The two magnetic layers act as polarizer and analyzer of electron spins, such that the relative orientation of the magnetization of the two layers determines the transmission of the base. The resulting salient feature of the SVT is that the collector current depends on the magnetic state of the base. This was first demonstrated by Monsma et al.[25] in a SVT showing a huge change of 390% of the collector current in an applied magnetic field at low temperature. In 1998, the first device operating at room temperature was achieved, having a 15% effect in fields of a few kilo-Oersteds.[28] More recently, we succeeded in the reproducible fabrication of SVTs that exhibit magnetocurrent (MC) effects up to 400% at room temperature, and in small magnetic fields of only a few Oersteds.[26,27]

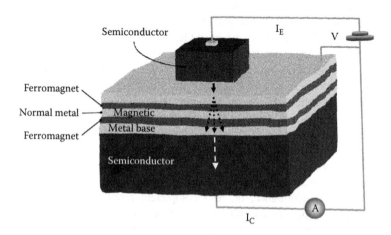

FIGURE 16.2 Basic layout of the SVT, showing the three-terminal arrangement with semiconductor emitter (top), semiconductor collector (bottom), and the metallic base comprising two ferromagnetic thin layers separated by normal metals (middle).

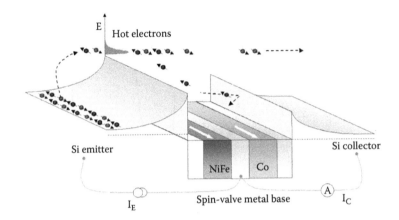

FIGURE 16.3 Schematic layout and energy band diagram of a SVT, showing the semiconductor emitter (left) and collector (right), and the metallic base comprising a spin valve (middle). Also depicted is the stream of electrons that is injected into the spin valve base above the Fermi energy.

Unlike other spintronic devices, the SVT is based on the spin-dependent transport of nonequilibrium, so-called hot electrons, rather than Fermi electrons. To illustrate this and explain the principle of operation, let us consider the specific structure that was used in Reference 26. That particular SVT uses silicon as the semiconductor for the emitter and collector and has a metallic base that contains a $Ni_{80}Fe_{20}$/Au/Co spin valve (see Figure 16.3). At the interfaces between the metal base and the semiconductors, energy barriers (Schottky barriers) are formed.[29] These energy barriers prevent electrons with the Fermi energy from traveling through the structure. To obtain the desired high-quality Schottky barrier with good rectifying behavior and thermionic emission dominating, low-doped Si (1 to 10 Ω cm) is used, and thin layers of, for example, Pt and Au are incorporated at the emitter and collector side, respectively. These also serve to separate the magnetic layers from direct contact with Si.

The operation of the SVT is as follows. A current is established between the emitter and the base (the emitter current I_E), such that electrons are injected into the base, perpendicular to the layers of the spin valve. Because the injected electrons have to go over the Si/Pt Schottky barrier, they enter the base as nonequilibrium, hot electrons. The hot-electron energy is determined by the emitter Schottky barrier height, which is typically between 0.5 and 1 eV, depending on the metal-semiconductor combination.[29] As the hot electrons traverse the base, they are subjected to inelastic and elastic scattering, which changes their energy as well as their momentum distribution. Electrons are only able to enter the collector if they have retained sufficient energy to overcome the energy barrier at the collector side, which is chosen to be somewhat lower than the emitter barrier. Equally important, a hot electron can only enter the collector if its momentum matches with that of one of the available states in the collector semiconductor. The fraction of electrons that is collected, and thus the collector current I_C, depends sensitively on the scattering in the base, which is spin-dependent when the base contains magnetic materials. The total scattering

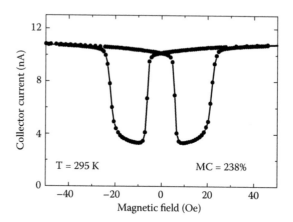

FIGURE 16.4 Collector current vs. applied magnetic field for a SVT with Si(100) emitter, Si(111) collector, and the following base: Pt (20 Å)/NiFe (30 Å)/Au (35 Å)/Co (30 Å)/Au (20 + 20 Å). I_E = 2 mA, V_{BC} = 0, and T = 295 K.

rate is controlled with an external applied magnetic field, which changes the relative magnetic alignment of the ferromagnetic $Ni_{80}Fe_{20}$ and Co layers in the base. This is illustrated in Figure 16.4, where I_C at room temperature is plotted as a function of magnetic field, for a transistor with a Si(100) emitter, a Si(111) collector and the following base: Pt (20 Å)/NiFe (30 Å)/Au (35 Å)/ Co (30 Å)/Au (20 + 20 Å). At large applied fields, the two magnetic layers have their magnetization directions aligned parallel. This gives the largest collector current (I_C^P = 11.2 nA). When the magnetic field is reversed, the difference in switching fields of Co (22 Oe) and NiFe (5 Oe) creates a field region where the NiFe and Co magnetizations are antiparallel. In this state the collector current is drastically reduced (I_C^{AP} = 3.3 nA). The magnetic response of the SVT, called the magneto-current (MC), is defined as the change in collector current normalized to the minimum value, that is,

$$MC = \frac{I_C^P - I_C^{AP}}{I_C^{AP}} \qquad (16.1)$$

where P and AP refer to the parallel and antiparallel magnetic state of the base spin valve, respectively. Thus, the relative magnetic response is MC = 240%, which is huge indeed.

It was shown[26,30] that the collector current and the MC are virtually independent of a reverse bias voltage applied across the collector Schottky barrier, provided that the intrinsic "leakage" current this induces in the collector diode is negligible compared to the hot-electron current. This is indeed the case, as can be seen in Figure 16.5. A voltage between base and collector does not afffect the hot-electron current because it does not significantly change the maximum of the Schottky barrier when measured with respect to the Fermi energy in the metal.[29] In other words, the energy barrier seen by hot electrons coming from the base is hardly changed. Similarly, a change

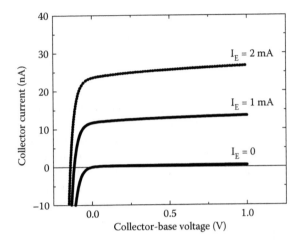

FIGURE 16.5 Transistor characteristics: collector current as a function of collector base voltage, for different values of the emitter current, at $T = 295$ K.

of the emitter to base voltage, or equivalently the emitter current, does not affect the energy at which hot electrons are injected into the base. Also here, the applied voltage hardly modifies the maximum of the emitter potential barrier measured with respect to the Fermi energy in the metal. The result is that the collector current is simply linearly proportional to the emitter current, as shown in Figure 16.6.

The SVT exhibits a huge relative magnetic effect that persists at room temperature, requiring only small magnetic fields of a few Oersteds. The combination of these three features has attracted significant attention. Furthermore, the results are reproducible and the properties of the device can be manipulated by controlling the

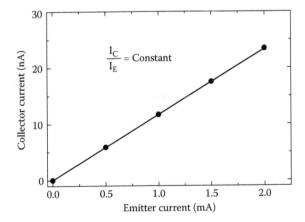

FIGURE 16.6 Collector current as a function of emitter current for the same SVT as in Figure 16.5. $V_{BC} = 0$, and $T = 295$ K.

thickness of layers, the type of materials, and so on. Nevertheless, a particular point that needs improvement is the absolute value of the output current. In the above example, an emitter current of 2 mA was used and the corresponding transfer ratio, defined as $\alpha = I_C / I_E$, is thus of the order of 10^{-6}. In this context it is important to remember that the term "transistor" should not be confused with "amplifier." The word transistor was chosen in analogy with the metal base transistor (MBT),[31-33] a device that is similar to the SVT except that the base contains only nonmagnetic metals. These structures were studied extensively in the 1960s and 1970s, from which it became clear that a metallic base has too little transmission to support amplification.[31-33] Hence, the SVT should not be judged on its (lack of) amplification. Rather, it should be viewed as a device with a magnetic field-dependent electrical output, which is the basic functionality one needs for a magnetic field sensor or a magnetic memory element. Although current gain is thus *not* required for most applications, a small absolute current is a disadvantage. This issue and the progress made so far are addressed in Section 16.2.7.

Before discussing the device in more detail, the essential role of the semiconductors in providing potential energy barriers must be emphasized. The energy barrier at the emitter side is needed to create injection of hot electrons, such that transport is governed by nonequilibrium processes. The collector energy barrier acts as an energy and momentum filter, allowing only a fraction of the hot electrons to pass into the collector and reflecting the rest. The collector Schottky barrier (Au/Si in the above example) selects only on the basis of energy and momentum, but not on the spin of the incoming hot electrons. In some sense, the role of spin is thus indirect as it is merely used to manipulate the energy and momentum distribution of the hot electrons during their motion through the base. This is essential though, as spin couples to an external applied magnetic field and is our handle to the outside world. As will be discussed in detail below, the origin of the large magnetic sensitivity of the SVT is the nonequilibrium nature of the transport that gives an exponential decay of base transmission, together with the strong spin dependence of the elastic and inelastic scattering parameters in ferromagnetic materials.

16.2.2 Fabrication: Vacuum Metal Bonding

For the correct operation of the SVT, it is crucial to have high-quality Schottky contacts that form defect-free energy barriers. This requires device-quality semiconductor material. Because there is no known method of growing crystalline Si with low defect density on top of a thin metal film, an alternative fabrication technology based on metal bonding was developed.[34] The basic idea, illustrated in Figure 16.7, is as amazing as it is essential for the SVT. In an ultrahigh vacuum, we first deposit part of the base metal layers onto the emitter Si wafer, using a shutter to prevent deposition on the collector Si wafer. Then the shutter is opened, metal is deposited onto both wafers simultaneously, during which the surfaces of the metal-coated wafers are brought into contact. A metal-metal bond is formed that bonds the two wafers together. The whole process of metal deposition and bonding is done *in situ* under ultrahigh vacuum conditions in a molecular beam epitaxy system. This yields clean and incredibly strong metallic bonding.

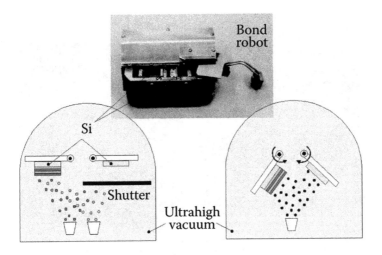

FIGURE 16.7 Concept of vacuum metal bonding, in which two Si wafers are joined together using freshly evaporated metal as the "glue."

As described previously,[34,35] a kind of recrystallization occurs when the two metal surfaces are brought into contact. This can repair structural defects at the bonding interface, provided the two surfaces are not too rough and atomic diffusion is sufficiently large. The driving force for the formation of the chemical bonds is the gain in free energy when two surfaces are combined to form a bulk crystal structure. The quality, strength, and reliability of the metal bond therefore depends on the type of metal used, as well as on the particular film topography (e.g., roughness).

To study the bonding interface, we performed metal bonding in structures that are simpler than the SVT in that they contain only one type of metal, instead of the complete spin valve. The bonding interfaces were then investigated by transmission electron microscopy (TEM). Two examples are shown in Figure 16.8, where two Si substrates are bonded together by two 5 nm thick Co layers (left panel) or two Au layers (right panel). For the Co structure, one can clearly recognize a lighter band that indicates a high density of structural defects at the bonded interface (denoted by the dashed arrow). In contrast, for the case of Au, one recognizes parallel atomic (111) planes running across the complete Au layer and no clear bonding interface is visible. Although we have mostly used Au, good results have also been obtained with other metals such as Cu and Ti.[34,35] For practical reasons related to the mechanical design of the ultrahigh vacuum bond robot, the bonding is typically done at room temperature with 1 cm² pieces of Si. Special care has to be taken to work in dust-free conditions and remove edge particulates after sawing of the wafers. We note that bonding is not limited to Si, but is a versatile technique that allows one to join many kinds of semiconductors. For example, SVTs with GaAs as the emitter and Si as collector have been successfully fabricated using vacuum metal bonding.[36]

After metal bonding, the structures are further processed into smaller devices using standard photolithography and a series of dry and wet etching steps. The first step is to pattern the emitters into squares of typically $350 \times 350\ \mu m^2$ using lithography

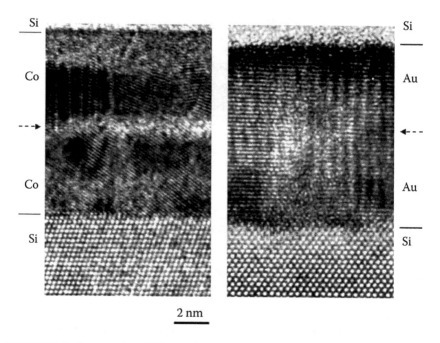

2 nm

FIGURE 16.8 Cross-sectional TEM images of two Si wafers after vacuum metal bonding by Co (left) and Au (right).

and wet etching. The most convenient and reproducible way to do this uses a silicon-on-insulator (SOI) wafer for the emitter. After bonding, the handle wafer of the SOI wafer is first etched away, using the buried oxide as the etch stop. Subsequently, the oxide is etched away, leaving a thin Si layer of homogeneous and well-defined thickness. This layer, 3 μm thick, has a doping profile already built in. The front side is low doped (concentration 10^{15} cm^{-3}, n type) to form a proper Schottky barrier with the metal base. The back side is highly doped such that a good ohmic contact is obtained by deposition of Cr and Au. The next step is patterning of the metal base into slightly larger rectangles using ion milling. A drawback is that this creates damage in the Si collector, which results in large leakage currents along the edges of the metal/collector Schottky contacts. This is detrimental as it reduces the MC by adding only to the total collector current (denominator in Equation 16.1). The edge leakage currents are found to depend strongly on temperature and, as shown previously,[26,30] may prevent a large magnetic response being obtained at room temperature. We therefore introduced an additional process step to remove the damaged collector Si with a last wet chemical etch. Collector Schottky diodes with reverse bias currents of the order of 0.1 nA can thus be obtained.

The final standard chip consists of a total of 52 SVTs with different sizes (see Figure 16.9). Electrical connections to these relatively large SVTs are made by ultrasonic wire bonding. This is however no longer feasible if smaller transistors are to be made. A modified fabrication process was therefore developed for the miniaturization of the device[37,38] for which the use of SOI emitter wafers is essential. Devices with interconnect

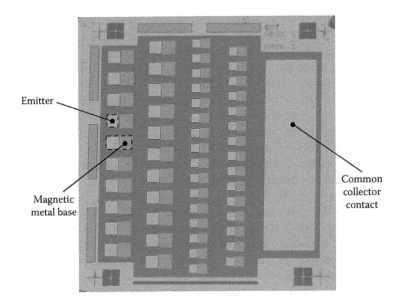

FIGURE 16.9 The final test chip with 52 SVTs of different sizes.

metallization and bond pads have been made so far with sizes down to $10 \times 10 \; \mu m^2$ and with properties comparable to the larger SVTs for which results are presented here.

16.2.3 PHYSICAL BASIS: HOT-ELECTRON SPIN TRANSPORT

Unlike other spintronic devices, the SVT is based on the spin-dependent transport of *hot* electrons, rather than Fermi electrons. Transport at the Fermi energy has been widely studied in connection with (giant) magnetoresistance effects and it is well-established that conduction in ferromagnets and their multilayers is dependent on the spin of the electrons.[39] However, hot-electron transport is distinctly different from ordinary transport at the Fermi energy. Therefore, to understand the operation of the SVT and avoid erroneous interpretation of results, it is best to "forget" about what we know about GMR and spin-dependent scattering in spin valves.

It has been known for quite some time that scattering rates of hot electrons in ferromagnetic materials are spin-dependent. Early work on hot-electron scattering at energies of about 5 eV employed spin-polarized photoemission from overlayer structures.[40,41] This and also later experiments[42,43] at energies as low as 1.5 eV showed unambiguously that the inelastic mean free path of hot electrons is spin-dependent in ferromagnets. More precisely, experiments carried out thus far have always found that the inelastic mean free path of hot electrons is shorter for the minority spin electrons.[44] A common interpretation is that this originates from the difference in the number of unoccupied states for the hot electron to scatter into, assuming electron-hole pair excitations to be the dominant scattering mechanism. Scattering processes have also been investigated using transmission through free-standing magnetic thin film foils,[45–48] as well as by a time-resolved, two photon photoemission (2PPE) experiment[49] that directly probed the inelastic lifetime for majority and minority spin.

Spin-dependent scattering of hot electrons is essential in a variety of spin-polarized electron spectroscopies that are widely used to examine magnetic materials. It leads to spin filtering and in magnetic multilayers it gives rise to phenomena such as the hot-electron spin-valve effect, first observed for secondary electrons.[50] This is employed in the magnetic version of ballistic electron emission microscopy (BEEM),[51] which enables magnetic imaging with nanometer resolution.[52,53] With the introduction of the SVT,[25,28] hot-electron spin transport was implemented in a solid-state electronic device. A thorough understanding of the spin-dependent scattering mechanisms of hot electrons is paramount to the further development of this type of device. Interestingly, the SVT itself has opened up a new route to study spin-dependent scattering processes of hot electrons, extending experiments to lower energy in the range between 0.5 and 1.5 eV. Moreover, for a solid-state device like the SVT it is rather easy to vary experimental parameters such as temperature or apply large magnetic fields. This has led to some new insights into the origin of the spin-dependent scattering processes, as discussed below.

It is now well-established that the difference between I_C^P and I_C^{AP} originates from the spin asymmetry in hot-electron attenuation length, leading to the dominant transmission of majority spins in each magnetic layer. This is schematically illustrated in Figure 16.10. We write

$$I_C^P \propto T_{NiFe}^M T_{Co}^M + T_{NiFe}^m T_{Co}^m \tag{16.2}$$

$$I_C^{AP} \propto T_{NiFe}^M T_{Co}^m + T_{NiFe}^m T_{Co}^M \tag{16.3}$$

where T^M and T^m are the transmission of majority (M) and minority (m) spin hot electrons, given by

$$T_{fm}^M = \Gamma_{in}^M \exp(-t_{fm} / \lambda_{fm}^M)\Gamma_{out}^M \tag{16.4}$$

$$T_{fm}^m = \Gamma_{in}^m \exp(-t_{fm} / \lambda_{fm}^m)\Gamma_{out}^m \tag{16.5}$$

Here, t_{fm} is the film thickness, λ_{fm}^M and λ_{fm}^m the hot-electron attenuation lengths for majority and minority spin, and fm denotes the ferromagnet. The factors Γ_{in}, Γ_{out} denote the transmission across the interface of the ferromagnet with the normal metal at each side, which can be spin-dependent due to the mismatch in the band structure of the magnetic and nonmagnetic metals. Thus, the interfacial attenuation is generally dependent on the combination of ferromagnet and normal metal. Now, an initially unpolarized current acquires, after transmission of a film fm, a spin polarization P given by

$$P_{fm} = \frac{T_{fm}^M - T_{fm}^m}{T_{fm}^M + T_{fm}^m} \tag{16.6}$$

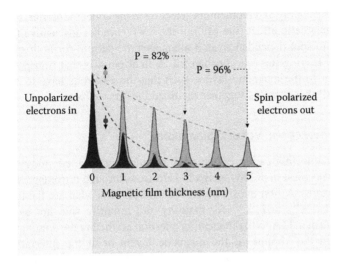

FIGURE 16.10 Schematic view on spin-dependent attenuation of hot electrons in a ferromagnetic thin film, causing preferential transmission of majority spin hot electrons.

This is related directly to the magnetocurrent MC via

$$MC = \frac{I_C^P - I_C^{AP}}{I_C^{AP}} = \frac{2P_{NiFe}P_{Co}}{1 - P_{NiFe}P_{Co}} \qquad (16.7)$$

In general, the attenuation is caused by a combination of inelastic and elastic scattering processes. It is important to realize that while the scattering parameters are uniquely defined for a given ferromagnet and hot-electron energy, the attenuation length is not.[54] The attenuation length depends not only on the scattering parameters, but to some extent also on the device geometry and the energy and momentum selection applied by the collector Schottky barrier. For example, in the *hypothetical* case that a collector is used which passes hot electrons regardless of their momentum, elastic scattering in the base metals would not directly cause attenuation, and the attenuation length would be governed mainly by inelastic processes.[55]

Let us address the origin of the large MC effect and consider the factors that distinguish the SVT from magnetoresistive structures such as a magnetic multilayer. First of all, transport is by hot electrons, which probe a different portion of the band structure than Fermi electrons, while the dominant scattering mechanisms are different and include inelastic scattering such as by electron-hole pair excitations. Second, the resistance of metals is governed by the mean free path, and the resistivity is *inversely* proportional to the mean free path. In contrast, the hot-electron transport in the SVT is a nonequilibrium phenomenon where the current transmission depends *exponentially* on the base thickness, and the parameter that controls the decay is the attenuation length. As stated above, this is related to the mean free path, but also depends on the energy and momentum selection at the collector barrier. Furthermore, the relative magnetic effect (the MC) of the SVT is rather insensitive to scattering that carries no spin dependence, as it attenuates the current for both spins with an equal factor, leaving the ratio unchanged. A final point is that transport is largely perpendicular to the magnetic layers, such that the electrons have to cross all the interfaces. All these factors together facilitate the large effects.

16.2.4 INTERFACE VS. VOLUME SCATTERING

The reproducible fabrication of the SVT permitted a detailed study of the spin-dependent transmission of low-energy hot electrons through ferromagnetic films and trilayer structures. A first question that was addressed is what are the typical attenuation lengths λ_{fm}^M and λ_{fm}^m for majority and minority spin hot electrons in a ferromagnetic metal? Another interesting question is whether the dominant scattering occurs within the volume of the magnetic layers or at the interfaces with the nonmagnetic layers. With respect to spin-dependent scattering, one may expect volume scattering to dominate as its contribution grows exponentially with ferromagnetic film thickness and will quickly outweigh the (constant) interfacial contribution (see Equation 16.4 and Equation 16.5). This is indeed what experiments with the SVT suggest.[56,57] Such experiments involve fabrication of a series of SVTs that are identical except for the thickness of one of the ferromagnetic layers of the base.

Let us discuss the particular example of a transistor with the following base structure Pt (30 Å)/$Ni_{80}Fe_{20}$ (x)/Au (44 Å)/Co (30 Å)/Au (44 Å), where the thickness x of the $Ni_{80}Fe_{20}$ layer was varied. Figure 16.11 shows I_C^P and I_C^{AP} (top panel) and MC (bottom panel) at $T = 100$ K, as a function of the thickness x of the $Ni_{80}Fe_{20}$ layer, keeping the Co layer thickness constant at 30 Å. For the parallel magnetic state of the spin valve, minority spins are strongly attenuated in both magnetic layers, and only majority spins contribute to I_C^P. From the exponential decay of I_C^P with thickness, we thus deduce a volume attenuation length of 43 ± 3 Å for majority hot spins in $Ni_{80}Fe_{20}$.

The interface attenuation for majority spins is extracted by comparing the I_C value of 169 nA for the device with $x = 0$, to the value of 78 nA obtained by extrapolation of the I_C^P data to zero thickness. The latter value is a factor of 2.2 lower, which is due to attenuation by the extra interface that is created when a $Ni_{80}Fe_{20}$ layer is inserted between the Pt and the Au layer of the base (more precisely, the factor of 2.2 represents the difference between the original Pt/Au interface, and the two new Pt/$Ni_{80}Fe_{20}$ and $Ni_{80}Fe_{20}$/Au interfaces). Note that this is the attenuation for the majority spins, as the minority spins are filtered out by the 30 Å Co layer, irrespective of their scattering at the $Ni_{80}Fe_{20}$ interfaces. The interfacial attenuation is a combination of the mismatch of the electronic states at both sides of the interface and elastic scattering due to interface disorder, defects, and so forth.

Information on the spin dependence of the volume and interface attenuation is contained in the difference between I_C^P and I_C^{AP}. Let us first compare the data with what is expected for two extreme cases, using Equation 16.2 to Equation 16.5 for the collector current. In the first case, represented by the dashed lines in Figure 16.11, all the spin dependence is assumed to arise from the interfacial scattering, while identical volume attenuation lengths are used for both spins. In that case, the MC

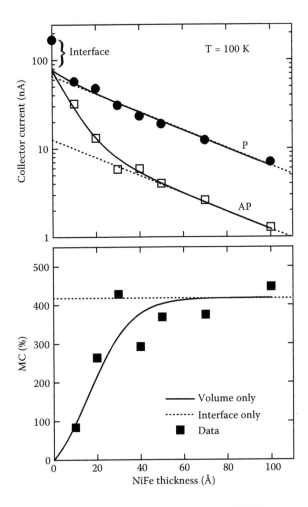

FIGURE 16.11 I_C^P and I_C^{AP} (top panel, labels P and AP) and MC (bottom panel) at 100 K, for SVTs with various $Ni_{80}Fe_{20}$ thickness. Solid and dashed lines, respectively, represent the extreme cases with only spin-dependent volume scattering or only spin-dependent interface scattering (see text).

would be independent of $Ni_{80}Fe_{20}$ thickness, and I_C^P and I_C^{AP} decay with x with the same slope. The experimental data clearly deviates from this behavior at small thicknesses. In contrast, the data agrees very well with the other extreme case (solid lines), in which the interface scattering is not spin-dependent, and the only spin asymmetry is that of the volume attenuation lengths. In this situation, we expect that I_C^P and I_C^{AP} approach each other as x goes to zero, such that the MC tends to zero. Note that for large x, filtering of minority spins is complete and the decay of both I_C^P and I_C^{AP} is determined by the majority spin attenuation length.

More precise analysis, taking into account the experimental error margins, shows that some weak spin dependence of the interface scattering cannot be excluded.[56]

For the ratio of the interface attenuation factor Γ for minority and majority spin, we obtain $\Gamma^{\downarrow} / \Gamma^{\uparrow} = 0.8 \pm 0.2$, where volume attenuation lengths are $\lambda_{NiFe}^{\downarrow} = 10 \pm 2$ Å for minority spin and $\lambda_{NiFe}^{\uparrow} = 43 \pm 3$ Å for the majority spin (at 0.9 eV). With these parameters, a 30 Å thick $Ni_{80}Fe_{20}$ film attenuates minority spins a factor of 10 more strongly than the majority spins. Volume scattering is thus by far the dominant contribution to the spin dependence of the hot-electron attenuation, mainly because it depends exponentially on the layer thickness.

16.2.5 ELASTIC VS. INELASTIC

In hot-electron scattering, it is obvious that inelastic scattering plays a role, as it leads to energy loss and thus makes the electrons "less hot." However, elastic scattering cannot be neglected *a priori*. The point is illustrated best by asking what we would expect if there were no elastic scattering whatsoever, neither in the volume of the layers nor at interfaces. Can we explain the basic electrical characteristics of the SVT using only inelastic scattering? A simple estimate based on exponential decay of the transmission of a particular layer is illuminating. It is well known from MBT$_s$[31–33] and BEEM[58,59] that scattering lengths for nonmagnetic materials such as Au or Cu are quite long, at least 100 Å. In the SVT, the thickness of these normal metal layers is typically only 30 to 60 Å such that inelastic attenuation in these layers can be neglected. To satisfy the experimental observations (a typical MC of 400% at low T and an overall transmission of 10^{-4} for SVTs with ferromagnetic layers of 30 Å, see Section 16.2.7), one would require inelastic attenuation lengths of 7 and 4.5 Å for majority and minority spins, respectively.

Such short inelastic scattering lengths appear unrealistic. Moreover, they are at odds with a direct measurement[49] of the inelastic lifetime of hot electrons by 2PPE, which yielded a lifetime of 8 fs and 4 fs for majority and minority spin hot electrons, respectively, in Co at an energy of about 1 eV. With a typical[60] electron velocity of 6×10^{5} m/s for 1 eV majority spins, this translates into an inelastic scattering length of about 50 Å in Co. This is much larger than the value quoted above. For minority spins, a significant density of d states is still present at energies of 1 eV above the Fermi energy, and one could argue that the electron velocity is much smaller. It is possible that there is a principle reason why inelastic lifetimes determined by the optical technique of 2PPE are much larger than those observed in electronic transport. This may be due to diffusion of electrons away from the optical spot,[61] due to matrix elements for optical transitions, or due to a different energy resolution of the techniques.

Attenuation lengths for several ferromagnetic materials in the SVT,[56] the MTT,[62] and in magnetic BEEM[53] all consistently give minority spin attenuation lengths around 10 Å and majority scattering lengths in the range of 25 to 60 Å. These measured values are significantly larger than values of 4.5 and 7 Å estimated above for the case in which only inelastic scattering is present, but smaller than what is derived from the 2PPE results. A consistent picture can only be made if elastic scattering is included. First, transport attenuation lengths that are determined by elastic and inelastic processes are naturally shorter than predicted from 2PPE, which probes only the inelastic scattering lifetime. Second, when *measured* attenuation

TABLE 16.1
Transfer Ratio α for Several Si/Pt/Au/Si Transistors with Different Crystal Orientations of the Collector. The Schottky Barrier Height Φ and Ideality Factor *n* Are also Given

Emitter	Φ_e (eV)	n	Base (Å)	Collector	Φ_c (eV)	n	α 10⁻⁴
Si(100)	0.91	1.01	40 Pt/40 Au	Si(111)	0.80	1.20	8.5
SOI(100)	0.85	1.07	40 Pt/40 Au	Si(111)	0.76	1.20	8.0
SOI(100)	0.88	1.02	40 Pt/40 Au	Si(100)	0.77	1.15	8.2

lengths are used to calculate the total volume attenuation of the base layers, one arrives at a transmission of about 0.01. Because observed base transfer ratios are at least two orders of magnitude smaller, a significant fraction of the attenuation must be attributed to scattering at interfaces, which is primarily elastic.

Finally, we mention that attributing the measured transport attenuation length solely to inelastic scattering is inconsistent with the general trend observed in SVT,[63] MTT,[62] and BEEM data[53] that the overall transmission *increases* with increasing hot-electron energy (see also Section 16.3 on the MTT). For inelastic scattering, the opposite should happen, because according to Fermi's golden rule the lifetime is reduced when at higher energy more final states become available to scatter into. This is a most compelling argument and we conclude that the experimental data cannot be explained if only inelastic processes are considered. The picture of purely ballistic transmission thus seems rather inappropriate. Experimental evidence for the importance of elastic scattering comes from the explicit observation of attenuation due to interfaces (Figure 16.11) and from the direct observation of spin-orbit scattering.[64] In addition, we have found that the magnitude of the collector current is insensitive to crystal orientation (100) or (111) of the Si collector, see data in Table 16.1.

An instructive way to examine the role of elastic scattering is to make use of phenomenological model calculations, in which scattering parameters can be systematically varied and the effect on magnetotransport can be studied. The model and the calculation procedure are described in detail in Reference 54. Here we merely discuss an example that illustrates the role of *interfacial* elastic scattering. For the case in which elastic interface scattering was set to zero, the top panel of Figure 16.12 shows how the calculated collectible current decays as the hot electrons traverse the metal base. Here, "collectible" refers to those electrons that still satisfy the energy and momentum constraints required for transmission into the Si collector.[54] We observe an exponential attenuation of the current in the volume of each layer, with the strongest decay in each of the ferromagnetic layers. The bottom panel of Figure 16.12 shows the results if the interface diffusivity is set to 0.9 for each interface in the base (including the metal semiconductor interfaces at the emitter and collector side). This corresponds to 90% of the electrons incident on an interface being scattered elastically, with scattered electrons distributed isotropically over all angular directions. One notes a pronounced drop in the calculated collectible current at each interface. This adds up to a large reduction of the overall base transmission.

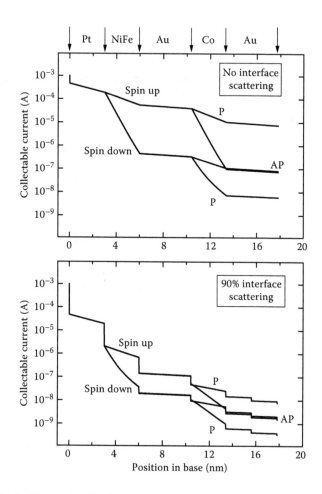

FIGURE 16.12 Calculated collectible current vs. position in the metal base, for zero and 90% interfacial scattering in the top and bottom panel, respectively.

For the emitter current of 2 mA used, the total transmitted current I_C^P is only 8×10^{-9} A for the strong interface scattering case, as opposed to 8×10^{-6} A without interface scattering. The difference is as much as three orders of magnitude. Thus, elastic scattering at interfaces, for example due to disorder, can cause a significant reduction of the collector current. Elastic scattering therefore plays a perhaps surprisingly important role in hot-electron transport.

16.2.6 TEMPERATURE AND SPIN WAVES

For ferromagnets, experiments carried out thus far have always found that the inelastic mean free path of hot electrons is shorter for the minority spin electrons.[44] A common interpretation is that this originates from the difference in the number of unoccupied states for the hot electron to scatter into, assuming excitation of electron-hole pairs to be the dominant scattering mechanism. However, recent experimental[65] and theoretical work[66–68] suggests that scattering by spin-wave excitations may also contribute to the spin dependence of the mean free path. The spin asymmetry is created by spontaneous spin-wave emission, which due to conservation of angular momentum is allowed only for minority spins. Spontaneous spin-wave emission is not dependent on temperature (T) occuring even at $T = 0$. It is calculated[66–68] to dominate at electron energies below ≈ 1 eV and is found to be the major source of small energy losses.[65] Besides spontaneous emission, there is also a thermal component that involves spin-wave absorption as well as emission. Information on these contributions is obtained most directly by varying T. Even though spectroscopic measurements have been performed, the T dependence had so far not been explored. Using the SVT, we have obtained the first measurements of spin-dependent hot-electron scattering as a function of temperature.[27] We found that thermal spin waves have a noticeable effect on spin-transport of hot electrons, in particular by introducing spin mixing.

Figure 16.13 shows the typical variation of the collector current with applied magnetic field at two temperatures (80 and 290 K). At high fields the magnetizations

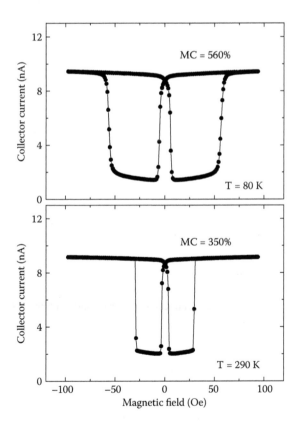

FIGURE 16.13 Collector current vs. magnetic field at $T = 80$ K (top) and $T = 290$ K (bottom). The SVT structure is n-type Si(100)/Pt (20 Å)/NiFe (60 Å)/Au (35 Å)/Co (30 Å)/Au (20 + 20 Å) /n-type Si(111). $I_E = 2$ mA and $V_{BC} = 0$.

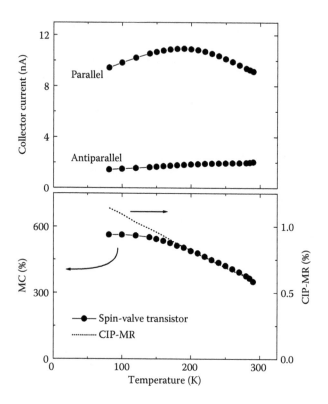

FIGURE 16.14 Temperature variation of I_C^P and I_C^{AP} (top panel) and the MC (bottom panel) for the same SVT as in Figure 16.13.

of the Co and $Ni_{80}Fe_{20}$ films in the spin-valve base are parallel. When one of the magnetizations is reversed, a clear antiparallel (AP) magnetization alignment is obtained. At $T = 80$ K, $MC = 560\%$, while a huge effect of 350% still remains at room temperature. Similar results were reproducibly obtained.

In Figure 16.14, the temperature variation of I_C^P and I_C^{AP} (top panel) and MC (bottom panel) are displayed. For the parallel case, the collector current first increases with T, but above 200 K it decreases. In contrast, I_C^{AP} increases over the whole temperature range. The resulting decay of the MC is relatively weak, especially below 200 K. The conventional current-in-plane magnetoresistance (CIP-MR) measured on an identical spin valve is also included. The CIP-MR is not only orders of magnitude smaller, but also exhibits a stronger thermal decrease that is linear, as was also found by others.[69]

To understand the data in Figure 16.14, we consider that thermal scattering may change the MC in two ways. First, scattering may attenuate, that is, remove electrons from the collected current in a spin-dependent fashion. The second possibility is spin mixing, in which electrons are scattered into the other spin channel by a spin-flip process, without being removed from the collected current. We have shown[27] that when we neglect possible thermally induced spin-mixing, the MC does not decay with T. Thus, the decay of MC is attributed to thermally induced spin-flip scattering, causing mixing of the two spin channels. This is consistent with the observation (Figure 16.2) that the current for the parallel state goes down above 200 K, while the current for the antiparallel case continues to go up. Becuase exchange scattering by paramagnetic impurities and spin-orbit scattering have negligible T dependence in our experimental range,[70] we consider spin-flip scattering by thermal spin waves.

We stress that spin mixing only results for those scattering events in which the energy or momentum transferred to the spin wave is such that after scattering, the hot electron is still able to enter the collector (recall that the electron needs a minimum energy to surmount the collector barrier and has to be incident at an angle smaller than the critical angle of acceptance[28,51]). We therefore consider the Bose–Einstein distribution function $N_q = 1/[\exp(Dq^2/kT) - 1]$, which determines the typical wave vector $\mathbf{q}$ and energy Dq^2 of thermal spin waves (D, the spin-wave stiffness, is typically 400 meV Å^2). We see that even at 300 K, virtually all thermal spin waves have low energies, which are smaller than the maximum allowed energy loss ($\approx$ 60 meV) given by the difference between emitter and collector barrier. Thus, the change of hot-electron energy will generally not remove the electron from the collected current. The situation is different when one considers the wave vector $\mathbf{q}$, which has magnitudes up to roughly 0.3 Å^{-1} at room temperature. For the largest $\mathbf{q}$, spin-wave scattering may deflect the hot electron enough to prevent collection, thus causing attenuation. However, for small $\mathbf{q}$ or when the $\mathbf{q}$ component perpendicular to the hot-electron direction of motion is small, the deflection is weak. In that case, the process contributes to spin mixing.

In a subsequent experimental study,[56] we have measured the actual attenuation length associated with thermal scattering by spin waves. Using Matthiessens's rule, we can write λ_{fm}^M and λ_{fm}^m in terms of the attenuation lengths for all the distinctive scattering processes

$$\frac{1}{\lambda_{fm}^M} = \frac{1}{\lambda_{e-h}^M} + \frac{1}{\lambda_{el}^M} + \frac{1}{\lambda_{ph}} + \frac{1}{\lambda_{TSW\,abs}^M} \tag{16.8}$$

$$\frac{1}{\lambda_{fm}^m} = \frac{1}{\lambda_{e-h}^m} + \frac{1}{\lambda_{el}^m} + \frac{1}{\lambda_{ph}} + \frac{1}{\lambda_{TS\,Wemis}^m} + \frac{1}{\lambda_{SS\,Wemis}^m} \tag{16.9}$$

where λ_{e-h}, λ_{el}, and λ_{ph} are the attenuation lengths associated with electron-hole pair excitations, elastic scattering, and phonon scattering, respectively. Also included are attenuation lengths λ_{TSWabs}^M and $\lambda_{TSWemis}^m$ due to absorption and emission of *thermal* spin waves, respectively, as well as a term $\lambda_{SSWemis}^m$ due to *spontaneous* emission of spin waves. Due to the conservation of angular momentum, only majority spin hot electrons can absorb spin waves, whereas (spontaneous and thermal) emission is allowed only for minority spins. Thus, the overall rate of spin-wave scattering has a spin asymmetry due to $\lambda_{SSWemis}^m$.

To extract the thermal component of the hot-electron attenuation, we have measured I_C^P as a function of temperature.[56] We divide out all temperature-independent attenuation factors by plotting the normalized collector current $I_C^N = I_C^P(T)/I_C^P(100K)$

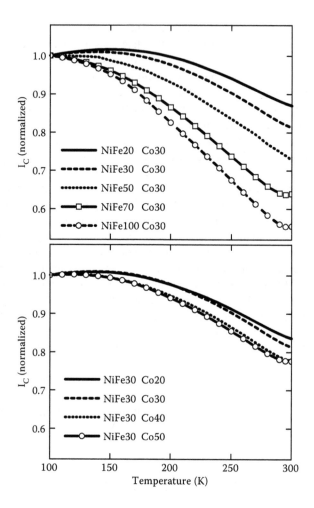

FIGURE 16.15 Top panel: normalized I_C^P vs. T for SVTs with a $Ni_{80}Fe_{20}$/Au/Co spin valve with different $Ni_{80}Fe_{20}$ thickness and a constant Co thickness of 30 Å. Labels indicate the thicknesses in Å. Bottom panel: similar for varying Co thickness and a constant $Ni_{80}Fe_{20}$ thickness of 30 Å.

(see Figure 16.15). The top panel shows data for transistors with different $Ni_{80}Fe_{20}$ thickness and constant Co thickness. Starting at 100 K, the collector current first goes up slightly as seen before, and then is reduced significantly toward room temperature. Moreover, the variation with T is more pronounced at higher $Ni_{80}Fe_{20}$ thickness, implying that the additional attenuation at higher T is due to a thermal volume scattering process. The same behavior is observed when the thickness of the Co layer is varied (bottom panel of Figure 16.15). However, the T variation for Co has weaker dependence on layer thickness than for $Ni_{80}Fe_{20}$, which shows that the thermal attenuation is stronger in $Ni_{80}Fe_{20}$ than it is in Co. This is consistent with attenuation due to scattering on thermal spin waves,[27] as the Curie temperature of Co (1388 K) is larger than that of $Ni_{80}Fe_{20}$ (873 K).

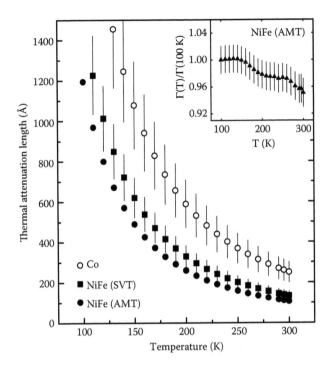

FIGURE 16.16 Thermal spin-wave attenuation lengths vs. T for Co and $Ni_{80}Fe_{20}$ (two data sets). The inset shows the thermal variation of the interface attenuation for $Ni_{80}Fe_{20}$.

Using the procedure described in Reference 56, one can extract the attenuation length $\lambda_{TSW}(T)$ due to thermal spin waves. The extracted thermal attenuation lengths for $Ni_{80}Fe_{20}$ and Co vs. T are shown in Figure 16.16. For $Ni_{80}Fe_{20}$, two sets of data are shown, one for the SVT, the other labeled AMT is extracted from a set of transistors with Pt (20 Å)/$Ni_{80}Fe_{20}$/Au (44 Å) base. As noted before, $\lambda_{TSW}(T)$ is shorter for $Ni_{80}Fe_{20}$ than for Co. Interestingly, $\lambda_{TSW}(T)$ is much shorter than hitherto assumed, with room temperature values of 130 ± 20 Å for $Ni_{80}Fe_{20}$ and 270 ± 40 Å for Co. Especially for $Ni_{80}Fe_{20}$, this is only three times larger than the majority spin attenuation length at low T. Hence, this shows convincingly that hot-electron attenuation lengths are dependent on T. For instance, the addition of thermal spin-wave scattering with a length scale of 130 Å reduces the attenuation length from 43 Å at 100 K, to a significantly lower value of $(1/43 + 1/130)^{-1} = 32$ Å at room temperature.

The inset of Figure 16.16 shows the thermal variation of the interface attenuation $\Gamma(T)$ for $Ni_{80}Fe_{20}$. Data is obtained by dividing the I_C^N curve for certain $Ni_{80}Fe_{20}$ thickness by the curve for $x = 0$ and using the attenuation lengths of Figure 16.16 to remove the volume scattering part. We find only a slight change of about 5% in the interface attenuation between 100 and 300 K. This shows that thermal spin-wave scattering is primarily a volume scattering process.

The above results suggest the importance of spin waves for the spin asymmetry of hot-electron transmission. As shown in Reference 66 and Reference 67, the asymmetry is created by the T-independent contribution of *spontaneous* spin-wave emission, which is only allowed for minority spins. It has so far not been possible to isolate spontaneous spin-wave emission and measure the corresponding scattering length. However, the above quantification of the thermal spin-wave contribution allows us to estimate the attenuation due to spontaneous emission, which is expected to be significantly stronger than the thermal scattering rate. This is because thermal spin waves occupy only a small fraction of the spin-wave phase space up to energies of the order of kilo-Tesla, while for spontaneous spin-wave emission, the complete phase space up to the hot-electron energy (≈ 0.9eV $>>$ kT) is available at all temperatures. The strength of the observed attenuation due to thermal spin waves is thus indirect evidence for the importance of *spontaneous* spin-wave emission. If we crudely estimate the spontaneous emission rate to be about an order of magnitude larger than the thermal emission rate, we obtain an attenuation length for spontaneous emission that is close to the measured minority spin attenuation length (10 Å for $Ni_{80}Fe_{20}$). This strongly suggests that the minority spin attenuation is dominated by spontaneous spin-wave emission, as theory predicts.[66–68] Because the process cannot contribute to attenuation of majority hot spins, we conclude that the spin asymmetry of the attenuation length may well be due to spontaneous spin-wave emission, instead of the spin-dependent rate of electron-hole pair generation that arises from the exchange split band structure.

The notion of spontaneous spin-wave emission as the dominant source of spin asymmetry explains the absence of band-structure features in spectroscopic data. In particular, it explains the puzzling observation[61] by time-resolved 2PPE that the inelastic lifetime for minority spins is shorter than the majority spin lifetime, not only in Co and Ni, but also in Fe. For Fe, a reversed spin asymmetry at low energy (< 1 eV) is expected from the band structure.[61] However, spontaneous spin-wave emission would, as it is forbidden for majority spins, always result in a larger lifetime for majority spins, as observed in the 2PPE experiment. In this context, it is of interest to note that from experiments on spin filtering of nonequilibrium holes,[71] it was recently suggested that there may be another source of spin asymmetry, namely, a spin-dependent group velocity. Subsequent calculations[72] for hot-electron mean free paths in ferromagnets have confirmed that this can indeed play a role, depending on the material.

16.2.7 Device Output and Transfer Ratio

In close harmony with the more fundamental research described in the preceeding sections, quite some effort has been devoted to improve one of the most important device parameters, namely, the output current level.[63] From the basic relation $I_C = \alpha I_E$, we see that we can increase I_C by either increasing the emitter current (I_E), or the transfer ratio ($\alpha = I_C/I_E$) of the transistor base. However, we know that the emitter current has an upper limit imposed by the breakdown of the device.[38] Therefore, we should optimize α. Parenthetically, the most fruitful way to achieve this is to change the nonmagnetic layers of the base. In a typical SVT structure of $Si(100)/NM_E/Ni_{82}Fe_{18}/Au/Co/NM_C/Si(100)$, thin normal metal (NM) layers NM_E and NM_C are incorporated at the emitter and collector side of the structure, respectively.

TABLE 16.2
Properties of SVTs with Different Combinations
of Emitter and Collector Barriers, where SV Denotes
a NiFe/Au/Co Spin Valve for T = 290 K

Base	Φ_e (eV)	Φ_c (eV)	$\Delta\Phi_B$ (eV)	α ($\times 10^{-4}$)	MC (%)
Pt/SV/Pt	0.88	0.86	0.02	0.01	213
Pt/SV/Au	0.88	0.83	0.05	0.07	260
Pt/SV/Cu	0.87	0.61	0.26	1.06	218
Au/SV/Au	0.81	0.80	0.01	0.10	204
Au/SV/Cu	0.82	0.69	0.13	1.18	230

The primary function of these layers is to create high-quality Schottky barrier diodes with thermionic emission dominating to ensure efficient hot-electron injection from the emitter and low reverse bias leakage of the collector diode. From a magnetic point of view, these layers are not active, providing some flexibility in device design. Below we show that this can be exploited[63] to enhance the transfer ratio in two ways: (i) controlling the difference in height of the emitter and collector Schottky barrier, and (ii) by carefully selecting materials with large scattering lengths for hot electrons.

We start by investigating the role of the Schottky barrier height of emitter (Φ_e) and collector (Φ_c) on the transfer ratio. The Schottky barrier heights in the SVT are determined by the choice of semiconductor, silicon in our case, and the nonmagnetic metal. We have systematically varied the Schottky barrier metal in two series of SVTs.[63] In the first series, denoted as Pt, series, NM_E is Pt, and NM_C is either Pt, Au, or Cu. In the second series, denoted as Au series, NM_E is Au, and NM_C is either Au, or Cu. The properties of the two series of transistors are summarized in Table 16.2.

In Figure 16.17, we show the parallel collector current vs. Schottky barrier height difference ($\Delta\Phi_B = \Phi_e - \Phi_c$) of the two series. Both the Pt and Au series show an increase in collector current for larger $\Delta\Phi_B$. The increase is not related to a change in hot-electron energy, as in each series the emitter Schottky barrier is fixed, and only the collector barrier is lowered as one goes from Pt to Au to Cu. The same studies have been performed on MBTs in which the spin valve has been omitted from the base. The structure is thus $Si(100)/NM_E/NM_C/Si(100)$, where NM_E and NM_C perform the same function and are made of the same materials as used for the SVTs in Table 16.1. The results (not shown) yield the same trend for the transfer ratio vs. $\Delta\Phi_B$ and material as for the SVT series.

The enhancement of I_C is due to the larger number of states available in the collector semiconductor. Electrons that are injected at an energy just above the collector barrier (small $\Delta\Phi_B$) arrive at the base/collector interface at energies just above the conduction band minimum of the collector Si. Near the bottom of the band, the density of states is small and only few electrons can enter the semiconductor. When the collector barrier is lowered or the emitter barrier is increased (large $\Delta\Phi_B$), the injected electrons can be transmitted to states higher up in the

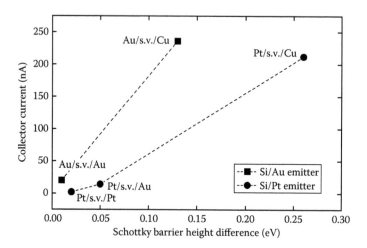

FIGURE 16.17 I_C^P for five types of SVTs vs. $\Delta\Phi_b$. The circles are the Pt series and the triangles are the Au series. I_E = 2 mA, T = 290 K and SV denotes a NiFe/Au/Co spin valve.

collector conduction band where the available density of states is larger. Thus, a larger fraction of the electrons can be collected. This is also seen in BEEM experiments,[51] as well as in transistors with a tunnel barrier emitter (as described in Section 16.3).

Although the transfer ratio improved by more than a factor of 100 from the Pt/SV/Pt base to the Au/SV/Cu base, the variation in MC is small and nonsystematic (see Table 16.1). This implies that the relatively small changes in the energy of the injected electrons have no large impact on the spin-dependent attenuation processes. Also the change of the collector barrier height enhances the collection of both the spin-up and spin-down electrons equally. We can thus enhance the transfer ratio and at the same time preserve the huge magnetic response. The maximum transfer ratio obtained so far was 1.2×10^{-4} for the Si/Au (20 Å)/NiFe (30 Å)/Au (70 Å)/Co (30 Å)/Cu (40 Å)/Si transistor. The corresponding MC at room temperature was 230% (See Figure 16.18) and nearly 400% at 100 K.

Besides the increase in transfer ratio with $\Delta\Phi_B$, we also observe that the transfer ratio is larger for the Au series than for the Pt series. The difference can be explained by the much shorter hot-electron attenuation length of Pt as compared to Au. BEEM studies yield an attenuation length (λ) of about 200 Å for Au.[58,59] Although no data is available for metallic Pt, a λ of about 40 Å is obtained in a BEEM study on PtSi.[73] Because the transfer ratio depends exponentially on the attenuation length, one expects a larger transfer ratio for the Au series than for the Pt series at comparable $\Delta\Phi_B$. Thus, nonmagnetic materials with large hot-electron scattering lengths, such as Au and also Cu, are preferred. Indeed, SVTs with Cu as the spacer between the two ferromagnetic layers show a larger transfer ratio.[74] Materials such as Pt should be avoided if possible, despite the fact that it produces the highest possible Schottky barrier on Si (which was the original motivation for its use in early devices).

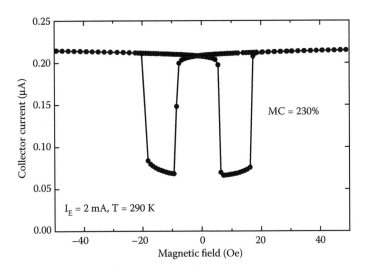

FIGURE 16.18 Collector current vs. applied magnetic field for a Si/Au (20 Å)/NiFe(30 Å)/ Au (70 Å)/Co (30 Å)/Cu (40 Å)/Si transistor. I_E = 2 mA and T = 290 K.

Note that by changing the Schottky barrier materials NM_E and NM_C, one may introduce changes in the structural properties of the SVT. For example the quality of the metal bond in NM_C may depend on the material used. With respect to layer NM_E, we have specifically examined its influence on the transfer ratio.[63] For that purpose, a series of transistors with a Pt/NiFe (30 Å)/Au (40 Å)/Co (30 Å)/Au (40 Å) base layer were prepared, with the Pt thickness varied from 20 to 60 Å. One would expect an exponential decay with layer thickness. Surprisingly, however, an initial *increase* in transfer ratio with Pt layer thickness was reproducibly obtained (see Figure 16.19). This is attributed to a structural change of the SVT with Pt thickness, although the precise cause has not yet been identified.

The above example nicely illustrates the effect of the structural quality of the SVT on the transfer ratio, and thereby points to a promising route for further improvement. As shown in Section 16.2.5, a significant part of the hot-electron attenuation is due to elastic scattering processes that likely originate from structural imperfections (defects, grain boundaries, stacking faults, disorder at interfaces, etc.). These are surely present in the polycrystalline, nonepitaxial structures we have studied so far. Going to epitaxial transistor structures is an approach we are currently exploring. Another option is to reduce the number of layers (and interfaces) in the base and with it the elastic scattering. This can be done in the slightly modified design of a MTT (see Section 16.3 on the MTT). Finally, we note that some optimization of the thickness of the ferromagnetic layers is needed, because there is a trade-off between transfer ratio and MC that yields a maximum in absolute change of collector current (ΔI_C) with magnetic field.[28,63]

In the preceding paragraphs we have focused on how to improve the transfer ratio, keeping the emitter current fixed at a somewhat arbitrary value of 2 mA. However, this is far below the breakdown current of the SVT.[38] Thus, the absolute

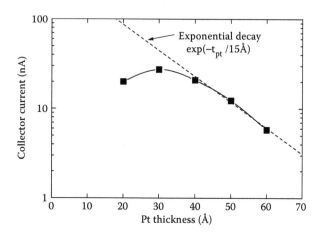

FIGURE 16.19 Parallel collector current vs. Pt thickness t_{pt} for a Si/Pt/NiFe(30 Å)/Au (40 Å)/Co(30 Å)/Au (40 Å)/Si transistor. $I_E = 2$ mA and T = 100 K.

value of the output collector current is by no means limited to the value of 0.2 μA shown in Figure 16.18. To illustrate this, Figure 16.20 shows how the collector current, transfer ratio, and MC vary with emitter current, for a SVT with similar structure as in Figure 16.18. We find that the output current of the SVT increases approximately linearly with emitter current, and large collector currents up to 50 μA are obtained.[75] Interestingly, the increase of output current by several orders of magnitude is accompanied by a rather weak reduction of the MC and even a slight increase of the transfer ratio. This convincingly demonstrates that the collector current is not intrinsically limited to small values in the nanoampere regime, but can approach 0.1 mA. In the end, the output current that can be achieved depends on the maximum emitter current that can be tolerated for a given application. This is determined by extrinsic requirements such as power consumption, device dimensions because the breakdown emitter current depends on device size,[38] and input impedance (for the maximum emitter current in Figure 16.20, the applied emitter voltage was about 4 V, corresponding to an input impedance of 16 Ω). This also implies that care has to be taken when comparing the properties of the SVT with similar devices such as the MTT, and correct conclusions can only be drawn if differences in the applied emitter current are properly considered.[76]

16.2.8 NOISE

An important parameter for any electronic device is the signal to noise ratio (SNR). Let us first examine the origin of the noise in the SVT. To characterize the noise behavior over a wide range of collector current values, measurements have been performed on three different types of transistors.[77] Besides the SVT with Pt/NiFe/Au/Co/Au spin-valve base, we also used a Pt/NiFe/Au MBT (with only a single magnetic layer), as well as the nonmagnetic MBT with a Pt/Au base.

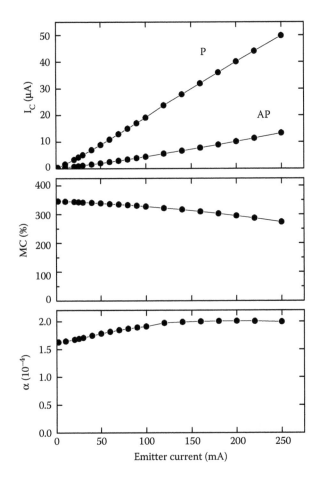

FIGURE 16.20 The dependence on emitter current of collector current (top panel), MC (middle), and transfer ratio α (bottom) for a Si/Au (20 Å)/NiFe (30 Å)/Au(70 Å)/Co (30 Å) /Cu(40 Å)/Si transistor.

The collector current noise spectrum showed only white noise in the frequency range of the measurement (10 Hz to 1 kHz). The noise current spectral density is plotted as function of I_C in the top panel of Figure 16.21, where the I_C for each of the three transistors was varied by changing I_E. The log-log plot yields a straight line with a slope equal to 1, which means a linear relationship between noise spectral density and collector current. This is expected when shot noise dominates. In fact, the experimental data are in excellent agreement with the calculated shot noise of $2qI_C$, as indicated by the solid line.

For the SVT, the relation between noise and the magnetic state of the base was studied. As shown in Figure 16.21, bottom panel, the noise varies with applied magnetic field very much like the collector current itself, and no extra noise appears in the field regions where the magnetization reversal occurs. The change of the noise

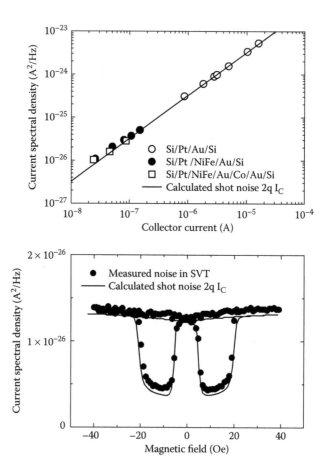

FIGURE 16.21 Top panel: Noise current spectral density vs. I_C for transistors with three different base structures: Pt/NiFe/Au/Co/Au (open squares), Pt/NiFe/Au (filled circles), Pt/Au (open circles). Bottom panel: Noise vs. magnetic field for the SVT. Solid lines in both panels are the shot noise calculated from the measured I_C.

with field is entirely due to the change of the collector current. The solid curve is the shot noise calculated from $2qI_C$, using the measured collector current vs. field curve. Again, good agreement is obtained with the data. We conclude that shot noise dominates in the investigated current and frequency regime.

We can now evaluate the SNR of the device, using as the signal the full difference between I_C^P and I_C^{AP}. The result is shown in Figure 16.22, where SNR is plotted against I_C^P for devices with different values of the MC percentage. It is evident that the SNR improves significantly at higher collector current, emphasizing the need for further increasing the transfer ratio. One also notes that there is a significant gain in SNR between a MC of 10 and 100%; however, beyond 100%, the gain levels off and only a marginal SNR increase is observed when the MC is further raised to 1000%. This may seem counterintuitive, but is simply because the absolute value of the

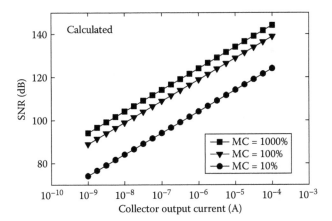

FIGURE 16.22 Calculated SNR in a 1 Hz bandwidth based on shot noise, as a function of collector current and for different values of the magnetic sensitivity.

signal $I_C^P - I_C^{AP}$ is already close to the maximum at a MC of several 100%. It is for this reason that we have never attempted to create a SVT with MC values above 1000%, although it is straightforward to do so, one simply increases the thickness of the ferromagnetic layers (to 50 Å or more). In fact, the net result would be a loss in SNR, because the collector current of such a structure would drop significantly.

16.3 MTT

The MTT is similar to the SVT and differs only in the structure of the emitter used to inject the hot electrons into the transistor base. Whereas the SVT uses a Schottky barrier, the MTT has a tunnel barrier as the emitter. In the first publication on the SVT in 1995 by Monsma et al.,[25] this alternative device geometry using a tunnel emitter was already proposed. In 1997, Mizushima et al. were the first to actually fabricate a MTT,[78] and they own the patent.[79]

The MTT device exists in two slightly different variations, as shown in Figure 16.23. The design on the left has a thin insulating layer to separate the metal base from a *nonmagnetic* metal emitter electrode. When a voltage V_{EB} is applied across the tunnel barrier between emitter and base, unpolarized electrons are injected by tunneling into the metal base, arriving at an energy eV_{EB} above the Fermi level in the metal base. The rest is the same as in the SVT: spin-dependent transmission of the hot electrons through the two ferromagnetic thin films in the metal base, and collection across a Schottky barrier with energy and momentum selection. The design on the right in Figure 16.23 also has a tunnel barrier, but uses a *ferromagnetic* emitter electrode. Because the tunnel probability is spin-dependent in such a structure, the injected hot electrons are already spin polarized as they enter the transistor base. Therefore, the first, polarizing magnetic layer in the base can now be omitted and only the second analyzer magnetic layer needs to be retained. Again, the collector current is determined by spin-dependent transmission of the hot electrons through

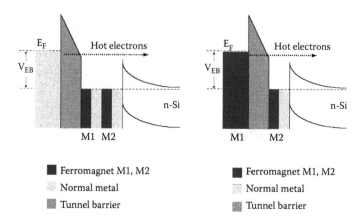

FIGURE 16.23 Energy diagrams of the MTT with a nonmagnetic tunnel injector (left) or a ferromagnetic tunnel injector (right) to generate the hot-electron current.

the ferromagnetic base and collection across a Schottky barrier with energy and momentum selection.

Because only the emitter part is different, we can directly apply much of the knowledge obtained from studies of the SVT described in Section 16.2.3 to Section 16.2.8. The exponential variation of the base transmission; the importance of elastic, inelastic, and thermal scattering processes in the layers and at interfaces; and the conditions for transmission into the collector are essentially the same for MTT and SVT. As an example, hot-electron attenuation lengths measured with the MTT[62] are in the same range as lengths previously found for the SVT[56] and magnetic BEEM.[53] Therefore, we will focus here only on those aspects that are different, referring the reader to the literature (see Reference 62, Reference 78, and Reference 80 to Reference 88) on the MTT for further information.

A first difference is that vacuum metal bonding, as used for the SVT, is not required. For the MTT, the base as well as the emitter structure can be created by vacuum deposition onto the collector semiconductor. For the fabrication of the thin tunnel barrier, methods well-established for fabricating magnetic tunnel junctions can be used.[7] Typically, this involves deposition of a 1 to 2 nm thin Al layer and subsequent oxidation. The emitter is then deposited on top and can be magnetically pinned using exchange bias from an antiferromagnet if desired.

A second notable difference is that in the MTT, the energy of the injected hot electrons is given by the voltage V_{EB} applied across the tunnel barrier. The hot-electron energy is thus tunable over a certain range of energies, typically up to 3 eV. This allows one to do spectroscopic studies. Also, as we have already seen in Section 16.2.7, raising the hot-electron energy is beneficial for the transfer ratio of the device, because more electrons can be transmitted into the larger density of states higher up in the conduction band of the collector semiconductor. For the MTT, this is illustrated in the example of Figure 16.24, which shows that above a certain threshold voltage set by the collector Schottky barrier, the transfer ratio grows more than linearly with V_{EB}. The same behavior is well known in BEEM. Thus, in the

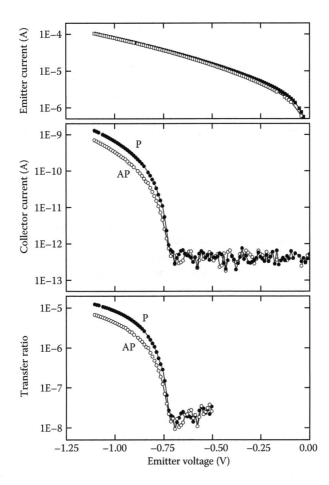

FIGURE 16.24 Emitter current, collector current, and transfer ratio of a MTT vs. emitter voltage V_{EB} applied across the tunnel barrier. (Data by O.M.J. van't Erve, MESA⁺, in collaboration with S.S.P. Parkin, IBM Almaden, unpublished.)

MTT, an increase of V_{EB} results in a larger collector current for two reasons, obviously the emitter current is increased, but at the same time the transfer ratio of the base is enhanced.

A third important aspect is that the MTT with ferromagnetic emitter has only a single magnetic layer in the base, as opposed to two magnetic layers and the nonmagnetic spacer for the SVT (and also for MTT with nonmagnetic emitter). From the work on the SVT, it became clear that each base layer and each interface is a source of scattering for the hot electrons. Therefore, we anticipate that the transfer ratio of a MTT with magnetic emitter should be significantly higher. A drawback of the MTT with a ferromagnetic emitter is that the relative magnetic response (MC) is limited by the tunneling spin polarization. This issue is discussed next.

16.3.1 MTT AND TUNNEL SPIN POLARIZATION

The expressions for the transmission and MC of the MTT with nonmagnetic tunnel emitter (left design of Figure 16.23) are the same as Equation 16.2 to Equation 16.7 for the SVT. However, the situation is different for the MTT with a ferromagnetic tunnel emitter, because the emitter current now becomes spin polarized, with the polarization determined by the tunneling process. For the parallel magnetization state, the emitter tunnel current is a sum of a majority and a minority spin contribution according to[7]

$$I_E^P \propto \delta_E^M \delta_B^M + \delta_E^m \delta_B^m \tag{16.10}$$

Here δ_E^M and δ_E^m are the fraction of majority and minority tunneling electrons associated with the emitter/insulator interface (label E), and δ_E^M and δ_E^m are the fraction of tunnel electrons associated with the insulator/base interface (label B). These are defined in the usual way with help of the tunneling spin polarization P_t

$$P_{t,E} = \frac{\delta_E^M - \delta_E^m}{\delta_E^M + \delta_E^m} \tag{16.11}$$

and

$$P_{t,B} = \frac{\delta_B^M - \delta_B^m}{\delta_B^M + \delta_B^m} \tag{16.12}$$

To obtain the collector current, we need to multiply each of the two emitter current components in Equation 16.10 with the majority or minority spin hot-electron transmission factor appropriate for the ferromagnetic layer in the base (M2 in Figure 16.23). We then have

$$I_C^P \propto \delta_E^M \delta_B^M T_B^M + \delta_E^m \delta_B^m T_B^m \tag{16.13}$$

where T^M and T^m are defined as before in Equation 16.4 and Equation 16.5. If we assume that the emitter is magnetically pinned, then for the antiparallel case, the magnetization of the base will be reversed and the resulting I_C^{AP} can be written as

$$I_C^{AP} \propto \delta_E^M \delta_B^m T_B^m + \delta_E^m \delta_B^M T_B^M \tag{16.14}$$

Using these equations, the MC can be cast into the familiar form

$$MC = \frac{I_C^P - I_C^{AP}}{I_C^{AP}} = \frac{2P_{t,E} P_B^*}{1 - P_{t,E} P_B^*} \tag{16.15}$$

where we have defined a *renormalized* polarization P_B^* of the base as

$$P_B^* = \frac{\delta_B^M T_B^M - \delta_B^m T_B^m}{\delta_B^M T_B^M + \delta_B^m T_B^m} \tag{16.16}$$

Note that this *renormalized* polarization is a combination of tunneling factors and hot-electron transmission factors and as such should not be confused with the regular tunneling spin polarization as defined in Equation 16.12. Thus, we see that the MC of a MTT with a ferromagnetic tunnel injector is determined by two completely unrelated physical processes. The polarization of the emitter current is due to spin-dependent tunneling, and the subsequent transmission of the hot-electrons in the base of the MTT is given by the spin-dependent scattering of hot electrons.

The fact that the MC depends on $P_{t,E}$ has an important consequence, because the tunnel polarization $P_{t,E}$ of the emitter/insulator interface is basically fixed, that is, not dependent on the thickness of the emitter electrode. Rather, it is determined by the choice of metal and tunnel insulator.[7] In practice, this means $P_{t,E}$ is limited to about 50% for typical ferromagnetic materials and Al_2O_3 tunnel barriers. Using the expressions given above, it is straightforward to calculate the expected MC for a given emitter tunnel polarization (see Figure 16.25, in which hot-electron transmission parameters for a Co base ferromagnet [$\lambda^M = 21$ Å and $\lambda^m = 8$ Å] were used). As can be seen, the MC grows with increasing thickness of the base ferromagnetic layer because the transmission polarization P_B^* of the base is dependent on the base layer thickness. Because the attenuation length of the majority spin electrons is considerably larger than that of minority spins, at large base thickness only majority spins can be transmitted ($P_B^* \sim 1$). The MC then saturates at a value of $2P_{t,E}/(1 - P_{t,E})$ and is dependent *only* on the tunnel spin polarization $P_{t,E}$. For a tunnel polarization of 50%, the MC cannot be larger than 200%. This constitutes a fundamental difference with the SVT, where the MC can be

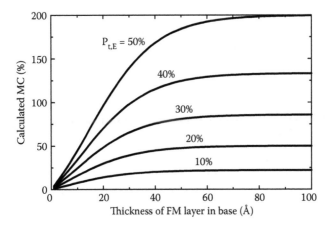

FIGURE 16.25 Calculated magnetic response of a MTT with ferromagnetic emitter, for different values of the tunnel spin polarization $P_{t,E}$ of the emitter.

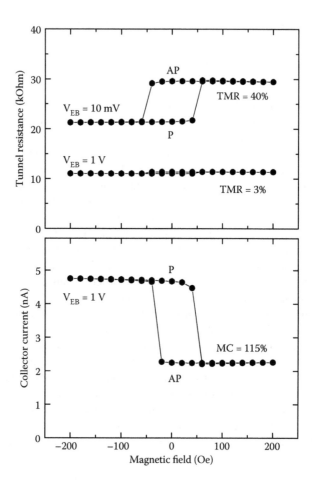

FIGURE 16.26 Bottom panel: Collector current vs. magnetic field for a $Co_{84}Fe_{16}/Al_2O_3/Co$ (10 nm)/Si MTT at 1 V bias and $T = 77$ K. The emitter was pinned by $Ir_{22}Mn_{78}$. Top panel: the TMR of the emitter-base tunnel junction, measured at 10 mV and 1 V bias, respectively. (Data by O.M.J. van't Erve, MESA+, in collaboration with S.S.P. Parkin, IBM Almaden, unpublished.)

made arbitrarily large by increasing the thickness of both the ferromagnetic layers of the base.

An interesting positive consequence is that the MTT can be used to *probe* the tunnel spin polarization at large bias. Recall that the MTT typically operates at a bias voltage of 1 V or larger. At such a voltage, the regular TMR of a ferromagnetic tunnel junction is usually negligibly small.[7] However, experimental MTTs with ferromagnetic emitter clearly display large MC. An example is shown in Figure 16.26 for a $Co_{84}Fe_{16}/Al_2O_3/Co$ (10nm)/Si transistor, fabricated by sputtering through shadow masks. A large difference in collector current between P and AP states is clearly observed, corresponding to a MC of about 115%. This demonstrates that highly spin-polarized electrons are injected into the base by tunneling from the $Co_{84}Fe_{16}$ emitter at an applied bias of 1.0 V. The corresponding tunnel spin polarization is about 36%. Note that the TMR for the same junction, measured between base and emitter, is only 3% at 1.0 V, whereas a TMR of about 40% was measured at a bias of 10 mV. We will return to this point after having described results on MTTs operating up to room temperature in Section 16.3.2.

16.3.2. MTT AT ROOM TEMPERATURE

So far, MTTs have been studied mostly at low temperature (77 K) because the rather large MTTs fabricated with shadow masks have a significant leakage current across the collector Schottky diode at room temperature.[62,78,85] However, to exploit the capability of the MTT to determine the tunnel spin polarization, it would be of interest to have MTTs operating up to room temperature. To achieve that, we have introduced predefined Si substrates and a photolithography process so as to be able to study the temperature dependence of the MC and the transfer ratio up to room temperature.[89] This allows us to probe the tunnel spin polarization up to room temperature, removing a limitation of the standard technique of tunneling into a superconductor.[90]

The MTTs were prepared by thermal evaporation in a molecular beam epitaxy system at a base pressure of 10^{-10} mbar. The structure of the MTT was n-type Si/Au (7 nm)/Co (8 nm)/Al_2O_3 (2.4 nm)/FM (10 nm)/Au (10 nm), where FM is $Ni_{80}Fe_{20}$ or Co (1 nm)/$Ni_{80}Fe_{20}$ (9 nm). The films were grown on a lithographically defined area of a Si wafer, surrounded by a thick SiO_2 to reduce device size and eliminate edge leakage currents across the collector diode (see Figure 16.27). The leakage current across the Schottky barrier in the patterned MTT is less than 0.1 nA at room temperature. The Al_2O_3 barriers were formed by plasma oxidation of a thin Al layer. MTT devices were processed using standard photolithography, ion beam etching, and lift-off techniques. The diameter of the active area of the devices varied from 10 to 100 μm and that of the base collector diode from 20 to 130 μm. Transport measurements were conducted using a four-point geometry for the emitter to base tunnel junction and a separate ohmic contact to the back of the Si collector.

Figure 16.28A shows the collector current at room temperature as a function of magnetic field for a MTT with a $Ni_{80}Fe_{20}$ emitter. The emitter bias voltage is 1 V. The largest collector current of 6.63 nA is obtained when both ferromagnetic layers

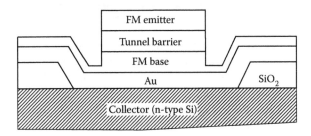

FIGURE 16.27 The layer structure of a lithographically defined MTT.

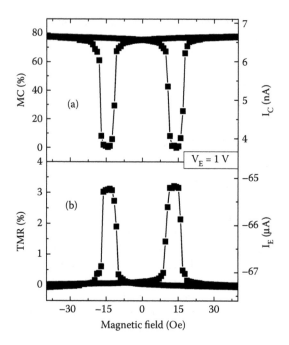

FIGURE 16.28 MC (A) and TMR (B) vs. magnetic field for a MTT with Si/Au (7 nm)/Co (8 nm)/Al$_2$O$_3$ (2.4 nm)/Ni$_{80}$Fe$_{20}$ (10 nm)/Au (10 nm) at an emitter bias of 1 V and at room temperature.

are aligned parallel to each other. In the antiparallel state, the collector current reduces to 3.7 nA, resulting in a MC of 79% at room temperature. The leakage current of less than 0.1 nA across the collector Schottky barrier is negligible compared to the output collector current.

For a 8 nm Co base virtually no minority spins are transmitted ($P_B^* \sim 1$), therefore we are in the regime where $MC = 2P_{t,E}/(1 - P_{t,E})$. A MC of 79% corresponds to a tunnel spin polarization of the Ni$_{80}$Fe$_{20}$/Al$_2$O$_3$ emitter interface of 29%, demonstrating that the tunnel current is still highly spin polarized at a high bias voltage of 1 V. However, the TMR at 1 V for the same junction is only 3.2% (Figure 16.28B); whereas the TMR is about 22% at a bias of 20 mV (not shown). This proves that the disappearance of TMR at high bias does not necessarily imply that the tunnel current is no longer spin polarized. Tunneling electrons originate mainly from states near the Fermi energy of the emitter at all bias voltages even though the electron energy distribution becomes broader with increasing bias voltage. However, empty states of the base into which electrons can tunnel depend on the bias voltage. These states are near the Fermi energy at low bias where the maximum TMR ratio is obtained and well above the Fermi energy at large bias voltage where the TMR ratio drops drastically. Thus, the low spin polarization of the states well above the Fermi energy is responsible for the drop of the TMR at high bias.

Figure 16.29A shows the temperature dependence of the MC for MTTs with Co (open squares) and Ni$_{80}$Fe$_{20}$ (solid circles) emitter. The MC at 100 K is around

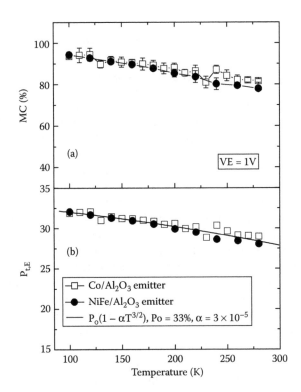

FIGURE 16.29 Temperature dependence of the MC (A) and tunnel spin polarization $P_{t,E}$ (B) for MTTs with a Co (open squares) and a $Ni_{80}Fe_{20}$ (solid circles) emitter. The line represents a fit to $P_{t,E}(T) = P_{t,E}(0)(1 - \alpha T^{3/2})$. The emitter bias voltage is 1 V.

94% for both emitter materials. As the temperature is increased to room temperature, the MC gradually decreases to 82% for a MTT with Co emitter and 79% for that with $Ni_{80}Fe_{20}$ emitter. The MTT with the Co emitter shows slightly weaker temperature dependence than that with the $Ni_{80}Fe_{20}$ emitter. Just like in the SVT, thermal spin-wave scattering can also affect the temperature dependence of the MC in the MTT. However, the thermal spin-wave attenuation length for Co is quite long compared to that of $Ni_{80}Fe_{20}$ as shown in Section 16.2.6. Consequently, the effect on the transmission polarization P_B^* of the base is too weak to explain the drop of the MC with temperature for the MTT. This implies that the temperature dependence of the MC is mainly due to that of the emitter tunnel spin polarization. Figure 16.29B shows the temperature dependence of the tunnel spin polarization extracted from the MC value in Figure 16.29A. The tunnel polarization decreases from 32 to 28% as the temperature is increased from 100 K to room temperature. The solid line in Figure 16.29B represents the fit obtained by using $P_{t,E}(T) = P_{t,E}(0)(1 - \alpha T^{3/2})$, where $P_{t,E}(0)$ is the tunnel spin polarization at T = 0 K and α is a material-dependent constant. This equation was previously used to describe the decay of TMR with temperature in magnetic tunnel junctions.[91] The fit gives a $P_{t,E}(0)$ of 33% and an α of 3×10^{-5} $K^{-3/2}$. These values agree well with the data in Reference 91.

This demonstrates that the MTT is a versatile tool to study tunneling spin polarization, not only at high bias, but also at elevated temperatures.

16.4 AVENUES FOR FURTHER RESEARCH

Perhaps the most influential aspect of the SVT is that it showed the value of functional integration of semiconductor and ferromagnetic materials into hybrid electronic devices employing spin. Room temperature effects as large as 400% in small magnetic fields are routinely obtained, a feature that has attracted significant attention and also stimulated research on other hybrid structures. For the MTT, the tunable hot-electron energy provides unique possibilities for spectroscopic studies. A number of device structures related to the SVT and MTT have been developed.[92–96] For the SVT, much progress has been made in optimizing device performance, focusing on the output current level and noise sources. The main limiting factors appear to be the inherently short hot-electron attenuation lengths of ferromagnetic metals, as well the significant transmission losses due to elastic scattering. Establishing the precise relation between scattering of nonequilibrium carriers and structural properties of the devices has remained largely unexplored. A next generation of devices may well use epitaxial growth techniques to obtain crystalline transistors. Moreover, it has been realized[1] that hot-electron filtering is a way to inject spins into a semiconductor that does not suffer from the well-known conductivity mismatch problem, with the ability to reach a spin polarization near 100% with conventional ferromagnets.

At the same time, the SVT opened up a new route to systematically study the fundamental physics of spin-dependent transport of hot electrons at energies in the range of 0.5 to 2 eV. New insights into the origin of spin-dependent hot-electron scattering have been obtained, including the dominance of volume effects over interfaces in the spin dependence of the transmission, the scattering by thermal spin waves, evidence for spontaneous spin-wave emission as a source of spin asymmetry, and the surprisingly important role of elastic scattering processes. Such insights can be further expanded by including studies on nonequilibrium holes below the Fermi energy, for which surprisingly strong spinfiltering was recently also reported.[71]

The relevance of understanding the fundamental scattering processes of nonequilibrium carriers extends beyond their use in SVT or MTT devices. For example, nonequilibrium carriers are created in ferromagnets when excited with ultrashort laser or magnetic field pulses. Moreover, excitation and switching of magnetization via spin-transfer torque involves injection of a spin-polarized current into a ferromagnet, where the exerted torque ultimately is determined by the microscopic mechanisms that provide the transfer of angular momentum. It is also of interest to explore nonequilibrium spin transport in novel materials such as half-metallic ferromagnets and oxides or ferromagnetic semiconductors, providing information about the materials supplementary to other transport techniques. A fruitful option is to employ BEEM and its magnetic counterpart,[52,53] offering a route to study these novel materials with nanoscale resolution. Last but not least, a particular feature of the MTT with ferromagnetic emitter, namely, the dependence of its magnetic sensitivity on spin-dependent tunneling, provides a unique opportunity to probe the tunnel spinpolarization of states away from the Fermi energy.

ACKNOWLEDGMENTS

The author is grateful to numerous colleagues at the MESA$^+$ Institute for Nanotechnology, University of Twente, for their contribution to the research described in this chapter. We acknowledge financial support from the Royal Netherlands Academy of Arts and Sciences, the Dutch Technology Foundation, the Dutch Foundation for Fundamental Research on Matter, and the European Commission.

REFERENCES

1. R. Jansen, *J. Phys. D: Appl. Phys.*, 36, R289, 2003; see www.iop.org/journals/jphysd.
2. M.N. Baibich, J.M. Broto, A. Fert, F. Nguyen Van Dau, F. Petroff, P. Etienne, G. Creuzet, A. Friederich, and J. Chazelas, *Phys. Rev. Lett.*, 61, 2472, 1988.
3. G.A. Prinz, *Science*, 282, 1660, 1998.
4. S.A. Wolf, D.D. Awschalom, R.A. Buhrman, J.M. Daughton, S. von Molnár, M.L. Roukes, A.Y. Chtchelkanova, and D.M. Treger, *Science*, 294, 1488, 2001.
5. S.S.P Parkin et al., *Proc. IEEE*, 91, 661, 2003.
6. J.S. Moodera, L.R. Kinder, T.M. Wong, and R. Meservey, *Phys. Rev. Lett.*, 74 3273, 1995.
7. J.S. Moodera and G. Mathon, *J. Magn. Magn. Mater.*, 200, 248, 1999.
8. S.S.P. Parkin et al., *J. Appl. Phys.*, 85, 5828, 1999.
9. J. de Boeck and G. Borghs, *Phys. World*, 12, 27, 1999.
10. S. Tehrani et al., *Proc. IEEE*, 91, 714, 2003.
11. J.M. Kikkawa and D.D. Awschalom, *Phys. Rev. Lett.*, 80, 4313, 1998.
12. J.M. Kikkawa and D.D. Awschalom, *Nature*, 397, 139, 1999.
13. H. Ohno, *Science*, 281, 951, 1998.
14. H. Ohno, D. Chiba, F. Matsukura, T. Omiya, E. Abe, T. Dietl, Y. Ohno, and K. Ohtani, *Nature*, 408, 944, 2000.
15. Y.D. Park, A.T. Hanbicki, S.C. Erwin, C.S. Hellberg, M. Sullivan, J.E. Mattson, T.F. Ambrose, A. Wilson, G. Spanos, and B.T. Jonker, *Science*, 295, 651, 2002.
16. S. Datta and B. Das, *Appl. Phys. Lett.*, 56, 665, 1990.
17. A. Fert and H. Jaffrès, *Phys. Rev. B*, 64, 184420, 2001.
18. S. Sugahara, *IEE Proc. — Circuits Devices Syst.*, 152, 355, 2005.
19. H.J. Zhu, M. Ramsteiner, H. Kostial, M. Wassermeier, H.-P. Schönherr, and K.H. Ploog, *Phys. Rev. Lett.*, 87, 016601, 2001.
20. A.T. Hanbicki, B.T. Jonker, G. Itskos, G. Kioseoglou, and A. Petrou, *Appl. Phys. Lett.*, 80, 1240, 2002.
21. A.T. Hanbicki, O.M.J. van 't Erve, R. Magno, G. Kioseoglou, C.H. Li, B.T. Jonker, G. Itskos, R. Mallory, M. Yasar, and A. Petrou, *Appl. Phys. Lett.*, 82, 4092, 2003.
22. V.F. Motsnyi, J. de Boeck, J. Das, W. van Roy, G. Borghs, E. Goovaerts, and V.I. Safarov, *Appl. Phys. Lett.*, 81, 265, 2002.
23. P. van Dorpe, V.F. Motsnyi, M. Nijboer, E. Goovaerts, V.I. Safarov, J. Das, W. van Roy, G. Borghs, and J. de Boeck, *Jpn. J. Appl. Phys. (Pt. 2)*, 42, L520, 2003.
24. S.A. Crooker, M. Furis, X. Lou, C. Adelmann, D.L. Smith, C.J. Palmstrøm, and P.A. Crowell, *Science*, 309, 2191, 2005.
25. D.J. Monsma, J.C. Lodder, Th.J.A. Popma, and B. Dieny, *Phys. Rev. Lett.*, 74, 5260, 1995.
26. P.S. Anil Kumar, R. Jansen, O.M.J. van 't Erve, R. Vlutters, P. de Haan, and J.C. Lodder, *J. Magn. Magn. Mater.*, 214, L1, 2000.
27. R. Jansen, P.S. Anil Kumar, O.M.J. van 't Erve, R. Vlutters, P. de Haan, and J.C. Lodder, *Phys. Rev. Lett.*, 85, 3277, 2000.

28. D.J. Monsma, R. Vlutters, and J.C. Lodder, *Science*, 281, 407, 1998.
29. S.M. Sze, *Physics of Semiconductor Devices*, 2nd ed., Wiley, New York, 1981.
30. P.S. Anil Kumar, R. Jansen, O.M.J. van 't Erve, R. Vlutters, S.D. Kim, and J.C. Lodder, *Physica C*, 350, 166, 2001.
31. S.M. Sze, C.R. Crowell, G.P. Carey, and E.E. LaBate, *J. Appl. Phys.*, 37, 2690, 1966.
32. C.R. Crowell and S.M. Sze, *J. Appl. Phys.*, 37, 2683, 1966.
33. S.M. Sze, *Physics of Semiconductor Devices*, 1st ed., Wiley, New York, 1969.
34. T. Shimatsu, R.H. Mollema, D.J. Monsma, E.G. Keim, and J.C. Lodder, *J. Vac. Sci. Technol. A*, 16, 2125, 1998.
35. J.C. Lodder, D.J. Monsma, R. Vlutters, and T. Shimatsu, *J. Magn. Magn. Mater.*, 198–199, 119, 1999.
36. K. Dessein, H. Boeve, P.S. Anil Kumar, J. de Boeck, J.C. Lodder, L. Delaey, and G, Borghs, *J. Appl. Phys.*, 87, 5155, 2000.
37. S.D. Kim, O.M.J. van 't Erve, R. Jansen, P.S. Anil Kumar, R. Vlutters, and J.C. Lodder, *Sensors and Actuators A*, 91, 166, 2001.
38. S.D. Kim, O.M.J. van 't Erve, R. Vlutters, R. Jansen, and J.C. Lodder, *IEEE Trans. Electron Devices*, 49, 847, 2002.
39. N.F. Mott, *Adv. Phys.*, 13, 325, 1964.
40. D.T. Pierce and H.C. Siegmann, *Phys. Rev. B*, 9, 4035, 1974.
41. D.P. Pappas, K.-P. Kämper, B.P. Miller, H. Hopster, D.E. Fowler, C.R. Brundle, A.C. Luntz, and Z.-X. Shen, *Phys. Rev. Lett.*, 66, 504, 1991.
42. J.C. Gröbli, D. Guarisco, S. Frank, and F. Meier, *Phys. Rev. B*, 51, 2945, 1995.
43. J.C. Gröbli, D. Oberli, and F. Meier, *Phys. Rev. B*, 52, R13095, 1995.
44. G. Schönhense and H.C. Siegmann, *Ann. Phys. (Leipzig)*, 2, 465, 1993.
45. H.-J. Drouhin, G. Lampel, Y. Lassailly, A.J. van der Sluijs, and C. Marlière, *J. Magn. Magn. Mater.*, 151, 417, 1995.
46. H.-J. Drouhin, A.J. van der Sluijs, Y. Lassailly, and G. Lampel, *J. Appl. Phys.*, 79, 4734, 1996.
47. D. Oberli, R. Burgermeister, S. Riesen, W. Weber. and H.C. Siegmann, *Phys. Rev. Lett.*, 81, 4228, 1998.
48. W. Weber, S. Riesen, and H.C. Siegmann, *Science*, 291, 1015, 2001.
49. M. Aeschlimann, M. Bauer, S. Pawlik, W. Weber, R. Burgermeister, D. Oberli, and H.C. Siegmann, *Phys. Rev. Lett.*, 79, 5158, 1997.
50. R.J. Celotta, J. Unguris, and D.T. Pierce, *J. Appl. Phys.*, 75, 6452, 1994.
51. L.D. Bell and W.J. Kaiser, *Phys. Rev. Lett.*, 61, 2368, 1988.
52. W.H. Rippard and R.A. Buhrman, *Appl. Phys. Lett.*, 75, 1001, 1999.
53. W.H. Rippard and R.A. Buhrman, *Phys. Rev. Lett.*, 84, 971, 2000.
54. R. Vlutters, R. Jansen, O.M.J. van 't Erve, S.D. Kim, J.C. Lodder, A. Vedyayev, and B. Dieny, *Phys. Rev. B*, 65, 024416, 2002.
55. Some indirect effect of elastic scattering still occurs because it changes the path length in the base and thereby increases the probability for an inelastic scattering event.
56. R. Vlutters, O.M.J. van 't Erve, S.D. Kim, R. Jansen, and J.C. Lodder, *Phys. Rev. Lett.*, 88, 027202, 2002.
57. R. Vlutters, R. Jansen, O.M.J. van 't Erve, S.D. Kim, and J.C. Lodder, *J. Appl. Phys.*, 89, 7305, 2001.
58. M.K. Weilmeier, W.H. Rippard, and R.A. Buhrman, *Phys. Rev. B*, 59, R2521, 1999.
59. L.D. Bell, *Phys. Rev. Lett.*, 77, 3893, 1996.
60. V.P. Zhukov and P.M. Echenique, private communication.
61. R. Knorren, K.H. Bennemann, R. Burgermeister, and M. Aeschlimann, *Phys. Rev. B*, 61, 9427, 2000.

62. S. van Dijken, X. Jiang, and S.S.P. Parkin, *Phys. Rev. B,* 66, 094417, 2002.
63. O.M.J. van 't Erve, R. Vlutters, P.S. Anil Kumar, S.D. Kim, F.M. Postma, R. Jansen, and J.C. Lodder, *Appl. Phys. Lett.,* 80, 3787, 2002.
64. R. Jansen, S.D. Kim, R. Vlutters, O.M.J. van 't Erve, and J.C. Lodder, *Phys. Rev. Lett.,* 87, 166601, 2001.
65. M. Plihal, D.L. Mills, and J. Kirschner, *Phys. Rev. Lett.,* 82, 2579, 1999.
66. J. Hong and D.L. Mills, *Phys. Rev. B,* 59, 13840, 1999.
67. J. Hong and D.L. Mills, *Phys. Rev. B,* 62, 5589, 2000.
68. V.P. Zhukov, E.V. Chulkov, and P.M. Echenique, *Phys. Rev. Lett.,* 93, 096401, 2004.
69. B. Dieny, V.S. Speriosu, and S. Metin, *Europhys. Lett.,* 15, 227, 1991; B. Dieny, P. Humbert, V.S. Speriosu, S. Metin, B.A. Gurney, P. Baumgart, and H. Lefakis, *Phys. Rev. B,* 45, 806, 1992.
70. A. Fert, J.L. Duvail, and T. Valet, *Phys. Rev. B,* 52, 6513, 1995.
71. T. Banerjee, E. Haq, M.H. Siekman, J.C. Lodder, and R. Jansen, *Phys. Rev. Lett.,* 94, 027204, 2005.
72. V.P. Zhukov, E.V. Chulkov, and P.M. Echenique, *Phys. Rev. B,* 73, 125105, 2006.
73. P. Niedermann, L. Quattropani, K. Solt, I. Maggio-Aprile, and O. Fischer, *Phys. Rev. B,* 48, 8833, 1993.
74. H. Gokcan, J.C. Lodder, and R. Jansen, *Mater. Sci. Eng. B,* 126, 129, 2005.
75. R. Jansen, H. Gokcan, O.M.J. van 't Erve, F.M. Postma, and J.C. Lodder, *J. Appl. Phys.,* 95, 6927, 2004.
76. S. van Dijken, X. Jiang, and S.S.P. Parkin, *Appl. Phys. Lett.,* 83, 951, 2003.
77. O.M.J. van 't Erve, P.S. Anil Kumar, R. Jansen, S.D. Kim, R. Vlutters, J.C. Lodder, A.A. Smits, and W.J.M. de Jonge, *Sensors and Actuators A,* 91, 192, 2001.
78. K. Mizushima, T. Kinno, K. Tanaka, and T. Yamauchi, *IEEE Trans. Magn.,* 33, 3500, 1997.
79. K. Mizushima, T. Kinno, T. Yamauchi, and K. Inomata, U.S. Patent 5,747,859, 1996; U.S. Patent 5,973,334, 1998.
80. K. Mizushima, T. Kinno, K. Tanaka, and T. Yamauchi, *Phys. Rev. B,* 58, 4660, 1998.
81. T. Yamauchi and K. Mizushima, *Phys. Rev. B,* 58, 1934, 1998; 61, 8242, 2000.
82. T. Kinno, K. Tanaka, and K. Mizushima, *Phys. Rev. B,* 56, R4391, 1997.
83. R. Sato and K. Mizushima, *Appl. Phys. Lett.,* 79, 1157, 2001.
84. S. van Dijken, X. Jiang, and S.S.P. Parkin, *Appl. Phys. Lett.,* 80, 3364, 2002.
85. S. van Dijken, X. Jiang, and S.S.P. Parkin, *Appl. Phys. Lett.,* 82, 775, 2003.
86. E.I. Rashba, *Phys. Rev. B,* 68, 241310(R), 2003.
87. S. van Dijken, X. Jiang, and S.S.P. Parkin, *Phys. Rev. Lett.,* 90, 197203, 2003.
88. T. Hagler, C. Bilzer, M. Dumm, W. Kipferl, and G. Bayreuther, *J. Appl. Phys.,* 97, 10D505, 2005.
89. B.G. Park, T. Banerjee, B.C. Min, J.S.M. Sanderink, J.C. Lodder, and R. Jansen, *J. Appl. Phys.,* 98, 103701, 2005.
90. R. Meservey and P.M. Tedrow, *Phys. Rep.,* 238, 173, 1994.
91. C.H. Shang, J. Nowak, R. Jansen, and J.S. Moodera, *Phys. Rev. B,* 58, R2917, 1998.
92. I. Appelbaum, K.J. Russell, D.J. Monsma, V. Narayanamurti, C.M. Marcus, M.P. Hanson, and A.C. Gossard, *Appl. Phys. Lett.,* 83, 4571, 2003.
93. K.J. Russell, I. Appelbaum, W. Yi, D.J. Monsma, F. Capasso, C.M. Marcus, V. Narayanamurti, M.P. Hanson, and A.C. Gossard, *Appl. Phys. Lett.,* 85, 4502, 2004.
94. G. Rodary, M. Hehn, T. Dimopoulos, D. Lacour, J. Bangert, H. Jaffrès, F. Montaigne, F. Nguyen van Dau, F. Petroff, A. Schuhl, and J. Wecker, *J. Magn. Magn. Mater.,* 290–291, 1097, 2005.
95. S. Ladak and R.J. Hicken, *Appl. Phys. Lett.,* 87, 232504, 2005.
96. P. LeMinh, H. Gokcan, J.C. Lodder, and R. Jansen, *J. Appl. Phys.,* 98, 076111, 2005.

Index

A

A and ß measurement, doped ZnO, 27
Absorption
  classical and quantum theory of, 19–23
  soft x-ray resonant magnetic scattering and,
    79
Alloys for spintronic devices, Heusler, 297–299,
  301–303
AlN and ion implantation, 42–45
Amplitude, scattering, 85–86
Andreev reflection (AR)
  probability, 290
  spin polarization and, 299–300
Angular dependence, 329–332
Anisotropic magnetoresistance (AMR),
  264
Anisotropic tensor scattering (ATS) reflections,
  79–80
Anisotropy, magnetic, 61–63
  in-plane, 68–73
  MFM images and, 82–84
  perpendicular (PMA), 61–63, 80, 82–84
Annealing, postimplantion, 42–45
Antiferromagnetic Cr/GaAs Schottky diodes,
  326–327
ARPES in e-ph interaction strength, 140

B

Ballistic magnetoresistance (BMR), 181
Ballistic spin transport
  optical magnetocurrent, 329–333
  sample preparation and characterization,
    328–329
  in spin-valve structures, 327–333
Band bending, 291
Bessel functions, 19
Bias dependence, 319–322
Bipolar spin LEDs, 281–283
Bir, Aronov, and Pikus (BAP) mechanism,
  248
  spin-diffusion length and, 254
Bloch theorem, 129
Boltzmann transport relationship, 249
Bragg angle, 89
Bragg peaks, 98–99
Bulk active regions, 271

C

Calculation
  results, *135t*
  spin transfer effect, 148–153
Cameras, charge-coupled device (CCD), 81, 92
Carrier electron mobility, 249–251
Carrier-induced ferromagnetism, 47–48, 57
CASTEP code, 301
CdTe cubic semiconductors, 23–24
Channel blocking and adiabaticity, 207–213
Charge-coupled device (CCD) cameras, 81, 92
CIMS
  for low switching field, 110
  in nanofabricated pillars, 115–119
Circuit model, magnetic nanowires, 213–219
Circular polarization, 87–88, 98
Co (Cobalt)
  doping of ZnO, 47–50, 51f
  ion implantation and, 42–45
  ZnO:, 29–30
Co2CrAL Heusler alloys, 301–303
CO/CU interface, 132–138
Co/GaAs, 348–349
  Schottky diodes, 324–326
Coherence of x-rays
  longitudinal, 90–91
  scattering from FePd nanowire, 93–96
  speckle pattern and, 91–92
  transverse, 90–91
Co/InAs, 349
Co2MnX Heusler alloys, 297
Complementary metal-oxide semiconductor
  (CMOS) technology, 269
Correlation length, magnetic, 89
Co2TiSN Heusler alloys, 301–303
Coupling, spin-orbit, 26
CPP-GMR ratio enhancement, 119–122
Cr (Chromium) doping of ZnO, 47–50, 51f
Crystal field splitting, 24–27
Cubic semiconductors, 23–24
Curie temperatures, 46, 47–48
  of ferromagnetic metals, 270
  Mn dopants and, 58–61
Current-controlled memory devices, CIMS, 110
Current-driven domain wall motion
  electron flow interaction, 234–235
  force and torque, 236–237

415